ENDOVASCULAR GRAFTING TECHNIQUES

ENDOVASCULAR GRAFTING TECHNIQUES

Juan Carlos Parodi, MD

Director and Chief
Department of Cardiovascular Surgery
Instituto Cardiovascular de Buenos Aires
Buenos Aires, Argentina
Vice Director
Post-Graduate Training Program in
Cardiovascular Surgery
University of Buenos Aires
Buenos Aires, Argentina

Frank J. Veith, MD

Professor and Chief
Vascular Surgical Services
Montefiore Medical Center
The University Hospital for the Albert Einstein
College of Medicine
New York, New York

Michael L. Marin, MD

Associate Professor of Surgery
Director
Endovascular Surgical Development
The Mount Sinai Medical Center/School of Medicine
New York, New York

SANS TACHE

Williams & Wilkins
A WAVERLY COMPANY

BALTIMORE • PHILADELPHIA • LONDON • PARIS • BANGKOK
BUENOS AIRES • HONG KONG • MUNICH • SYDNEY • TOKYO • WROCLAW

Editor: Carroll C. Cann
Managing Editor: Danielle Hagan
Marketing Manager: Adam Glazer
Project Editor: Jeffrey S. Myers

351 West Camden Street
Baltimore, Maryland 21201-2436 USA

Rose Tree Corporate Center
1400 North Providence Road
Building II, Suite 5025
Media, Pennsylvania 19063-2043 USA

Printed in the United States of America

Library of Congress Cataloging-in-Publication Data

Endovascular grafting techniques / [edited by] Juan Carlos Parodi, Frank J. Veith, Michael L. Marin.
p. cm.
Includes bibliographical references and index.
ISBN 0-683-30209-4
1. Blood vessel prosthesis. 2. Blood vessels—Endoscopic surgery. I. Parodi, Juan Carlos. II. Veith, Frank J., 1931– . III. Marin, Michael L.
[DNLM: 1. Aortic Aneurysm—surgery. 2. Blood Vessel Prosthesis Implantation—methods. 3. Arterial Occlusive Diseases—surgery. 4. Blood Vessel Prosthesis. 5. Angiography—methods. WG 410E56 1999]
RD598.55.E534 1999
617.4′13—dc21
DNLM/DLC
for Library of Congress 98-28509
CIP

99 00 01 02 03
1 2 3 4 5 6 7 8 9 10

Preface

In 1990, a transluminally placed endovascular graft was first used successfully to treat a high-risk patient with an abdominal aortic aneurysm. Since then, worldwide interest in these devices has ignited, and several thousand endovascular grafts have been used to treat a variety of aneurysmal, traumatic, and occlusive arterial lesions.

The explosive interest in these devices stems from the facts that they may constitute a safer, better way to treat a variety of arterial lesions and that their less invasive nature is extremely attractive to both patients and referring physicians. As a result, a large number of surgeon-made and, subsequently, industry-made endovascular grafts have been developed and are in various stages of clinical trials. However, in many patients, insertion procedures have been long and complicated, significant early and late morbidity has occurred, and the comparative performance of endovascular grafts relative to prior standard therapies in the treatment of common garden variety aneurysms and occlusive lesions has yet to be demonstrated. Accordingly, the ultimate and precise role of endovascular grafts in the treatment of various lesions remains unknown.

On the other hand, endovascular graft devices have already been effectively used to treat patients with limb- and life-threatening arterial lesions, which would otherwise be difficult or impossible to treat. Examples include injuries that produce false aneurysms and arteriovenous fistulas of central blood vessels, recurrent or para-anastomotic aneurysms after aortoiliac surgery, and thoracic and abdominal aneurysms in patients with serious medical or surgical comorbidities that would render standard operations extremely high-risk.

This volume provides a snapshot view of the present state of the art in endovascular grafting. We have tried to accomplish this goal by asking pioneers in the field to summarize their experience to date with these grafts. We recognize the limitations imposed by writing a book in a rapidly developing field in which new grafts are introduced on a monthly basis, new investigators enter the field on a weekly basis, and new knowledge is produced on a daily basis. However, we believe our volume will have value as a starting point for all, to be supplemented and updated by articles in journals, presentations, and discussions among those interested in endovascular grafts.

Despite the as-yet unfulfilled requirement for long-term proof of safety, efficacy, and effectiveness of endovascular grafting compared with standard treatments, we believe that these new devices hold great future promise for altering the way that vascular disease is treated. Improvements in endovascular graft devices and methods to insert them will make them safer and easier to use for presently justified indications.

We also believe that indications will expand as long-term effectiveness and durability are demonstrated and as modified devices are developed to treat more complex arterial lesions, such as those involving the visceral segment, the arch of the aorta, and aortic dissections. We believe that the potential usefulness and advantages of these devices are almost without limit and that we are at the dawn of a new era in vascular disease treatment. This belief is reinforced by the intense interest of industry and engineers in these new devices. However, we admit that our belief is still largely speculation and that the use of these devices may be restricted to a small niche. Only time and good science will tell.

Endovascular grafts, because they are based on interventional, catheter-based techniques and on vascular grafts, have given rise to some conflict between interventional specialists and vascular surgeons, with both specialties having a basis for believing they should control and use these devices. In fact, most initial efforts in endovascular grafting have been spearheaded by vascular surgeons with full collaborative support from their interventional colleagues, although in some instances, the reverse relationship has existed.

It is our sincere hope that these new grafting devices will lead to closer collaboration and cooperation between surgeons and interventionalists, as we have enjoyed in the past. It is also our hope that this book will be valuable to interventionalists as well as to vascular surgeons, and it has been written with this in mind. However, we recognize that endovascular grafts could provoke turf battles, and the potential for such conflicts are real. Again, only time will tell what the outcome will be.

FJV and JCP

Contributors

George L. Berdejo, BA, RVT
Technical Director
Vascular Diagnostic Laboratories
Montefiore Medical Center
The University Hospital for the Albert Einstein College of Medicine
and Jack D. Weiler Hospital
Bronx, New York

John J. Bergan, MD
Clinical Professor of Surgery
University of California
San Diego, California

Jacob Cynamon, MD
Vascular and Interventional Radiology
Montefiore Medical Center
The University Hospital for the Albert Einstein College of Medicine
Bronx, New York

Michael D. Dake, MD
Associate Professor of Radiology and Medicine (Pulmonary)
Stanford University School of Medicine
Chief
Cardiovascular and Interventional Radiology
Stanford University Hospital
Stanford, California

Edward B. Diethrich, MD
Medical Director
Arizona Heart Institute and
Arizona Heart Hospital
Phoenix, Arizona

Gerald Dorros, MD
President
Arizona Heart Institute Foundation
Phoenix, Arizona

James I. Fann, MD
Clinical Assistant Professor
Cardiothoracic Surgery
Stanford University School of Medicine
Stanford, California
Staff Cardiovascular Surgeon
Veterans Affairs Health Care System
Palo Alto, California

Richard M. Green, MD
Chief
Section of Vascular Surgery
Professor of Surgery
University of Rochester School of Medicine and Dentistry
Rochester, New York

Brian R. Hopkinson ChM, FRCS
Professor of Vascular and Endovascular Surgery
University of Nottingham
Nottingham, United Kingdom

Michael Jaff, DO
Peripheral Vascular Medicine Specialist
Milwaukee Heart and Vascular Clinic
Milwaukee, Wisconsin

Hector Lardani, MD, FACC
Head
Section of Ultrasound
Instituto Cardiovascular de Buenos Aires
Buenos Aires, Argentina

George Latus
The William Dorros-Isadore Feuer Interventional Cardiovascular Disease Foundation
Milwaukee, Wisconsin

Yu Liu, PhD
The William Dorros-Isadore Feuer Interventional Cardiovascular Disease Foundation
Milwaukee, Wisconsin
Cardiovascular Magnetic Resonance Research Unit
St. Luke's Medical Center
Milwaukee, Wisconsin

Ross T. Lyon, MD, RVT
Director
Vascular Laboratory
Assistant Professor of Surgery
Montefiore Medical Center
The University Hospital for the Albert Einstein College of Medicine
Bronx, New York

Shekker Maddineni, MD
Fellow
Vascular and Interventional Radiology
Montefiore Medical Center
The University Hospital for the Albert Einstein College of Medicine
Bronx, New York

Michael L. Marin, MD
Associate Professor of Surgery
Director
Endovascular Surgical Development
The Mount Sinai Medical Center/School of Medicine
New York, New York

James May, MS, FRACS, FACS
Bosch Professor of Surgery
University of Sydney
Sydney Australia

D. Craig Miller, MD
Professor of Cardiovascular Surgery
Stanford University School of Medicine
Stanford, California

R. Scott Mitchell, MD
Associate Professor
Cardiovascular Surgery
Stanford University School of Medicine
Stanford California

Wesley S. Moore, MD
Professor of Surgery/Vascular
UCLA School of Medicine
Los Angeles, California

Takao Ohki, MD
Chief
Endovascular Program
Assistant Professor of Surgery
Montefiore Medical Center
The University Hospital for the Albert Einstein College of Medicine
Bronx, New York

Juan C. Parodi, MD
Director and Chief
Department of Cardiovascular Surgery
Instituto Cardiovascular de Buenos Aires
Buenos Aires, Argentina
Vice Director
Post-Graduate Training Program in Cardiovascular Surgery
University of Buenos Aires
Buenos Aires, Argentina

Alla Rozenblit, MD
Director
Abdominal Imaging
Associate Professor of Radiology
Montefiore Medical Center
The University Hospital for the Albert Einstein College of Medicine
Bronx, New York

Grigory Rozenblit, MD
Director
Vascular and Interventional Radiology
Associate Professor of Radiology
New York Medical College
New York, New York

Cynthia Sampson, MD
The William Dorros-Isadore Feuer Interventional Cardiovascular Disease Foundation
Milwaukee, Wisconsin
Cardiovascular Magnetic Resonance Research Unit
St. Luke's Medical Center
Milwaukee, Wisconsin

Luis A. Sanchez, MD
Assistant Professor of Surgery
Montefiore Medical Center
The University Hospital for the Albert Einstein College of Medicine
Bronx, New York

Claudio J. Schönholz, MD
Clinica La Sagrada Familia
Vascular and Interventional Radiology
Instituto Cardiovascular de
Buenos Aires
Buenos Aires, Argentina

Charles P. Semba, MD
Associate Director
Cardiovascular-Interventional
Radiology
Stanford University Medical Center
Stanford, California

Wolf-J. Stelter, MD, PhD
Professor of Surgery
J.W. von Goethe - Universität Frankfurt
Chief of Surgery (General, Thoracic, Vascular,
Surgical Intensive Care)
Teaching Hospital
Frankfurt Germany

Frank J. Veith, MD
Professor and Chief
Vascular Surgical Services
Montefiore Medical Center
The University Hospital for the Albert
Einstein College of Medicine
Bronx, New York

Reese A. Wain, MD
Assistant Professor of Surgery
Montefiore Medical Center
The University Hospital for the Albert Einstein
College of Medicine
Bronx, New York

L. Samuel Wann, MD
Clinical Professor in Medicine (Cardiology)
University of Wisconsin-Madison
Madison, Wisconsin
Cardiovascular Magnetic Resonance Research Unit
St. Luke's Medical Center
Milwaukee, Wisconsin

Geoffrey H. White, MB, BS, FRACS
Vascular Surgeon
Royal Prince Alfred Hospital
Associate Professor of Surgery
University of Sydney
Sydney Australia

Weiyun Yu, BSC, MB, BS
Endovascular Research Fellow
Department of Surgery
University of Sydney
Sydney Australia

SW Yusuf, MB, BS, DM, FRCS
Division of Vascular Surgery
University Hospital Nottingham
Nottingham, United Kingdom

Contents

Chapter 1

History, Potential Influence, and Facility Requirements of Transluminally Placed Endovascular Grafts

Frank J. Veith

The treatment of arterial and venous disease has changed over the last 15 years and will change even more dramatically in years to come. Catheter-guidewire–based techniques have already made significant contributions to the manipulative or surgical management of aortoiliac, superficial femoral, and renal artery stenoses; arterial thromboembolism; iliofemoral venous thrombosis, arteriovenous malformations; and some visceral artery aneurysms. Some of these lesions can now be effectively treated with percutaneous balloon angioplasty (PTA), intravascular stents, catheter-directed lytic drugs and aspiration, intracaval filters, and coil embolization. These techniques all have in common the use of endovascular pathways accessed from remote sites to treat the offending vascular lesion, usually under fluoroscopic guidance.

Because these endovascular techniques have proved to be a simpler, safer way of treating some patients with vascular disease, they have been embraced by vascular surgeons and increasingly used by them. In some cases, these catheter-based treatments have been directly used by the vascular surgeon; in other cases, management has been by referral to a collaborating interventional radiologist or cardiologist.

Transluminally placed endovascular grafts (or TPEGs) are a new treatment modality that builds on some of the endovascular techniques already mentioned and combines them with standard vascular methods. This combination makes it possible to place vascular grafts within the arterial tree by inserting them at a remote access site, guiding them intraluminally to the appropriate location within the vascular tree, and fixing them there with a variety of attachment systems, such as expandable stents.

TPEGs repair the vascular pathology in a less invasive fashion than would characterize a standard open, operative approach. The TPEG terminology refers to the concept as well as the technology and devices used to carry it out. TPEGs can be used in the arterial or venous systems, and they can use autogenous, biologic, or prosthetic conduits.

In the following discussion, we will consider only TPEGs that are used in the arterial tree. Because TPEG repairs can be performed less invasively, their risks and costs may be lower than those of standard vascular graft operations. They are therefore extremely attractive to both patients and physicians and may have a major impact on the way patients with vascular diseases are treated. We will outline the history of arterial TPEGs and then consider their potential impact on the practice of vascular surgery and radiology. In addition, we will describe the facilities currently required to safely perform TPEG insertions.

HISTORY OF TPEGs IN THE ARTERIAL SYSTEM

The TPEG concept was first expressed in 1969 by Charles Dotter (1), the pioneer of fluoroscopically controlled angioplasty and other catheter-directed techniques. Throughout the 1980s, several investigators explored the use of TPEGs to treat experimental aneurysms (2–5). In 1985, the Russian surgeon Nicholas Volodos (6) first used a TPEG device with a self-expanding stent to treat iliac arteriosclerotic occlusive disease. However, his work, published only in Russian, received scant attention until Juan Parodi, an Argentinean surgeon, performed the first TPEG repair of an aortic aneurysm in a high-risk patient in September 1990. This original successful case along with five others was reported in 1991 by Parodi and his two associates (7), Julio Palmaz, a radiologist and developer of a balloon-expandable metal stent, and Hector Barone, an Argentinean developer-manufacturer of vascular grafts and other related devices.

More recently, Parodi (8) reported his overall experience with 57 TPEGs, of which 50 were inserted to treat aortic or aortoiliac aneurysms and 7 were used to treat a variety of other traumatic lesions, mostly arteriovenous fistulas.

The clinical TPEG program at Montefiore and Einstein hospitals in New York began in November 1992 under the leadership of vascular surgeons with collaboration from interventional radiologists and support from the Argentinean group. To date, the Montefiore-Einstein group has inserted 191 TPEGs in 148 patients with a variety of aneurysmal, occlusive, and traumatic lesions at different sites within the arterial tree (9–25) (Table 1.1).

Almost all the patients with aneurysmal and occlusive disease and many of the patients with traumatic lesions had some major medical comorbidity or a surgical problem (scarring, infection, or other organ disease) that precluded standard surgical or interventional treatment. Many were not candidates for general or regional anesthesia. All but three patients had some open vascular surgical component to their procedure.

The techniques for performing the TPEG procedures and the early results extending now to 56 months are reported in detail elsewhere (9–25) and in later chapters of this book. Even though many of these patients had end-stage disease with many previous therapeutic failures, our results to date have been encouraging.

All but 5 of the 46 trauma and nonaortic aneurysm patients have effective, functioning TPEGs from 1 to 56 months after their procedure. Many of the 44 aortic or aortoiliac aneurysm TPEG procedures were and continue to be successful, although five of the patients died after their procedure—one from multiorgan failure, two from microembolization, and two from cardiac causes. Of the 54 patients with aortofemoral or iliofemoral grafts, most still have patent grafts, although several patients have died of unrelated causes with patent grafts, five grafts have failed, and eight grafts required relatively minor reinterventions. In this group of late-stage limb salvage patients, the life table secondary patency rate at 3½ years is 82%, and the limb salvage rate is 92%.

Table 1.1. Montefiore/Einstein Experience With Transluminally Placed Endovascular Stented Grafts

Lesion	Number of Patients	Number of Grafts
Abdominal aortic or aortoiliac aneurysm	44	74
Iliac aneurysm	22	24
Popliteal aneurysm	3	2
Subclavian/innominate aneurysm	2	2
Traumatic false aneurysm	14	16
Arteriovenous fistula	5	5
Occlusive disease	54	61
Aortoiliac femoral (3 bilateral; 8 to superficial femoral or popliteal artery)		
Femoropopliteal	4	7
Total	148	191

See references 9–25.

This experience is still too scant to make strong recommendations about widespread use of the endovascular methods or to suggest that our indications be broadened to include any but these otherwise difficult-to-treat patients. Central vascular trauma, anastomotic aortoiliac aneurysms, and iliac aneurysms are probably exceptions in which TPEG treatment already appears to be advantageous. In addition, our overall experience to date allows us to emphasize a few generalities, which those wishing to enter this alluring new field should consider.

Many of the TPEG procedures were complex and difficult to perform. Most required and taxed the combined skills of those who regard themselves as expert in the techniques of open vascular surgery and the catheter-guidewire-imaging skills of interventional radiology. Successful attainment of a satisfactory endpoint sometimes required a major vascular surgical or catheter-guidewire "rescue" component to the procedure. Almost all the procedures were performed with an anesthesiologist in attendance and full invasive pressure monitoring in an operating room equipped with a movable, radiolucent operating table and a digital fluoroscope. In several instances, this venue and staffing were critical to achieving a successful outcome because some of the procedures proved to be long and complicated.

The problems with TPEGs are likely to be obviated by improvements in our relatively primitive devices and technical refinements that come with experience. This is particularly likely with some of the new second-generation, commercially fabricated devices that are becoming more widely available.

INFLUENCES OF TPEGs ON VASCULAR SURGERY AND RADIOLOGY

Despite the latter reservations, our preliminary observations in addition to those of Parodi and others (8–30) suggest that TPEGs may prove to be effective. If so, they will constitute a potentially better method to treat a major proportion of current vascular surgical cases. Because they offer the advantage of being less invasive to insert, TPEGs will be inherently attractive to patients and physicians. These devices also offer the potential advantages of lower mortality, fewer complications, shorter hospital stays, and consequent cost reductions when compared with current grafting procedures.

If the advantages of TPEGs—some of which have already been observed (12,29)—are further documented, these devices will certainly replace a substantial percentage of current standard vascular grafts. We estimate this percentage to be from 30% to 70% of the standard prosthetic grafts currently used to treat aneurysmal, traumatic, and occlusive arterial lesions. Because prosthetic grafts are used in approximately 60% of major arterial operations, TPEG procedures could replace from 20% to 40% of all major vascular operations. Adding this figure to the 25% of major vascular operations that may be replaced by other endovascular technologies means that from 45% to 65% of major vascular operations could be performed by an endovascular technique (31).[1] In addition, many of the remaining operations, such as vein bypasses and graft or native artery thromboembolectomy, will be greatly facilitated by endovascular fluoroscopically guided adjuncts such as PTA of residual inflow and outflow lesions and guidewire-assisted passage of balloon thromboembolectomy catheters (32).

The potential influence of these possibilities on vascular surgery is substantial and has caused vascular surgeons to have concerns and some anxieties. Clearly, the effective use of TPEGs and other endovascular techniques, although sometimes requiring vascular surgical skills, always requires catheter-guidewire-imaging skills, which some vascular surgeons do not possess. How will they cope with these endovascular techniques, which could provide better treatment options than standard open operations in most of their patients? How can they avoid becoming obsolete and being replaced by other interventional specialists? Several possible courses of action exist for vascular surgeons.

Approaches of Vascular Surgeons

One approach is to adopt an attitude of skepticism and denial: "No interventional endovascular technique has been any good and none will be." This *ostrich approach* is dangerous and wrong. It will result in vascular surgeons being largely excluded from a leadership role in the use of endovascular devices that are likely to work increasingly well in coming years as technology improves and experience is gained. The ostrich approach should be shunned by vascular surgeons. So also should the *wait-and-see* approach, which will allow other interventional specialists in radiology and cardiology to gain control of a technology that may be required to treat most of the patients whose vascular lesions justify intervention. These other interventional specialists sincerely believe their catheter-guidewire-imaging skills better qualify them to use TPEGs and other endovascular techniques than do our vascular surgical skills and knowledge of vascular disease.

Another approach that provides vascular surgical and interventional skills is to have patients with vascular disease cared for by a *team comprising collaborating vascular surgeons and interventional radiologists or cardiologists.* Such teams, largely under the leadership of vascular surgeons, have been responsible for most of the successful TPEG developmental efforts to date. Indeed, because vascular surgeons usually supervise the care of most vascular disease patients, most current, collaborative, multispecialty groups dealing with vascular disease patients are led by vascular surgeons. However, other interventional specialists have advocated changing current practice and assuming major responsibility for all care of vascular disease patients (33–37). Although a case can be made for these changes, such major patient care responsibilities should be based on all the knowledge and judgment currently used by most vascular surgeons in the care of their patients. This is not presently the case because many nonsurgical, vascular interventional specialists are largely trained in and oriented toward treating lesions and not patients.

A final approach to deal with emerging endovascular technologies such as TPEGs is for *vascular surgeons to become sufficiently competent in catheter-guidewire-imaging-stent techniques* so that they can perform these procedures independently. This creates a number of ancillary problems for many vascular surgeons. Obtaining appropriate training in these basic endovascular techniques is a fundamental requirement. Fortunately, adequate numbers of vascular surgeons possess these skills to make such training available with appropriate organization and an effective administrative structure.

Alternatively, many vascular surgeons have adequate, mutually beneficial relationships with other specialty colleagues so that adequate training could be obtained to perform endovascular procedures as part of healthy collaboration. This does not imply that vascular surgeons should become interventional radiologists or attempt to replace them. It simply means that, as vascular surgeons develop improved endovascular techniques for managing vascular lesions, they must also develop sufficient proficiency with catheter-guidewire-imaging-stent methods. This will enable them to safely and effectively use the TPEGs and other endovascular technologies that may replace standard arterial repairs and bypasses.

[1] Portions of this chapter were based on articles published in the *Journal of Vascular Surgery* (Veith FJ. Presidential Address: Transluminally placed endovascular stented grafts and their impact on vascular surgery. J Vasc Surg 1994;20:855–860) and the *American Journal of Surgery* (Veith FJ, Marin ML, David M. Hume Memorial Lecture: Impact of endovascular technology on the practice of vascular surgery. Am J Surg 1996;172:100–104).

Relationship of Vascular Surgery and Interventional Radiology

How should the relationship evolve between two previously distinct specialties as a new overlapping technology is introduced. Some of the newer endovascular technologies, particularly TPEGs, are cases in point. Both vascular surgeons and interventional radiologists are justified in feeling that their specialties should control these new developments. Accordingly, conflict can and has resulted. Many believe serious turf battles are inevitable. However, these are costly and will result in casualties at both the physician level and, more important, at the patient level. Such battles must be avoided. How to do this remains the challenge.

Both vascular surgeons and interventional radiologists fear displacement by the other group, and this fear is not unfounded. If vascular surgeons develop catheter-guidewire skills, some interventional radiologists fear they will become redundant. On the other hand, if interventional radiologists can insert TPEGs independently and the TPEGs work, a major portion of vascular surgery will be eliminated. Some vascular surgeons therefore worry that they will be displaced.

The crucial step in avoiding the development of turf battles is for both groups to remove the fear factor and reassure each other by word and deed that mutual replacement is not a goal. Interventional radiologists must be reassured that vascular surgeons do not intend to take over diagnostic angiography, balloon angioplasty, and the many other interventional techniques currently within their domain. Vascular surgeons must be assured that they will be able to perform newer endovascular procedures including TPEG insertions as they evolve to replace open, vascular, operative procedures. However, these reassurances will not be enough.

Combined Services

Interventional radiologists and vascular surgeons have to become confederated at the highest levels and be willing to represent each other's interests. In the United States, this requires creation of a combined executive committee of the Society for Vascular Surgery/International Society for Cardiovascular Surgery, North American Chapter, and the Society for Cardiovascular and Interventional Radiology. Similar collaborative arrangements will have to evolve in other countries.

The new committees will have to deal with such complex issues as defining acceptable practice guidelines across the specialties, training each other's members and trainees, and limiting the number of trainees so that excessive numbers of specialists are not produced in both fields.

The long-term goal of collaborative leadership functions would be to produce adequate numbers of optimally trained specialists so that the needs for treating vascular disease could be appropriately met. This would work best if vascular surgeons and interventional radiologists of the future worked in combined services or departments of vascular disease treatment. Such a radical solution cannot easily happen. There will be resistance, and it will take time. However, this solution is right because it will provide optimal patient care and treatment and because it will reduce costs. This combined treatment services approach can do this by avoiding the splintering of care, financially motivated procedures, and choice of procedure based on an individual physician's training or skills or what resources he or she controls. Combined services will also facilitate the more rapid evaluation of different treatments and prevent painful, costly turf battles, which are the worst possible effect that these better endovascular technologies such as TPEGs can exert on vascular surgical and radiologic practices.

FACILITY REQUIREMENTS FOR SAFE USE OF TPEGs

Although some of the TPEG devices may probably be ultimately suitable for safe and reliable percutaneous introduction (9,27), experience indicates that open vascular operations are required or may be required as part of many TPEG insertion procedures or for treatment of complications or adverse events (7,10–26,28–30,38). Accordingly, facilities in which these procedures will be performed must be suitably equipped for open arterial operations as well as for catheter-directed, fluoroscopically guided techniques. Such facilities can take one of two forms.

First, a *standard operating room* or comparable location may be equipped to perform open arterial operations *and* digital, fluoroscopically guided arteriography and catheter-guidewire manipulations at several levels of the arterial tree. A number of expensive ceiling- and floor-mounted digital fluoroscopy/arteriography units are available in addition to movable, radiolucent operating tables that are ideally adapted to the performance of TEPG procedures in an operating room setting. However, it is possible to adequately perform them with a radiolucent, electrically operated, movable operating table and a C-arm portable, digital fluoroscope with last-image hold, hard copy and videotape recording capabilities, and cine playback, digital subtraction, and road mapping capabilities.

For most TPEG insertions, the procedural room should be equipped with 525- or 1024-line television integrated with a high-quality video chain capable of resolving 1.8-line pairs per mm in the fluoroscopy mode. The image intensifier should have a minimum diameter of 9 inches (preferably 12 inches) and should be able to resolve 3-line pairs per mm. Movable collimators are essential, and fluoroscopic equipment with an over-table x-ray tube is not appropriate for these studies. The room should be equipped with a power injector for adequate digital angiography.

Alternatively, for most TPEG procedures a *standard arteriography suite* equipped for digital fluoroscopy and catheter-guidewire technologies can be modified to permit sterile, open operations. For procedures in which devices will be placed in the abdominal or thoracic aorta or iliac arteries, patient safety considerations and the possible need to deal with emergent complications mandate that they be performed in a room that is fully equipped and can be staffed with appropriate nursing and anesthesia personnel. This is so that major vascular operations can be safely and expeditiously performed if required. For thoracic aortic procedures, perfusionists and equipment for instituting total cardiopulmonary bypass should be available.

In the future, different combinations of various imaging modalities other than fluoroscopy may prove equally effective and safe for the deployment of TPEGs. Once proved safe and effective, these other modalities may be substituted for the previously described methods to facilitate TPEG device insertion and evaluation in different treatment environments.

References

1. Dotter CT. Transluminally-placed coilspring endarterial tube grafts: Long-term patency in canine popliteal artery. Invest Radiol 1969;4:329–332.
2. Balko A, Piasecki GS, Shah, DM, et al. Transfemoral placement of intraluminal polyurethane prosthesis for abdominal aortic aneurysm. J Surg Res 1986;40:305–309.
3. Mirich D, Wright KC, Wallace S, et al. Percutaneously placed endovascular grafts for aortic aneurysms: Feasibility study. Radiology 1989;170:1033–1037.
4. Laborde JC, Parodi JC, Clem MF, et al. Intraluminal bypass of abdominal aortic aneurysm: Feasibility study. Radiology 1992;184:185–190.
5. Chuter TAM, Green RM, Ouriel K, et al. Transfemoral endovascular aortic graft placement. J Vasc Surg 1993;18:185–197.
6. Volodos NL, Shekhanin VE, Karpovich IP, et al. Self-fixing synthetic prosthesis for endoprosthetics of the vessels. Vestn Khir Im I I Grek 1986;137:123–125.
7. Parodi JC, Palmaz JC, Barone HD. Transfemoral intraluminal graft implantation for abdominal aortic aneurysms. Ann Vasc Surg 1991;5:491–499.
8. Parodi JC. Endovascular repair of abdominal aortic aneurysms and other arterial lesions. J Vasc Surg 1995;21:549–557.
9. Marin ML, Veith FJ, Panetta TF, et al. Percutaneous transfemoral stented graft repair of a traumatic femoral arteriovenous fistula. J Vasc Surg 1993;18:298–301.
10. Marin ML, Veith FJ, Panetta TF, et al. Transfemoral endoluminal stented graft repair to popliteal artery aneurysm. J Vasc Surg 1994;19:754–757.
11. Marin ML, Veith FJ, Panetta TF, et al. Transfemoral stented graft treatment of occlusive arterial disease for limb salvage: A preliminary report. Circulation 1993;88(4):I-11 (Abstract).
12. Marin ML, Veith FJ, Panetta TF, et al. Transluminally placed endovascular stented graft repair for arterial trauma. J Vasc Surg 1994;20:466–473.
13. Marin ML, Veith FJ, Cynamon J, et al. Transfemoral endovascular stented graft treatment of aortoiliac and femoropopliteal occlusive disease for limb salvage. Am J Surg 1994;168:156–162.
14. Marin ML, Veith FJ. Endoluminal stented graft aorto-bifemoral reconstruction. In: Greenhalgh RM, ed. Vascular and endovascular surgical techniques: An atlas. 3rd ed. London: WB Saunders, 1994.
15. Marin ML, Veith FJ, Cynamon J, et al. Transfemoral endoluminal repair of a penetrating vascular injury. J Vasc Intervent Radiol 1994;5:592–594.
16. Marin ML, Veith FJ, Cynamon J, et al. Initial experience with transluminally placed endovascular grafts for the treatment of complex vascular lesions. Ann Surg 1995;222:449–469.
17. Marin ML, Veith FJ. Clinical application of aortoiliac occlusive disease and vascular trauma. Cardiovasc Surg 1995;3:115–120.
18. Sanchez LA, Marin ML, Veith FJ, et al. Placement of endovascular stented grafts via remote access sites: A new approach to the treatment of failed aortoiliofemoral reconstructions. Ann Vasc Surg 1995;9:1–8.
19. Marin ML, Veith FJ, Sanchez LA, et al. Endovascular aortoiliac grafts in combination with standard infrainguinal arterial bypasses in the management of limb-threatening ischemia: Preliminary report. J Vasc Surg 1995;22:316–325.
20. Marin ML, Veith FJ, Lyon RT, et al. Transfemoral, endovascular repair of iliac artery aneurysms. Am J Surg 1995;170:179–182.
21. Sanchez LA, Marin ML, Veith FJ, et al. Combined use of endovascular and standard operative procedures for the treatment of multiple aortoiliac pseudoaneurysms. Application for emerging surgical techniques. Vasc Surg 1996;30:399–406.
22. Patel AV, Marin ML, Veith FJ, et al. Endovascular graft repair of penetrating subclavian artery injuries. J Endovasc Surg 1996;3:382–388.
23. Ohki, T, Marin ML, Veith FJ, et al. Endovascular aortounifemoral grafts and femorofemoral bypass for bilateral limb-threatening ischemia. J Vasc Surg 1996;24:984–997.
24. Marin ML, Veith FJ, Sanchez LA, et al. Endovascular repair of aortoiliac occlusive disease. World J Surg 1996;20:679–686.
25. Yuan JG, Marin ML, Veith FJ, et al. Endovascular grafts for noninfected aortoiliac anastomotic aneurysms. J Vasc Surg 1997; 26:210–221.
26. Parodi JC. Endovascular repair of abdominal aortic aneurysms. Adv Vasc Surg 1993;1:85–106.
27. Cragg AH, Dake MD. Percutaneous femoropopliteal graft placement. Radiology 1993;187:643–648.
28. May J, White GH, Waugh R, et al. Transluminal placement of a prosthetic graft-stent device for treatment of subclavian aneurysm. J Vasc Surg 1993;18:1056–1059.
29. Moore WS, Vescera CL. Repair of abdominal aortic aneurysm by transfemoral endovascular graft placement. Ann Surg 1994; 220:331–341.
30. May J, White G, Waugh R, et al. Treatment of complex abdominal aortic aneurysms by a combination of endoluminal and extraluminal aortofemoral grafts. J Vasc Surg 1994;19:924–933.
31. Veith FJ, Marin ML. David M. Hume Memorial Lecture: Impact of endovascular technology on the practice of vascular surgery. Am J Surg 1996;172:100–104.
32. Parsons RE, Marin ML, Veith FJ, et al. Fluoroscopically assisted thromboembolectomy: An improved method for treating acute arterial occlusions. Ann Vasc Surg 1996;10:201–210.
33. Kinnison ML, White RI, Auster M, et al. Inpatient admissions for interventional radiology: Philosophy of patient management. Radiology 1985;154:349–351.
34. DeMaria AN. Peripheral vascular disease and the cardiovascular specialist. J Am Coll Cardiol 1988;12:869–870.
35. Katzen BT, van Breda A. Developing an interventional radiology practice. Semin Intervent Radiol 1988;5:99–102.
36. Kerelan RK, Marone T, Ring EJ. The clinical role of the interventional radiologist. Semin Intervent Radiol 1988;5:103–104.
37. Cook JP, Dzau VJ. The time has come for vascular medicine. Ann Intern Med 1990;112:138–139.
38. Veith FJ, Abbott WM, Yao JST, et al. Endovascular Graft Committee. Guidelines for development and use of transluminally placed endovascular prosthetic grafts in the arterial system. J Vasc Surg 1995;21:670–685.

Chapter 2

The Role of Arteriography in Endovascular Grafting Techniques

Jacob Cynamon

Vascular surgeons have always considered angiography an important diagnostic tool. It is the best method for obtaining a vascular map to guide therapy. Patients have benefited from the percutaneous therapies developed by interventional radiologists. Although newer modalities such as computed tomography angiography (CTA), magnetic resonance angiography (MRA), and vascular ultrasonography play a significant role and at times even supersede arteriography, arteriography remains the examination of choice for patients undergoing reconstruction for peripheral vascular disease (1–8).

On the other hand, many vascular surgeons argue that patients with abdominal aortic and some nonaortic aneurysms can be diagnosed and managed based on the findings of one or more of the above noninvasive imaging modalities (9–18). This may be true when treating aneurysms by conventional surgery. However, when stent-grafts are used, additional information must be gathered, which is not obtainable from these other modalities. Therefore, all patients at our institution who are being considered for endoluminal reconstruction undergo preprocedure arteriography.

Angiography can be performed using conventional cut films or digital subtraction techniques (19). Digital subtraction has many advantages, including the fact that it uses less contrast material, multiple views can be rapidly obtained, and images can be observed in rapid sequence so that the direction of flow can be more easily assessed. Finally, original images can be easily reproduced, magnified, and manipulated so that the maximum amount of information is obtained from a single injection.

Because the degree of magnification cannot be determined when digital subtraction techniques are used, a catheter with radiopaque markers must be used to provide accurate measurements.

Patients with elevated creatinine or significant contrast allergies can still undergo diagnostic angiography using the digital subtraction technique with carbon dioxide (CO_2) contrast (20,21). CO_2 angiography requires a carbon dioxide tank and an appropriate three-way valve so that the gas can be aspirated into the injector without the addition of any room air. Although the images obtained are somewhat degraded compared with images obtained in iodinated contrast angiography, they are frequently diagnostic. With digital fluoroscopic equipment, a roadmap image can be made, which would help to complete diagnostic and therapeutic procedures (i.e., stenoses can be crossed, tortuous vessels can be negotiated, and stents and stent-grafts can be accurately positioned).

ARTERIOGRAPHIC ROUTES

Diagnostic arteriography is usually performed by a transfemoral route using the Seldinger technique (22). In most cases, a single-wall, 19-gauge needle is used. Care must be taken to make the puncture below the inguinal ligament. To be certain that an intraluminal passage has occurred, the guidewire must exit the needle and travel freely without resistance. Alternatively, a 21-gauge needle with a 0.018-inch guidewire can be used for access and then converted to a 4F or 5F catheter.

If the femoral approach is not possible because of bilateral occlusions, pseudoaneurysms, or recent operative incisions, an axillary or brachial artery puncture can be performed (23). We prefer to use the brachial approach because it has a lower risk of producing a hematoma, which would result in significant neurovascular compromise.

Finally, a translumbar arteriographic route can also be used; however, it may not be possible to obtain a complete examination, and percutaneous intervention may not be possible.

PREOPERATIVE ASSESSMENT AND INTERVENTION

The indications for performing the arteriogram should guide the study. Evaluation of an abdominal aortic aneurysm for possible stent-graft treatment requires obtaining the same information that is needed for a routine repair as well as additional data, as described in the following text. Coexisting pathology of the celiac axis and superior mesenteric artery should be identified, and any vascular variant should be noted. The number and quality of the renal arteries should be determined. The lumbar and inferior mesenteric arteries should be identified, if patent. The quality of the iliac vessels should be evaluated, inasmuch as tortuous or stenotic iliac vessels may add to the difficulty encountered in performing these diagnostic and therapeutic procedures.

Because accurate measurements must be obtained, it is advisable to use a pigtail catheter that has several radiopaque markers placed 1 cm apart. The diameter and length of the proximal and distal necks can then be determined. The length of the aneurysm, which is not crucial in a routine repair, may be important in stent grafting and can also be measured using a marker catheter (Fig. 2.1) (24–30).

In addition to the diagnostic evaluation, the angiogra-

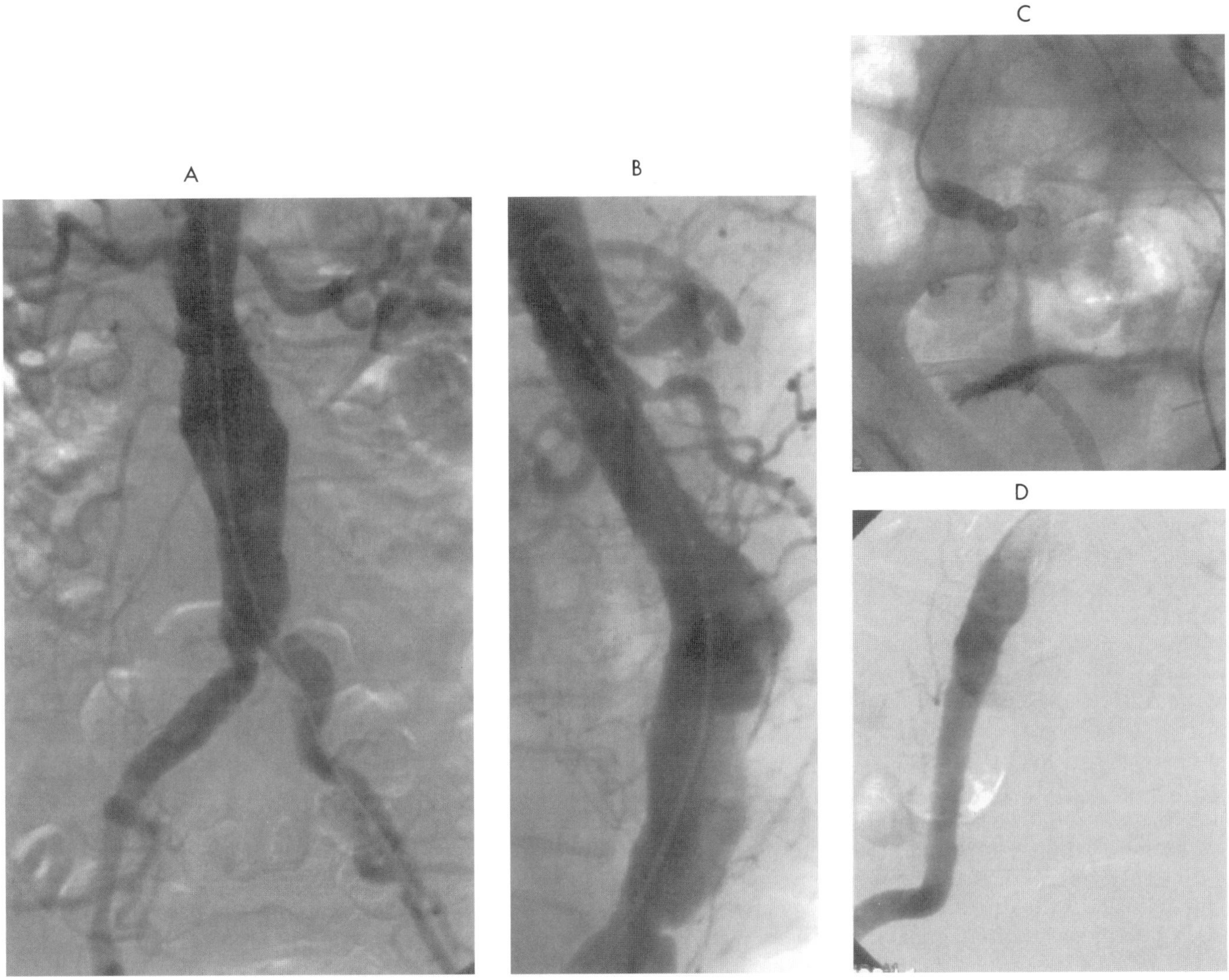

Figure 2.1. Angiogram of a 73-year-old man with an abdominal aortic aneurysm evaluated in the posteroanterior (**A**) and lateral (**B**) projections. A pigtail catheter with radiopaque markers was used. Celiac and left common iliac artery stenoses were identified. In addition, the inferior mesenteric and lumbar arteries were not visualized and were presumed to be occluded at their origins. The proximal infrarenal neck can be measured. **C** and **D**. The right hypogastric artery is embolized via a left common femoral artery approach so that retrograde filling of the iliac artery is no longer possible after an aorto-to-right external iliac artery stent-graft is placed.

A B

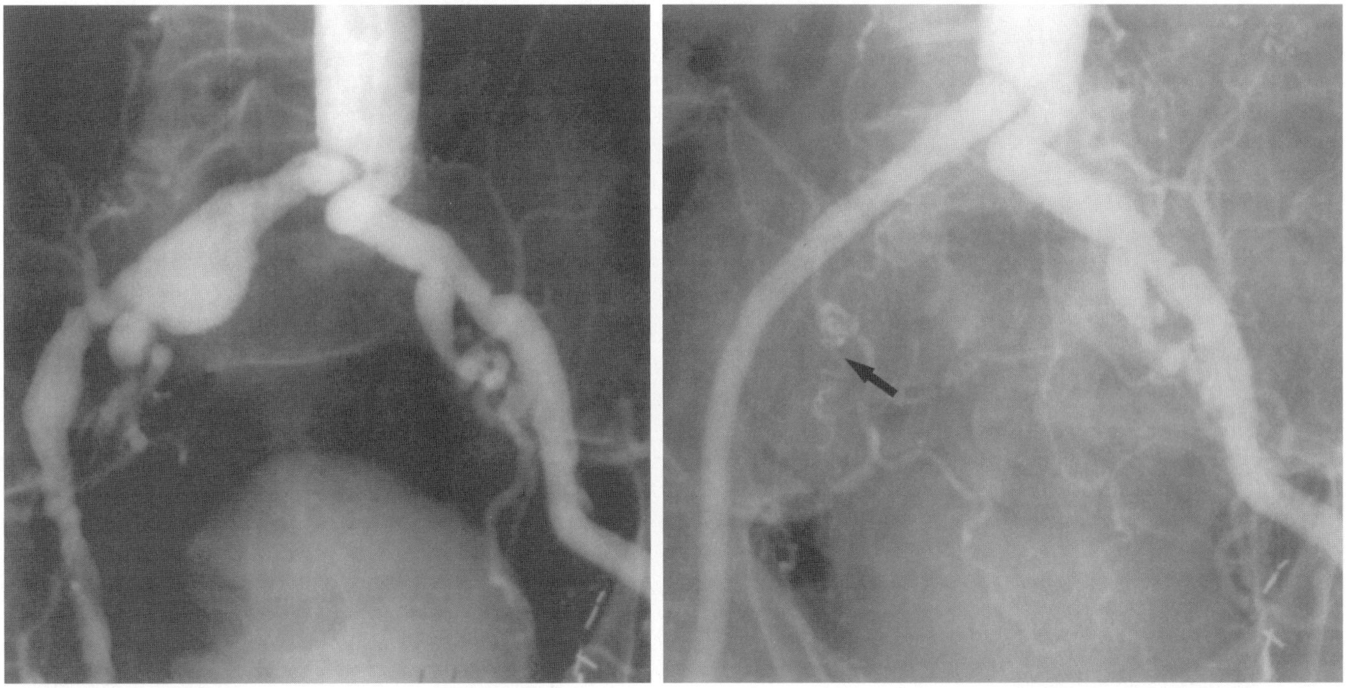

Figure 2.2. **A.** Angiogram of a 62-year-old man with a right common iliac artery aneurysm (**B**) treated with a stent-graft. Before placement of the stent-graft, the hypogastric artery was occluded with coils proximally (*arrow*) to avoid retrograde filling of the aneurysm after the stent-graft is in

pher can treat certain entities before the stent-graft procedure. Aneurysmal disease may coexist with atherosclerotic disease. Access vessel stenoses can be treated at the time of the diagnostic arteriogram, facilitating the introduction of a large delivery device for the stent-graft. Persistent filling of the aneurysm must be prevented when treating aortic aneurysms with an aortofemoral stent-graft in conjunction with a cross-femoral bypass. This is accomplished in two steps.

During the stent-graft procedure and after the cross-femoral bypass is in place, an occlusion device is placed in the contralateral common iliac artery, which obstructs retrograde flow into the aneurysm but maintains the patency of the hypogastric artery. Pelvic collaterals through the hypogastric artery would maintain patency of that artery on the side of the stent-graft and allow retrograde flow into the iliac artery, which could pressurize the aneurysm sac. It would be difficult to manage this type of leak postoperatively. Therefore, preoperative coiling of the hypogastric artery on the side of the stent-graft is essential (see Fig. 2.1C). Hypogastric artery coiling is easily accomplished during the diagnostic arteriogram using the side opposite the access vessel for the arteriogram.

If necessary for anatomic reasons, the ipsilateral hypogastric artery can be embolized. We use Gianturco coils (Cook Inc, Bloomington, IN) for this type of embolization. Common iliac and hypogastric artery aneurysms also require pre–stent-graft embolization,

A B

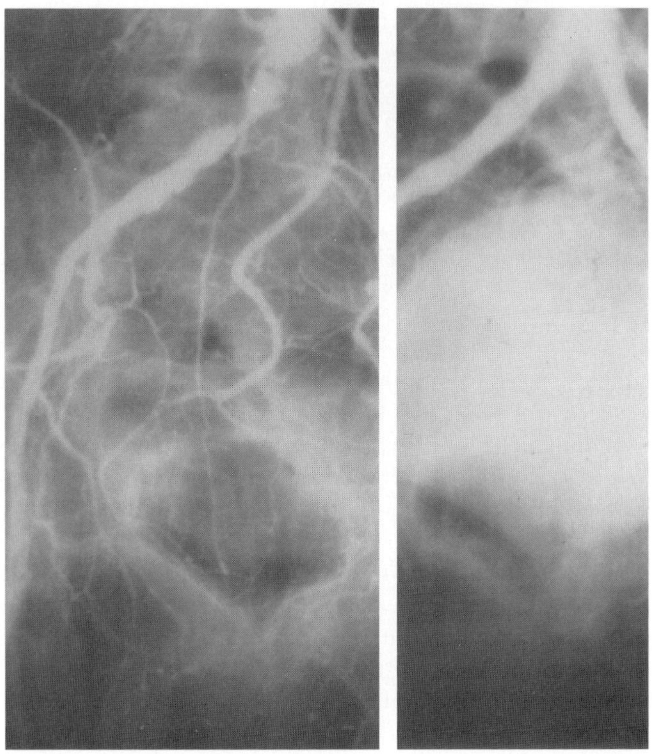

Figure 2.3. Angiogram of a 65-year-old man with gangrene in the left foot and left common and external iliac artery occlusions. **A.** The right common iliac artery has a high-grade stenosis. **B.** A right common iliac stent was placed during the initial diagnostic study to facilitate access for the left-sided stent-graft and to permit a cross-femoral bypass if treatment with a stent-graft is not possible.

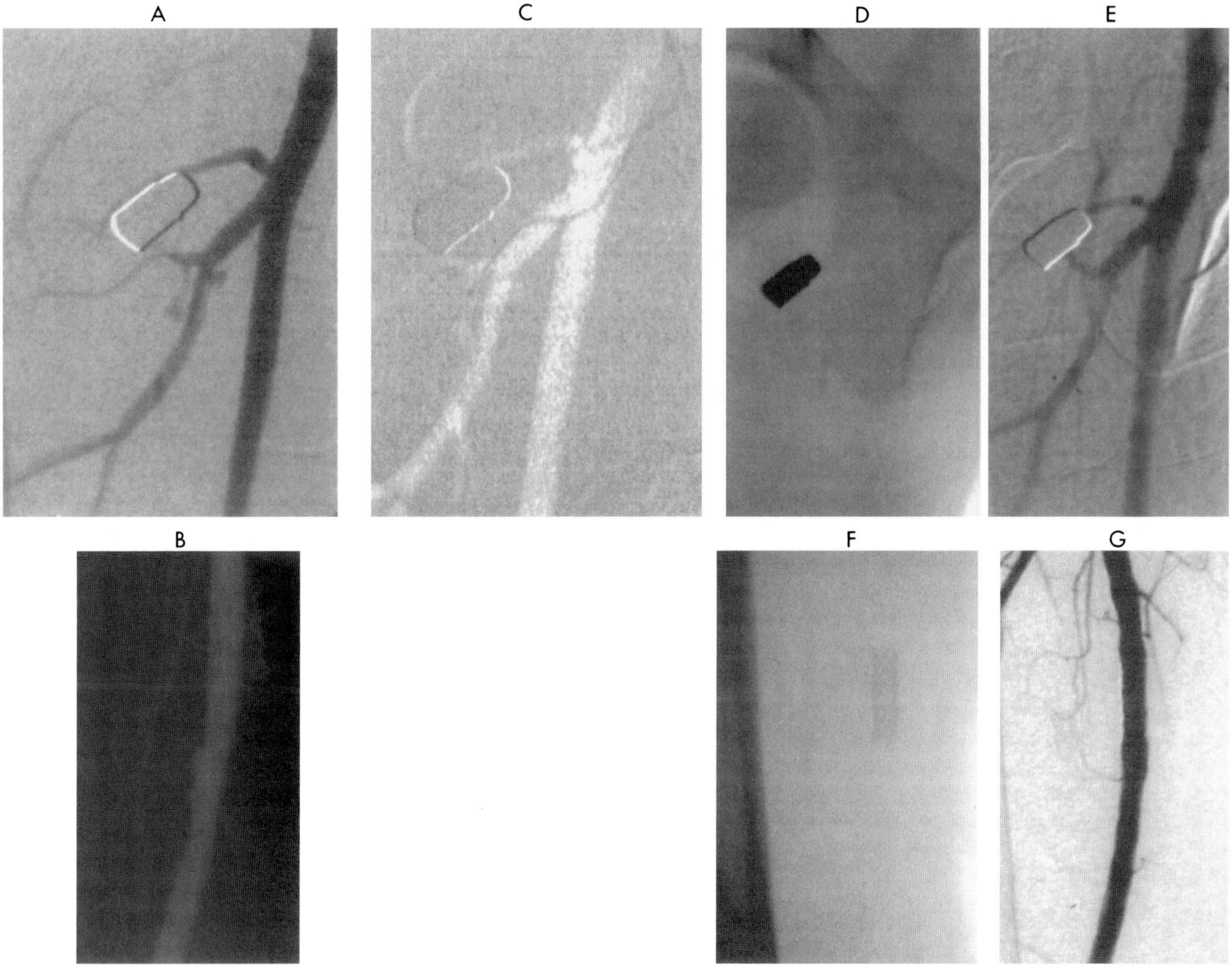

Figure 2.4. Angiogram of a 25-year-old man after a gunshot wound to the right thigh. **A.** Pseudoaneurysms of the profunda and a branch of the profunda are noted. **B.** A significant intimal injury of the superficial femoral artery is also identified. **C** and **D.** The pseudoaneurysm of the profunda branch vessel was treated with a microcoil. **E** through **G.** The injuries of the superficial femoral and profunda femoral arteries were treated with covered stents.

which can be accomplished at the time of the diagnostic arteriogram (Fig. 2.2) (28–30).

When occlusive disease is being treated, the iliac artery in the asymptomatic extremity can be treated, if necessary, to prepare the patient for a cross-femoral bypass if stent grafting is unsuccessful (Fig. 2.3) (30–33). In addition, a thoracic arteriogram with evaluation of the axillary arteries should be performed to determine whether these arteries can be used as donor vessels for an axillofemoral bypass if the stent-graft procedure fails.

When treating vascular injuries, concomitant injuries must be searched for (34–39). If several injuries are present and one or more are amenable to percutaneous embolization, treatment should be carried out before the stent-graft procedure. The injury requiring a stent-graft can then be addressed without concern for the other lesions (Fig. 2.4).

INTRAPROCEDURE ASSESSMENT AND INTERVENTION

If necessary, stent-graft treatment can be carried out in an angiography suite that has been modified to perform operative intervention and open repair (40). An operating room with excellent fluoroscopic capabilities can also be used. An angiography suite is preferred because of its overall improved capabilities for obtaining excellent pre-, intra-, and postprocedure images using multiple projections, increased sensitivity to contrast, and ability to visualize catheters, guidewires, stents, angioplasty balloons, and other equipment. This equipment may not be very radiopaque and may be difficult to see using an operating room fluoroscopy unit. Accuracy is essential when placing stents and stent-grafts.

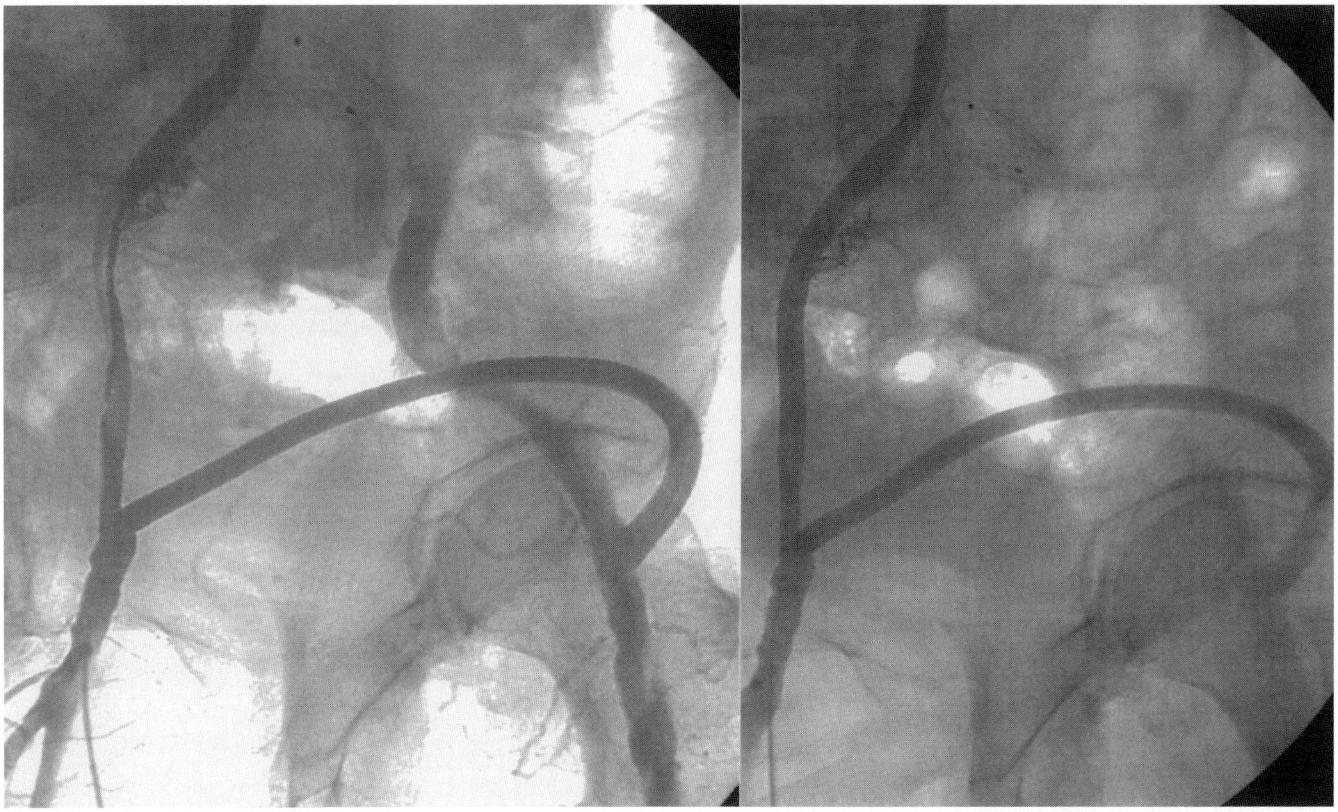

A

B

Figure 2.5. A postprocedure angiogram was performed because femoral pulses were absent after an aortoiliac stent-graft was placed to treat an abdominal aortic aneurysm. **A.** A long flow-limiting defect is seen in the iliac limb. **B.** Status after a percutaneously placed Wallstent. There is no residual obstruction.

Road mapping of vessels can be helpful; however, it can also be misleading because of fluctuation in the position of the vessels with the placement of stiff guidewires and respiratory and cardiac motions. Therefore, a confirmation arteriogram must be performed before deployment to ensure accurate placement. This additional arteriogram can be carried out via a second, percutaneously placed, catheter. Alternatively, a port in the stent-graft delivery system can be made available for contrast injection before the delivery of the stent-graft.

Radiopaque rulers, marker boards, surgical clips, and anatomic landmarks all have been used to aid in the positioning of the stent-graft. After the target for stent-graft deployment has been determined, the table and the fluoroscope unit must not be moved, because parallax can significantly alter the relative location of a surface marker and the true target.

It is necessary to be familiar with rescue techniques, such as snaring foreign bodies, repositioning partially deployed stents, and selective catheterization of small, tortuous arteries, the renal arteries, and the contralateral iliac and hypogastric arteries.

Identifying and Correcting Leaks

A completion angiogram should be performed to search for leaks at the proximal and distal anastomoses, for kinks in the graft, for extrinsic compression, or for any other lesion causing a significant hemodynamic gradient. All surgical anastomoses should be studied. If there is any suspicion of peripheral embolization or thrombosis, the runoff should also be evaluated. The best method for avoiding a proximal leak is accurate placement of the proximal stent.

After a leak is identified, it must be corrected, or the procedure will be deemed a failure. If a leak is due to inaccurate positioning, an additional covered stent can be placed. If a leak is due to inadequate expansion of the proximal stent, a larger balloon can be used to further expand the stent. It is unclear whether placing coils or other embolic agents in the leak to encourage thrombus formation is sufficient to consider the procedure a success.

Coiling the leak can be technically challenging, because it may be difficult to discern where the leak is coming from. Multiple oblique projections and the use of a directional catheter may help to identify the exact location of a leak so that embolization can be performed. Distal leaks can be a significant problem when tube grafts are used to treat abdominal aortic aneurysms with short distal necks, because it may be difficult to anchor the distal graft. In most of our cases, we elected to bring the graft down to the external iliac artery where it could be easily anchored. A distal leak is unusual with this procedure. However, the proximal ipsilateral hypogastric artery must be sufficiently coiled to prevent back bleeding into the iliac artery and the aneurysm.

Graft Defects

Kinks and other defects within the graft can be managed either by balloon angioplasty alone or usually with either a balloon-expandable or self-expanding stent. These additional procedures should be performed before completion of the case.

POSTPROCEDURE ASSESSMENT

CT scanning and noninvasive studies have proved to be excellent tools for postprocedure evaluation. For example, aneurysm exclusion and patency of grafts can be confirmed. Duplex studies can identify failing grafts. If questions arise regarding the integrity of the graft—whether a hemodynamic issue or a question of a leak—an angiogram should be performed. Many of the issues already addressed that can be managed during the procedure can also be managed in a similar fashion after the procedure with percutaneous techniques. Leaks can be embolized, and stenoses or kinks can be dilated or stented (Fig. 2.5).

References

1. Zierler RE. Duplex and color-flow imaging of the lower extremity arterial circulation. Semin Ultrasound CT MR 1990;11:168–179.
2. Cossman DV, Ellison JE, Wagner WM, et al. Comparison of contrast arteriography to arterial mapping with color-flow duplex imaging in the lower extremities. J Vasc Surg 1989;10:522–528.
3. Jager KA, Phillips DJ, Martin RL, et al. Noninvasive mapping of lower limb arterial lesions. Ultrasound Med Biol 1985;11:515–521.
4. Edwards JM, Coldwell DM, Goldman ML, Strandness DE Jr. The role of duplex scanning in the selection of patients for transluminal angioplasty. J Vasc Surg 1991;13:69–74.
5. Kohler TR, Nance DR, Crammer MM. Duplex scanning for diagnosis of aortoiliac and femoropopliteal disease: A prospective study. Circulation 1987;76:1074–1080.
6. Yucel EK, Kaufman JA, Geller SC, Waltman AC. Atherosclerotic occlusive disease of the lower extremity: Prospective evaluation with two-dimensional time-of-flight MR angiography. Radiology 1993;187:637–641.
7. Borrello JA. MR angiography versus conventional x-ray angiography in the lower extremities: Everyone wins. Radiology 1993;187:615–617.
8. Mulligan SA, Matsuda T, Lanzer P, et al. Peripheral arterial occlusive disease: Prospective comparison of MR angiography and color duplex US with conventional angiography. Radiology 1991;178: 695–700.
9. Zeman RK, Fox SH, Silverman PM, et al. Helical (spiral) CT of the abdomen. Am J Radiol 1993;160:719–725.
10. Rubin GD, Walker PJ, Dake MD, et al. Three-dimensional spiral computed tomographic angiography: An alternative imaging modality for the abdominal aorta and its branches. J Vasc Surg 1993;18:656–665.
11. Rubin GD, Dake MD, Napel SA, et al. Three-dimensional spiral CT angiography of the abdomen: Initial clinical experience. Radiology 1993;186:147–152.
12. Pavone P, Di Cesare E, Di Renzi P, et al. Abdominal aortic aneurysm evaluation: Comparison of US, CT, MRI, and angiography. Magn Reson Imaging 1990;8:199–204.
13. Couch NP, O'Mahony J, McIrvine A, et al. The place of abdominal aortography in abdominal aortic aneurysm resection. Arch Surg 1983;118:1029–1034.
14. Godwin JD. Conventional CT of the aorta. J Thorac Imaging 1990;5:18–31.
15. Torres WE, Maurer DE, Steinberg HV, et al. CT of aortic aneurysms: The distinction between mural and thrombus calcification. Am J Radiol 1988;150:1317–1319.
16. Narnke MD, Gould HR, Goldman MH. Computed tomography in the evaluation of the patient with symptomatic abdominal aortic aneurysm. Surgery 1988;103:638–642.
17. Collin J. The value of screening for abdominal aortic aneurysm by ultrasound. In: Greenhalgh RM, Mannick JA, eds. The cause and management of aneurysms. London: WB Saunders, 1990.
18. Ellis M, Powell JT, Greenhalgh RM. Limitations of ultrasonography in surveillance of small abdominal aortic aneurysms. Br J Surg 1991;78:614–616.
19. Malden E, Picus D, Vesely T. Peripheral vascular disease: Evaluation with stepping DSA and conventional screen-film angiography. Radiology 1994;191:149–152.
20. Seeger JM, Self S, Harward TR, et al. Carbon dioxide gas as an arterial contrast agent. Ann Surg 1993;217:688–698.
21. Weaver FA, Pentecost MJ, Yellin AE, et al. Clinical applications of carbon dioxide/digital subtraction arteriography. J Vasc Surg 1991;13:266–273.
22. Lang EK. A survey of the complications of percutaneous retrograde arteriography: Seldinger technique. Radiology 1963;81: 257–260.
23. Grollman JH, Marcus R. Transbrachial arteriography: Technique and complications. Cardiovasc Intervent Radiol 1988;11:32–35.
24. Parodi JC, Palmaz JC, Barone HD. Transfemoral intraluminal graft implantation for abdominal aortic aneurysms. Ann Vasc Surg 1991;5:491–499.
25. Volodos NL, Karpovich IP, Shekhanin VE. A case report of thoracic aortic pseudoaneurysm repair by means of a synthetic self-fixing endoprosthesis. Grudnaya Khir (Russia) 1988;6:84–86.
26. May J, White GH, Yu W, Waugh RC. Endoluminal grafting of abdominal aortic aneurysms: Causes of failure and their prevention. J Endovasc Surg 1994;1:44–52.
27. Parodi JC. Endovascular repair of abdominal aortic aneurysms and other arterial lesions. J Vasc Surg 1995;21:549–557.
28. Marin ML, Veith FJ, Lyon RT, et al. Transfemoral endovascular repair of iliac artery aneurysms. Am J Surg 1995;170:179–182.
29. Cynamon J, Marin ML, Veith FJ, et al. Endovascular repair of an internal iliac artery aneurysm with use of a stented graft and embolization coils. J Vasc Interv Radiol 1995;6:509–512.
30. Marin ML, Veith FJ, Cynamon J, et al. Initial experience with transluminally placed endovascular grafts for the treatment of complex vascular lesions. Ann Surg 1995;222:449–469.
31. Sanchez LA, Marin ML, Veith FJ, et al. Placement of endovascular stented grafts via remote access sites: A new approach to the treatment of failed aortoiliofemoral reconstructions. Ann Vasc Surg 1995;9:1–8.
32. Marin ML, Veith FJ, Sanchez LA, et al. Endovascular aortoiliac grafts in combination with standard infrainguinal arterial bypasses in the management of limb-threatening ischemia: Preliminary report. J Vasc Surg 1995;22:316–322.
33. Marin ML, Veith FJ, Cynamon J, et al: Transfemoral endovascular stented graft treatment of aorto-iliac and femoropopliteal occlusive disease for limb salvage. Am J Surg 1994;168:156–162.
34. Marin ML, Veith FJ, Panetta TF, et al. Transluminally placed endovascular stented graft repair for arterial trauma. J Vasc Surg 1994;20:466–473.

35. Becker GJ, Benenati JF, Zemel G, et al. Percutaneous placement of a balloon-expandable intraluminal graft for life-threatening subclavian arterial hemorrhage. J Vasc Interv Radiol 1991;2:225–229.
36. Marin ML, Veith FJ, Panetta TF. Percutaneous transfemoral insertion of a stented graft to repair a traumatic femoral arteriovenous fistula. J Vasc Surg 1993;18:299–302.
37. Marin ML, Veith FJ, Cynamon J, et al. Transfemoral endoluminal repair of a penetrating vascular injury. J Vasc Interv Radiol 1994; 5:592–594.
38. Ruse SC, Moore EE. Angiography in patients with arterial trauma: Correlation between angiographic abnormalities, operative findings, and clinical outcome. Am J Radiol 1987;149:613–616.
39. Coughlin BF, Paushter DM. Peripheral pseudoaneurysms: Evaluation with duplex US. Radiology 1988;168:339–342.
40. Katzen BT. Endovascular stent grafts: The beginning of the future or the beginning of the end? J Vasc Interv Radiol 1996;7: 469–476.

Chapter 3

The Role of Helical CT-Angiography in Endovascular Grafting Procedures

Alla Rozenblit, Grigory Rozenblit, and Shekker Maddineni

Various vascular abnormalities, including aneurysms, dissections, arteriovenous malformations, occlusive disease, and trauma, are amenable to treatment with endovascular grafting procedures (1,2). However, in recent years most of the experience has been gained in the treatment of abdominal aortic and pelvic aneurysms (3–7).

The technical and clinical success of these complex procedures depends on numerous factors, the most important of which is proper patient selection. Candidates for endovascular treatment should be selected not only on clinical grounds, but also by their individual anatomy, which must be suitable for the chosen procedure. Preoperative anatomy evaluation can be carried out with angiography as well as with various cross-sectional techniques, including computed tomography (CT), magnetic resonance imaging (MRI), and Doppler sonography. These same techniques can be used to assess ultimate procedural success and to detect complications in the postoperative period and during follow-up. An ideal imaging modality must provide maximum anatomic and diagnostic information with minimum invasiveness and cost.

Angiography has been the "gold standard" for evaluation of the vasculature because it is an accurate method of showing the vascular anatomy and regional hemodynamics, which are presented in an easily understandable format. However, angiography shows only intraluminal structures and often cannot provide essential information regarding the vascular wall, thrombus, precise vascular dimensions, and periaortic abnormalities. In addition, angiography is more invasive and expensive than cross-sectional imaging.

The complementary use of angiography and noninvasive imaging methods has been a common practice that results in increased cost, but these tests do not necessarily provide additional information. Recently, angiography has been challenged by less invasive techniques that are capable of producing angiographic-like images using either MRI or CT—known as MR- or CT-angiography. Both methods are effective for the preoperative assessment of abdominal aortic aneurysms (AAAs) and in many cases have the potential for replacing angiography. Since its introduction in 1989, spiral CT with CT-angiography (CTA) has grown to be a leading modality for the evaluation of major vessels. Using this method, vascular aneurysms, dissections, stenoses, and occlusions can be demonstrated. CTA is invaluable for postoperative evaluation and follow-up of patients treated with endovascular grafts (8–10). Unlike CTA, MR-angiography (MRA) cannot be used for imaging the currently used metallic endovascular devices because of a significant susceptibility artifact that interferes with diagnostic information.

PRINCIPLES OF SPIRAL CT

In every CT scanner, the information is collected during the rotation of the x-ray tube in the scanning gantry (Fig. 3.1). There are presently two methods of CT scanning — conventional or incremental and volumetric, also known as spiral or helical CT—which have different principles of x-ray data acquisition and image reconstruction.

Conventional (slice-by-slice) CT is performed by sequential scanning at multiple predetermined levels (Fig. 3.2). The incremental data are obtained when the patient on the CT table is in a stationary position and suspending respiration at each level. This often results in

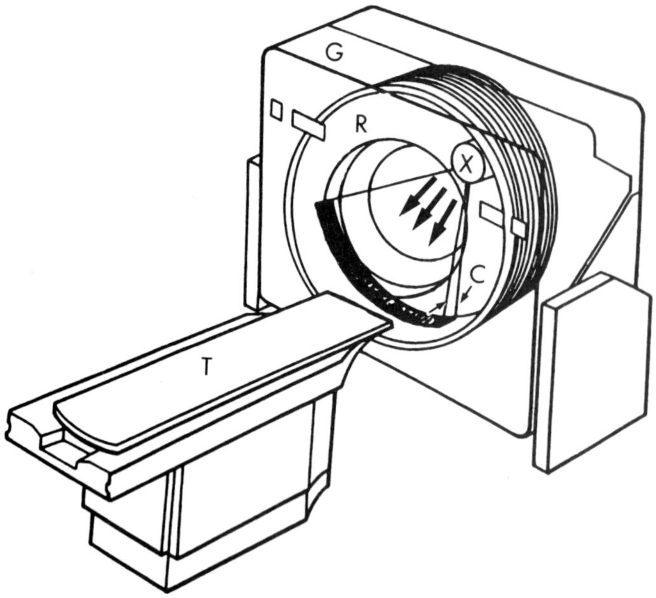

Figure 3.1. Schematic drawing of a CT scanner. An x-ray tube (*X*) is mounted on a ring (*R*), which rotates inside the scanning gantry (*G*). Collimation (*C*) of the x-ray beam (*arrows*) determines the slice thickness. The transition of the patient's table (*T*) through the opening in the gantry is synchronized with the tube motion.

misregistration artifacts that are related to the difference in the position of the diaphragm and other organs during multiple breath-holds. The patient is allowed to breathe between scans while the table moves from one fixed position to the next. Some patients find the breathing instructions difficult to follow during the entire scan, and consequently motion artifacts are common.

Another disadvantage of the incremental technique is that data are not acquired during the table transition from one slice to the next (interscan delay), thereby significantly prolonging the examination time. With the modern scanners, each slice can be obtained in 1 to 3 seconds, and a typical interscan delay varies between 1 and 5 seconds.

Optimal demonstration of the vascular lumen requires opacification with iodinated contrast medium injected intravenously (IV) during the scan. Evaluation of the vasculature is limited with conventional techniques because a long scan time would require a prolonged IV contrast injection for adequate vascular opacification, leading to an unacceptably high dose of iodinated contrast material.

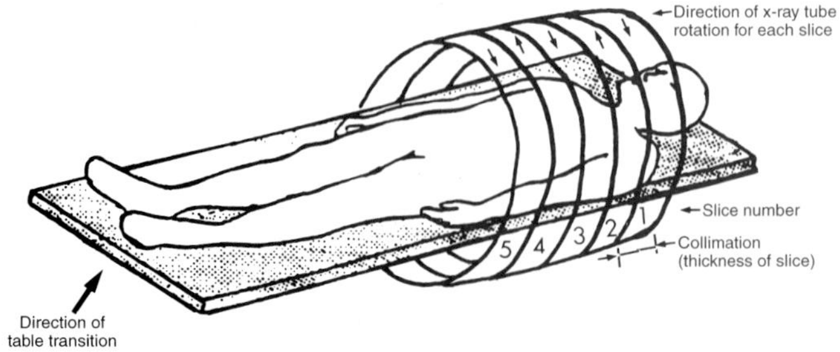

Figure 3.2. Conventional CT. Images are acquired in a slice-by-slice fashion. The table moves in a step-like motion between (but not during) the acquisition of two neighboring slices when the x-ray tube is "off." The trajectory of the x-ray tube for each slice represents a ring. Contiguous neighboring rings form a "stack" of images for the entire scan.

Several technical advances contributed to the progress of the conventional CT, including improvement of the x-ray tube, detectors, generators, and computers. However, the breakthrough came with the introduction of a slip-ring design in the construction of the scanning gantry. The latter innovation coupled with the development of an unconventional mathematical interpolation algorithm for image reconstruction became essential for spiral CT.

In a conventional CT, the x-ray tube has to change the direction of motion after each full rotation (see Fig. 3.2) because the mobile tube is connected to the stationary generator and processor by electrical cables. A pause for change of direction coincides with the table transition, and both contribute to the interscan delay.

A slip-ring gantry is designed to scan without the use of cables. An x-ray tube is mounted on a ring with special slips and brushes that provide electrical contacts. Thus, it may rotate continuously, eliminating the need to pause for change of direction. Yet, a pause is still unavoidable in conventional CT when the acquisition must be interrupted for the patient's transition between slices.

In the volumetric method, the need for a pause is completely eliminated because the x-ray data are continuously collected during uninterrupted, unidirectional rotation of the x-ray tube This is all done while the patient is being transported through the gantry without stopping (11,12). The resulting data are gathered in a spiral-like fashion corresponding to the trajectory of the x-ray tube around the patient (Fig. 3.3). Using the spiral mode, the entire upper abdomen can be scanned within 30 seconds compared with the 2 to 3 minutes needed for the conventional technique. With shortened acquisition, only a brief contrast injection time is needed for optimal vascular opacification. Moreover, an intravenously injected bolus of contrast material can be captured with the peak concentration in the arterial system, thus simulating angiography (11).

With the helical acquisition, the distance covered by the CT table during the full (360-degree) revolution of the x-ray tube is expressed in so-called "pitch." The pitch represents the ratio of this distance to the thickness of the x-ray beam collimation (12). For example, if the table covers a 5-mm distance and the collimation is 5 mm, the pitch is one (see Fig. 3.3*A*); if the table covers 10 mm with the same slice thickness, the pitch is two (see Fig. 3.3*B*). With contemporary scanners the use of a pitch higher than two results in a loss of information.

The length of anatomic coverage is determined by both the collimator width and the pitch. In most current scanners, the full gantry revolution is accomplished in 1 second. Thus, using 10-mm collimation and a pitch of

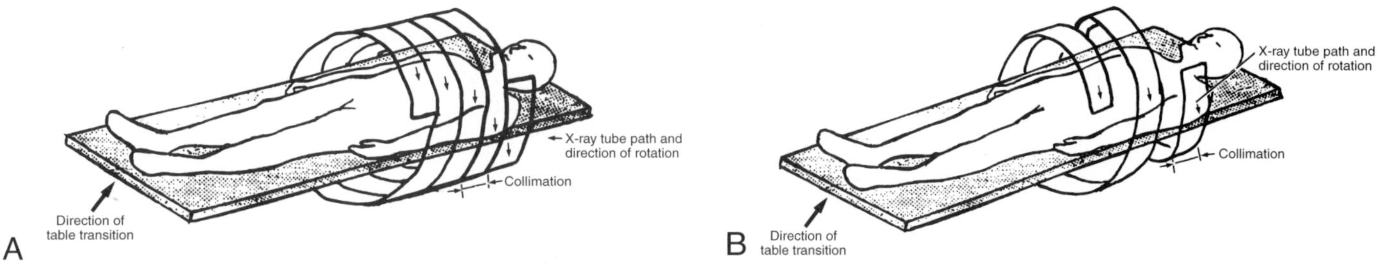

Figure 3.3. Helical CT. Images are acquired volumetrically. The table moves continuously during the acquisition while the x-ray tube is "on." The information is collected between the cranial and caudal limits of the volume of interest. The trajectory of the x-ray tube over the scanned volume represents a helix. Axial images can be reconstructed at any level of the imaging volume using the slice thickness prescribed for the acquisition. **A.** Pitch 1. During one full x-ray tube rotation (360 degrees), the table covers the distance equal to collimation. **B.** Pitch 2. During one full x-ray tube rotation (360 degrees), the table covers the distance equal to two collimations.

one, 30 cm of anatomy can be scanned during a 30-second exposure. With 1-mm slice thickness and a pitch of one, only 3 cm of anatomy can be covered in the same 30 seconds. If a longer anatomic segment needs to be imaged, it can be done with a wider collimation, greater pitch, or both. Either maneuver, however, decreases the spatial resolution (12–16). The narrower the collimation, the better the spatial resolution along the helical path; therefore, smaller details of anatomy can be visualized in the direction of the table transition (*Z* axis).

However, because of the limitations in power of the current x-ray tubes, collimation below 3 mm is only occasionally used in abdominal CT in adults. This low collimation is avoided because thinner slices have unacceptably high image noise due to the limited amount of x-ray available to form an image (17).

Helical CT data do not represent any single slab of tissue but reflect the entire three-dimensional volume of the examined part of the body. Using special mathematical algorithms, the acquired volumetric CT data can be recalculated into axial planar images similar to the images of a conventional CT. Almost all contemporary CT scanners use a 180-degree interpolation algorithm, which allows optimal performance (13).

The slice thickness of reconstructed images cannot be smaller than the collimation of the x-ray beam used during the acquisition. Yet, the images can be reconstructed at any point along the Z axis. To obtain contiguous images without gaps, the intervals between reconstruction levels should not exceed the slice thickness. In addition to contiguous sections, an indefinite number of overlapping axial images can be reconstructed. For example, 5-mm thick slices can be reconstructed at 3-mm intervals, equivalent to a 40% overlap; or reconstruction can be done at 1-mm intervals equal to an 80% overlap. A typical overlap for an abdominal CTA is between 33% and 60% (18,19).

Overlapping reconstruction is unique for helical scanning. It improves visualization of small vascular structures by diminishing partial volume averaging without an additional x-ray exposure. Also, it results in a better quality three-dimensional rendering. Image reconstruction, however, is a lengthy process, which requires 20 to 60 minutes of additional time after the scan is completed and the patient leaves the CT suite (9,15,19).

Reconstructed slices can be viewed and photographed as axial images. The stack of images can be stored in the computer workstation for subsequent rendering with either two- or three-dimensional techniques. The former can be obtained with the use of either multiplanar or curved reformation, and the latter requires the initial generation of a three-dimensional model. Several currently available three-dimensional rendering techniques can then be applied for creating vascular images. These include shaded surface display, maximum-intensity projection, ray-sum projection, and volume rendering techniques.

Shaded surface display depicts the surface of an intravascular contrast column with attenuation values above the chosen threshold (usually between 100 HU [hound-filled units] and 200 HU), whereas the image of low-attenuation tissues is eliminated. This is performed by assigning an attenuation threshold to the three-dimensional, computer-generated model. The resulting image is presented as a reflection of the vascular surface illuminated by an imaginary light source.

The surfaces of the contrast column and overlapping vessels are well depicted with the latter technique (Fig. 3.4*A* and *B*). However, calcified plaques and metallic devices often cannot be differentiated from intraluminal contrast material, small vessels with a relatively low attenuation may be deleted, and stenosis may be exaggerated (9,15,18,20).

Both maximum-intensity projection and ray-sum projection are techniques that display two-dimensional projections of a three-dimensional model. In contrast to shaded surface display, both can differentiate calcium and metal from intraluminal contrast material. Images are generated in a selected plane by mathematically projected multiple rays through the volumetric model. The resulting two-dimensional image is similar to an angiogram, but, unlike angiography, it can be displayed in any projection.

Maximum-intensity projection is designed to depict only the highest attenuation structures within a three-dimensional model along the chosen projection; therefore, the image is formed only by maximum-attenuation

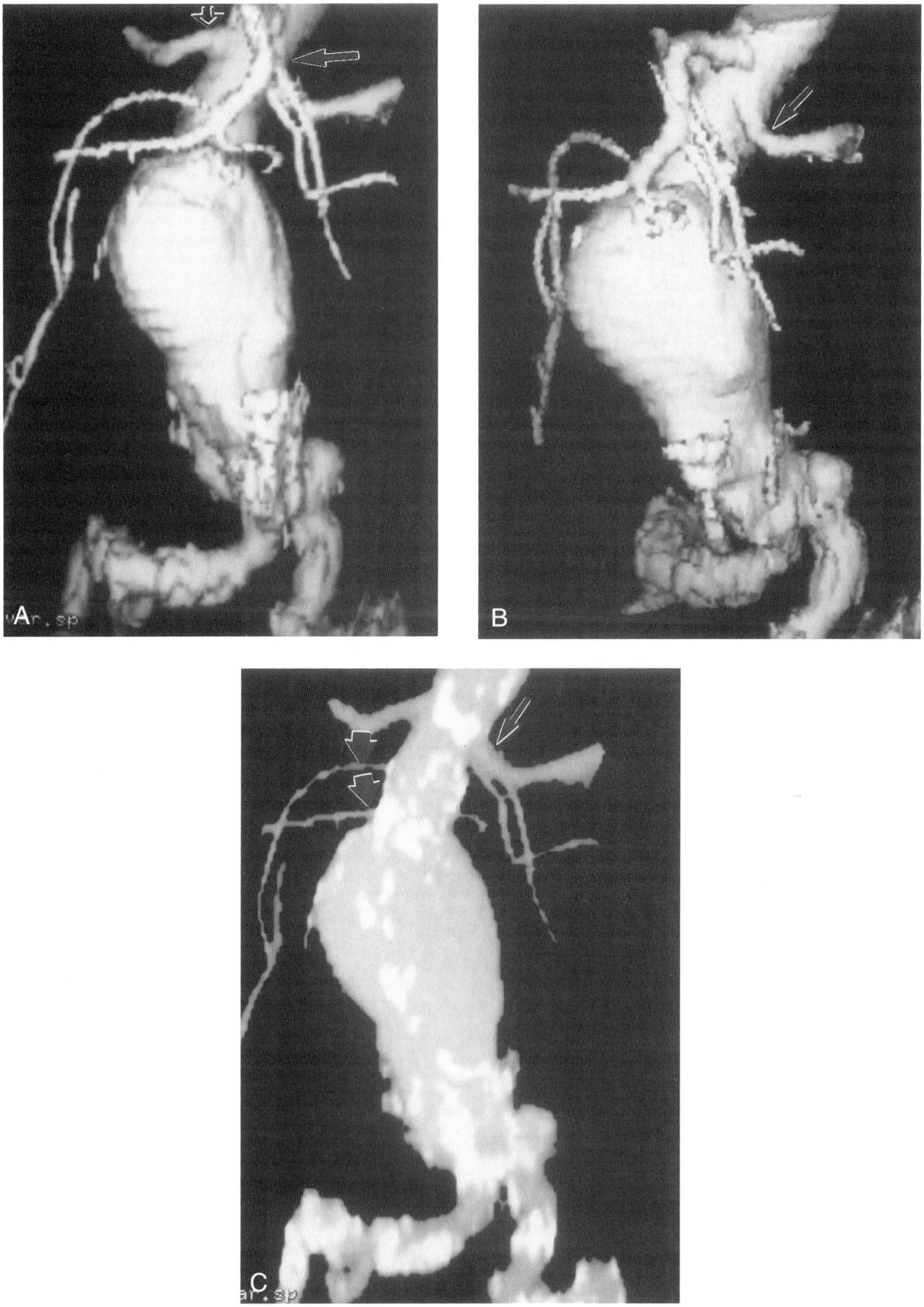

Figure 3.4. CTA with three-dimensional rendering. **A.** Shaded surface display. The surface of the contrast column, including the right renal (*arrowhead*) and superior mesenteric arteries (SMA), is well seen. SMA branches overlap and obscure the origin of the left renal artery (*arrow*). Note that calcification cannot be distinguished from intravascular contrast. **B.** Shaded surface display. An oblique view of the three-dimensional model is necessary to show the origin of the left renal artery (arrow). **C.** Maximum-intensity projection in the same view as in **A** shows calcification throughout the aorta and iliac arteries. Overlapping branches of the SMA (*arrow*) cannot be distinguished from and may be mistaken for the left renal artery. Other SMA branches (*arrowheads*) may be misinterpreted as accessory renal arteries.

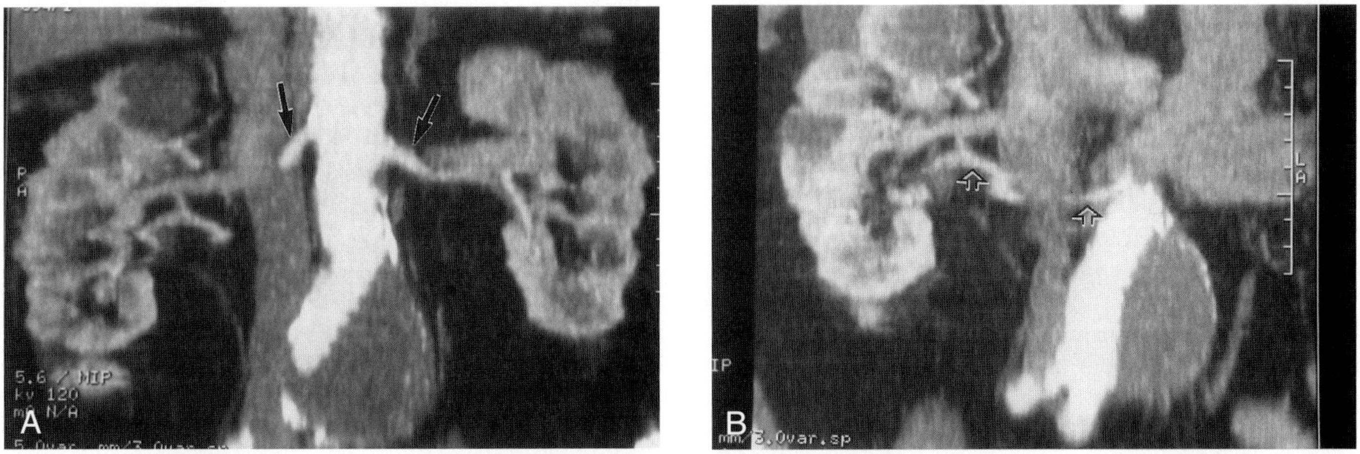

Figure 3.5. CTA with two-dimensional multiplanar reformation. **A.** Planar reformation in the coronal plane shows a juxtarenal AAA. Mural calcification, the proximal neck, luminal thrombus, and the proximal renal arteries (*arrows*) are well depicted. **B.** Oblique view shows the distal extent of the aneurysm. In addition, an accessory renal artery originating just above the aneurysm is visualized (*arrowheads*).

pixels among all other pixels (8,15). Maximum-intensity projection images are more accurate than shaded surface display for the assessment of vascular stenosis (19–21). A drawback of this technique is that frequently superimposed vessels, such as the origin of the aortic branches, cannot be distinguished (see Fig. 3.4C). To alleviate this problem, multiple maximum-intensity projection images are usually displayed in a cine loop, creating an illusion of a three-dimensional depth.

In ray-sum projection, the image is generated by the sum of several different attenuation values (rather than only the maximum value used in maximum-intensity projection) along a projectional ray through the three-dimensional model. The resultant angiographic-like image is translucent with the potential to distinguish multiple overlapping vessels (16,18). At present, a high degree of image transparency is difficult to display without limiting the image sharpness and overall resolution.

Image editing is recommended before generating three-dimensional images, particularly maximum-intensity projection (9,15,18,19). This involves the removal of bones from the three-dimensional model by either manual drawing or semiautomated methods.

Volume rendering for CTA using modern, fast computers has recently been reported with excellent results (22). With this technique, the entire volume of available data is included in the three-dimensional model. Consequently, the full range of densities within the model can be used for the display (unlike other rendering techniques in which only a fraction of the data is displayed).

The resulting images after image editing can depict all levels of intravascular and perivascular attenuation, optimally showing both the vascular surface and the lumen, including thrombi and dissections as well as calcification in the vascular wall. The volume rendering system is not readily available and not yet refined and still requires considerable editing time, whereas the spatial resolution is somewhat degraded.

Two-dimensional rendering includes generation of multiplanar views through a stack of axial images reconstructed from the original raw data. The plane of reformation may have any direction and may be flat (Fig. 3.5) or curved (Fig. 3.6). It is determined by selecting a straight or curved line through the axial, coronal, or sagittal plane. The resultant planar reformatted image is perpendicular to the chosen plane.

The overall accuracy of CTA depends on the scanning parameters (collimation and pitch), reconstruction interval, the degree and uniformity of vascular enhancement, and the available software for generation of three-dimensional images.

Advantages

The major advantage of helical scanning is the ability to rapidly acquire information. For CTA, the speed of the acquisition is very important for two reasons. First, it permits the optimal timing for imaging of the vasculature.

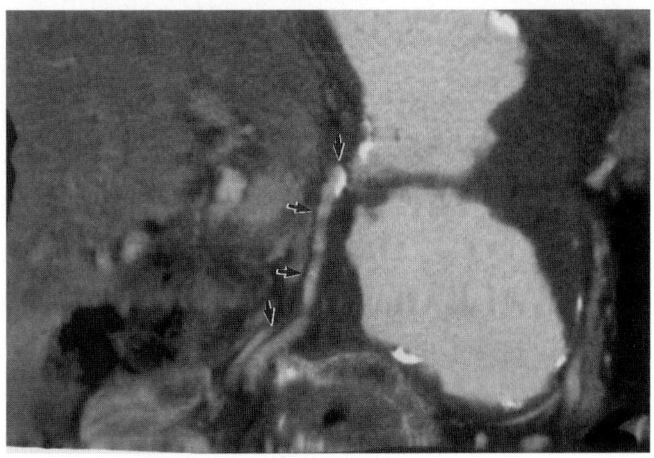

Figure 3.6. CTA with two-dimensional curved reformation. This image was reformatted along the curved course of the right renal artery (*arrowheads*) so that the entire artery is visible in a patient with a suprarenal AAA. The artery is narrowed and originates from the aneurysm.

Therefore, images obtained with a peripheral IV injection are, in many respects, similar to those produced in angiography. Second, the entire volume of interest could be examined in a single breath-hold, thus eliminating misregistration artifact and resulting in motion-free, high-quality images. Also, compared with the slice-by-slice method, a smaller volume of intravascular contrast material can be used in some cases, leading to improved safety and decreased cost (23). Because of its speed, helical acquisition can be repeated in the venous phase, providing additional information regarding body anatomy, pathology, late distribution of IV contrast, and renal function without added contrast material.

Overlapping reconstruction allows better visualization of small inconspicuous structures and more accurate characterization of otherwise indeterminate lesions. Volumetric acquisition is also intrinsically better suited for three-dimensional and multiplanar reformation.

Compared with angiography, CTA is less invasive and costly. It has higher contrast but lower spatial resolution than angiography. The latter is compensated for by the ability to image the vasculature in innumerable three-dimensional and planar views. Several recent publications have favorably compared spiral CT with conventional angiography (19,24,25). Technically adequate spiral CT with CTA should obviate the need for angiography before routine aortic surgery for aneurysms or dissections, with angiography used only as a problem-solving modality in a small number of cases (8,26). CTA may prove to have a similar role in the planning of endovascular treatment of AAAs. Several studies have indicated the usefulness of CTA for renal and mesenteric vascular assessment, both before and after endovascular intervention (20,26). These applications, however, will require further prospective analysis with angiographic and surgical correlation.

Disadvantages

One of the problematic areas of CTA is the limited coverage of the craniocaudal anatomy. It is hindered by both the short period of optimal vascular contrast enhancement after IV injection and the insufficient maximum allowable heat capacity of the x-ray tube, restricting the exposure time. An increase in anatomic coverage can be achieved with thicker collimation and higher pitch, which result, however, in decreased spatial resolution (9,12–16).

The spatial resolution of spiral CT with pitch one is comparable to conventional CT (27). With 3-mm collimation it is approximately 0.5 mm in the axial plane, or 1 line pair (lp) per mm. However, the resolution along the Z axis may be only 2 to 4 mm (depending on pitch), equivalent to 0.16 to 0.13 lp/mm. This is much lower than the spatial resolution of screen-film systems used in conventional angiography and is capable of resolving 2 to 7 lp/mm. Therefore, small axially oriented vessels and vascular stenosis may be difficult to detect on reformatted images of spiral CT. Specifically, renal artery stenosis, thin intraluminal webs of fibromuscular dysplasia, and small accessory renal arteries may be inaccurately demonstrated (26,28).

Image noise, a serious problem in spiral CT, is related to the limited output of x-ray tubes that results in photon-deficient (noisy) images (13–17). Therefore, patients with large body habitus (weighing more than 250 pounds) cannot be adequately imaged with 3-mm and even 5-mm collimation using the current equipment (Fig. 3.7).

Image quality is also degraded in patients with large metallic devices, such as cardiac defibrillators or pacers, spinal fixation devices, and hip prostheses, within the scanning volume. Image quality is not significantly affected with the presence of small metallic surgical clips or staples (26).

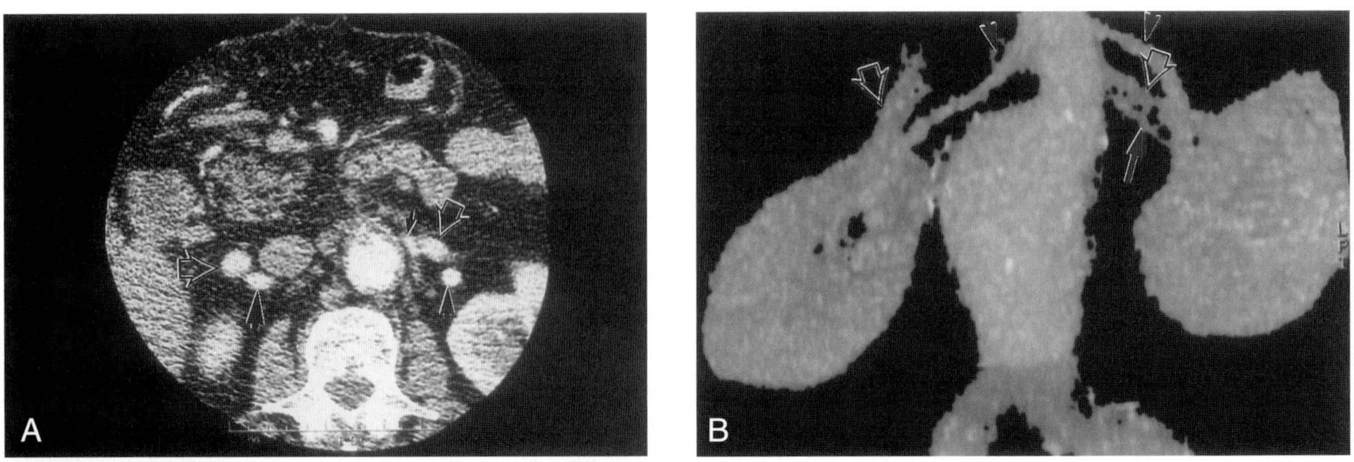

Figure 3.7. CTA image noise in a patient with a large body habitus. Both axial (**A**) and three-dimensional images with maximum-intensity projection rendering (**B**) are photon-deficient, or "noisy," owing to the inability of the x-ray tube to deliver an adequate technique for patients weighing over 250 pounds. A surgically confirmed left accessory renal artery (*arrow*) just above the aneurysm is barely visible. The main renal arteries (*arrowheads*) and both renal veins (*open arrows*) are reasonably well seen.

In addition, partial volume artifact, inherent in CT scanning, affects spiral CT as well. Because of partial volume averaging, CTA tends to underestimate the residual vascular lumen in a stenotic vessel. Therefore, accurate measurements of a vascular lumen and grading of an arterial stenosis cannot be obtained with CTA using the current software (26).

Failure to detect small vessels is also attributed to partial volume effect as well as to lower intravascular iodine concentration, achieved with a peripheral IV injection compared with the intra-arterial angiographic technique.

Although spiral CT can be completed quickly, image reconstruction takes additional and considerable machine time. Image rendering is often both time-consuming and tedious, requiring valuable physician time (9,15,19). With any rendering technique, a truly volumetric analysis of helical CT is still not possible owing to the limitations of the currently available computers.

Spiral CTA requires IV contrast medium with its potential problems, including hypersensitivity and immune-mediated reactions, nephrotoxicity, local extravasation, and cardiac failure. High osmolar contrast materials are very inexpensive and have a time-proven safety record. Yet, they almost invariably induce nausea during and immediately after IV administration. This inevitably leads to the patient's motions deteriorating the quality and diagnostic accuracy of the scan.

In addition, inadvertent extravasation of high osmolar contrast causes severe pain and occasionally local tissue loss (skin sloughing). Low osmolar and especially the newest iso-osmolar contrast media are costly but are probably safer to use in patients with cardiac disease. They are much better tolerated by patients and practically never induce nausea, thereby resulting in overall superior quality images. In addition, if incidentally injected into the soft tissues, they cause only minor local discomfort. Consequently, all our patients currently receive only low- or iso-osmolar IV contrast.

Patients with an increased risk of immune-mediated reactions to IV contrast are premedicated with 32 mg of methylprednisolone (sometimes in combination with 25 mg of hydroxyzine [Atarax] and 300 mg of cimetidine [Tagamet]) given orally in at least two doses, 12 hours and 2 hours before the CT scan (29). Patients with a history of anaphylactic reactions undergo a noncontrast CT and, if necessary, an alternative imaging modality (MRI or sonography). The risk of nephrotoxicity is generally low and is usually more of a concern in patients with preexisting renal insufficiency (30). Preprocedural patient hydration is aimed at increasing urine output and is generally believed to reduce the risk.

PREOPERATIVE ASSESSMENT OF PATIENTS WITH ABDOMINAL AORTIC ANEURYSMS

The following are goals for preoperative assessment of abdominal aortic aneurysms:

1. To define the anatomy of the aneurysm in regard to the size, extent, and relationship with the aortic branch vessels as well as calcification of the wall and configuration of the thrombus
2. To identify and estimate the dimensions of the proximal and distal aneurysmal necks, when present
3. To detect pertinent vascular variations, such as accessory renal arteries and circumaortic and retroaortic renal veins
4. To diagnose associated vascular and extravascular abnormalities that may alter the management of the patient (18,19,31–33).

The size of an aneurysm can be accurately measured on transaxial sections without IV contrast. Usually, two orthogonal dimensions of the aneurysm are obtained. In patients having a follow-up CT for assessment of the stability of the aneurysm, all measurements must be taken at identical levels to accurately compare any interval change in size (5).

On conventional CT, the aneurysmal size can be overestimated because the aneurysm is often erroneously measured in the oblique plane relative to the course of the aorta (10). Thus, an anteroposterior diameter can be overestimated when the aorta deviates anteriorly from the spine. A similar overestimation of the transverse diameter occurs with tortuosity of the aorta in the coronal plane (Fig. 3.8). This common problem is easily overcome by using spiral CT and obtaining multiplanar reformation. The course of the aorta and tortuosity become apparent, and appropriate measurements can be obtained in the planes perpendicular to the vascular course.

The cross-section of the proximal and distal aneurysmal necks can also be accurately measured so that suitable candidates are selected for endovascular treatment. Different requirements regarding the maximal diameter of the neck may be applicable for various types of endovascular devices (5–7). The length of the neck can be approximately measured either along its surface on shaded surface display or in the center of the lumen using curved planar reformation. When necessary, similar measurements of the distal aneurysmal neck can be obtained. In many cases, accurate measurement of the length of the aneurysmal neck is still problematic with the current software. This is expected to improve in the near future. A longer aneurysmal neck is required when a covered stent is used, whereas an uncovered stent may extend to the renal artery level without any vascular compromise (7).

The angulation between the aneurysmal neck and the axis of the aneurysm is apparent, and it can be measured on both three-dimensional and multiplanar reformation images.

On conventional CT the longitudinal dimension of a vascular segment cannot be accurately assessed other

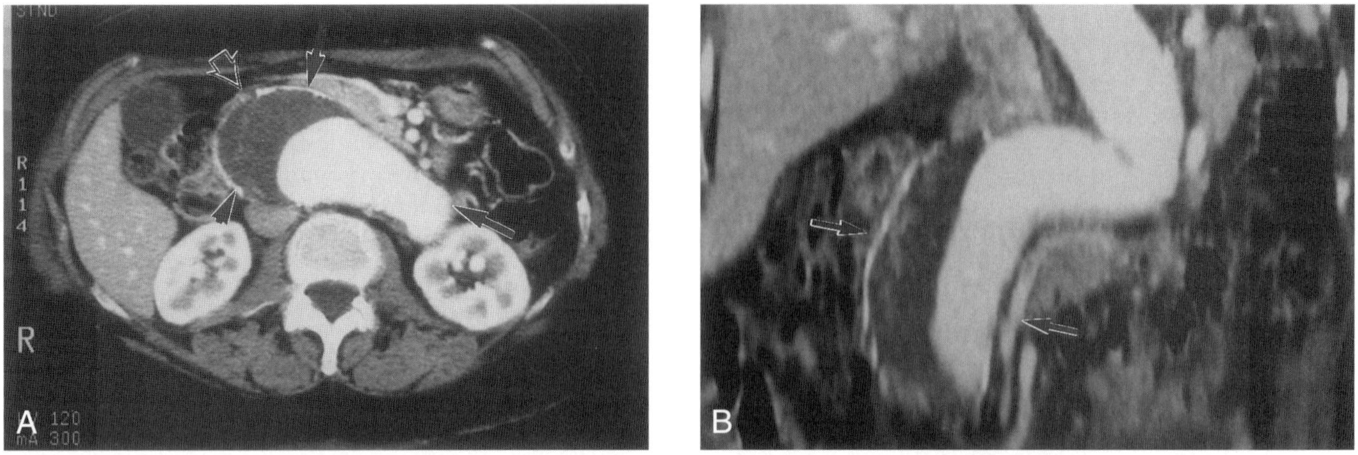

Figure 3.8. Aneurysm size estimation by CT. **A.** On axial CT this 6-cm AAA (*arrowheads*) may be overestimated if it is erroneously measured between the proximal (*arrow*) and the distal (*open arrow*) portions of the aorta, both imaged on the same slice. **B.** Both the tortuosity of the aorta and the aneurysm's true size (*arrows*) are apparent on oblique multiplanar reformation. The plane between the arrows corresponds to the axial image in **A.**

than by calculating the number of slices. With the spiral technique, all cross-sectional and most longitudinal measurements can be accurately obtained, even on an unenhanced CT when it is supplemented with multiplanar reformation. This is especially beneficial in patients who have contraindications to IV contrast.

Spiral or conventional noncontrast CT is also helpful in the demonstration of mural calcification. All other details of the vascular anatomy, specifically visualization of the renal arteries and other branch vessels, require IV contrast.

Irregularity of an aneurysmal thrombus or the presence of ulcerated plaques along the femoroiliac access may be a significant risk factor for the embolic complications described with the use of endovascular grafting procedures. Although these features have not been prospectively studied, their presence may influence the treatment decision in favor of an open repair.

Endovascular grafting may become a preferred treatment in patients with inflammatory aneurysms. The diagnosis can be suggested by a CT that demonstrates a homogeneous soft tissue rind around the aneurysm, with typical anterior and lateral distribution. Perianeurysmal inflammation or fibrosis may show strong enhancement on delayed images (32). It should be differentiated from a retroperitoneal hemorrhage, mycotic aneurysm, and periaortic lymphadenopathy. Inflammatory aneurysms may present with very subtle CT changes, and the sensitivity of CT for the detection of this abnormality varies between 25% and 47% (33,34).

The cranial extent of an AAA can be identified on conventional CT with an accuracy between 76% and 100% (32,33,35,36), with a lower percentage reported in more recent and larger series. Incremental CT tends to overestimate the extent of the aneurysm even with thin sections. Volume averaging and a tortuous nature of a juxtarenal aneurysm may produce an appearance falsely suggesting involvement of the main renal arteries, whereas these vessels are actually cranial to the aneurysm (Fig. 3.9). Vowden and associates (36) reported that only one of seven patients studied with thin-section conventional CT with suspected involvement of the renal arteries had an actual suprarenal aneurysm found at surgery, and the remaining cases were false-positive.

Spiral CT is capable of depicting the cranial extent of an aneurysm (see Fig. 3.9) with an accuracy between 82.6% (37) and 97.4% (38). Van Hoe and colleagues (38) showed that helical CT was more accurate than digital subtraction angiography for diagnosing supra- and juxtarenal AAAs. In our series of 33 patients, CTA was as accurate as angiography for the evaluation of both the proximal and the distal extent of the aneurysm (Fig. 3.10) (25). The accuracy of conventional CT in predicting the distal extent of an AAA ranges from 76% to 100% (32,35,39,40). False-positive results may occur owing to volume averaging of a dilated aorta and normal-caliber iliac arteries.

Although the origins of the renal arteries are often difficult to visualize on conventional CT (41), they are usually apparent on a helical scan (Fig. 3.11) (9,18–20). Rubin and associates (20) reported a 92% sensitivity and an 83% specificity of CTA in the detection of a renal artery stenosis of at least 70%. Van Hoe and associates (38) demonstrated a 95% accuracy in the assessment of renal artery stenosis using 2-mm collimation spiral CT.

Conventional CT almost never demonstrates the accessory renal vessels that are present in 10% to 20% of patients with AAAs (32,35). These vessels may originate from the aneurysm and therefore may require reimplantation. With CTA, accessory renal arteries can be identified with an accuracy of 75% to 100% (see Fig. 3.5) (19,26,37,38).

The origins of the celiac and superior mesenteric arteries are usually visualized on conventional CT; how-

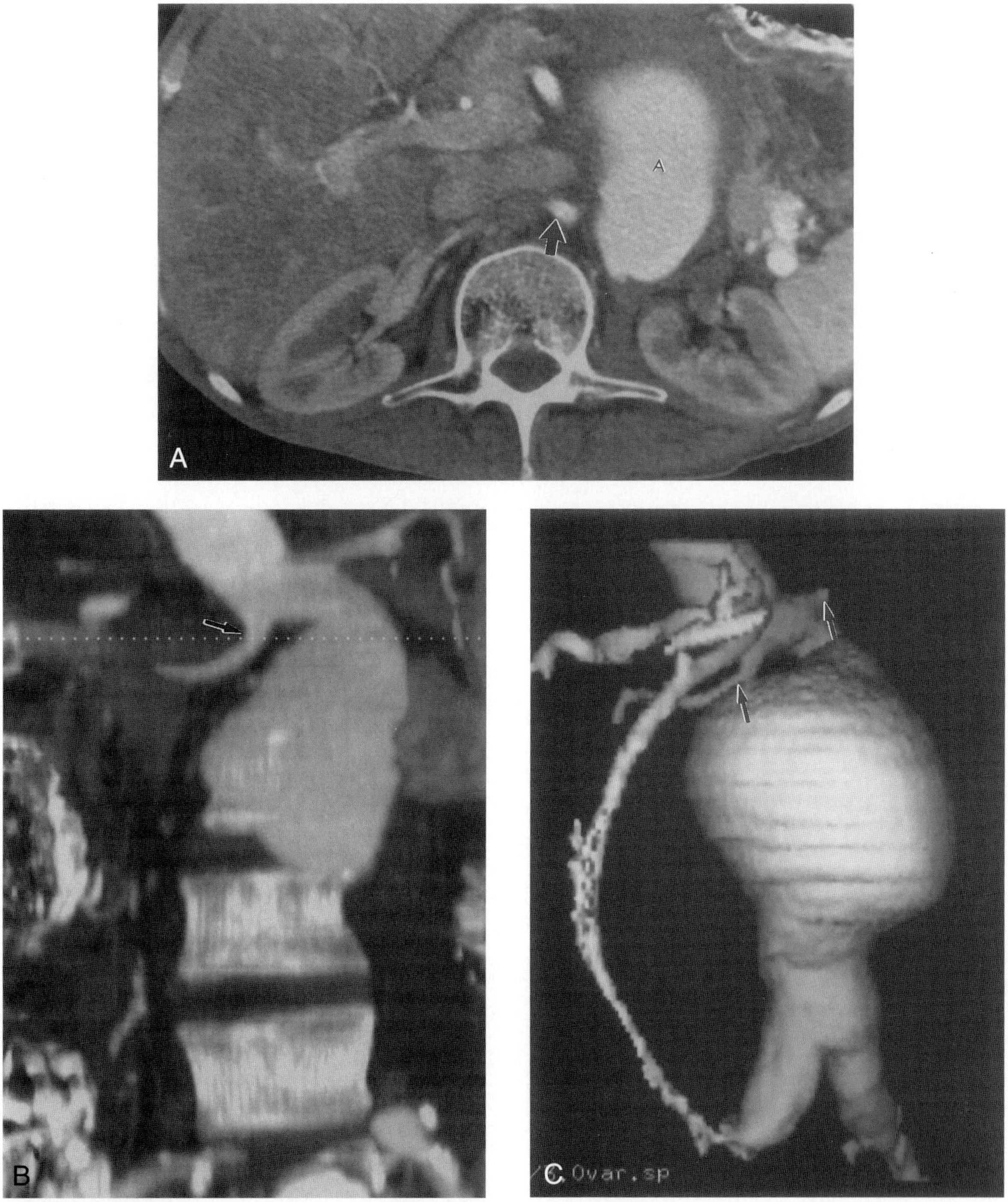

Figure 3.9. Demonstration of the proximal extent of an AAA. **A.** Axial CT at the proximal right renal artery (*arrow*) displays the aneurysm (*A*) at the same level, suggesting a suprarenal extension. **B.** Multiplanar reformation clearly shows that the AAA is infrarenal, with a tortuous proximal neck that is about 2 cm long. The dotted line indicates the level of the axial image in **A. C.** Three-dimensional reformat with shaded surface display rendering confirms the location of the renal arteries (arrows) cranial to the AAA. The distal extent of the aneurysm to the aortic bifurcation as well as involvement of both common iliac arteries is also apparent.

ever, arterial patency cannot be accurately judged by this modality. A proximal vascular obstruction involving the superior mesenteric and celiac arteries can be demonstrated on CTA with an accuracy of 75% to 100%. An inferior mesenteric artery obstruction can also be detected, but with a lower success rate owing to the small caliber of this vessel (25).

Large lumbar arteries can be well seen on CTA; small ones are often difficult to assess (25). Some investigators have suggested that the patency of the inferior mesenteric and lumbar arteries represents a contraindication to endovascular therapy, because patent branch vessels may form the backflow pathway around the graft (3,42). As of this writing, CTA cannot provide sufficient diag-

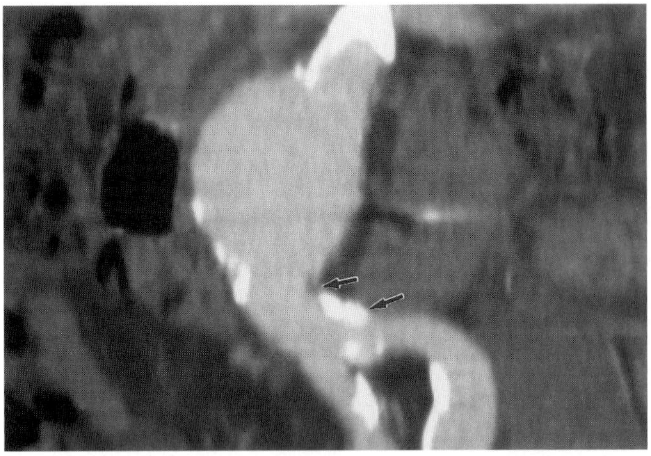

Figure 3.10. Distal extent of an AAA. Multiplanar reformation shows a 2-cm distal neck (*arrows*) between the aneurysm and the aortic bifurcation. The left common iliac artery is tortuous and dilated distally.

nostic information regarding all small aortic branches, and angiography may still be required to answer some specific questions.

Although venous anatomic variants such as circumaortic and retroaortic renal veins can be easily visualized with either conventional or spiral CT, they may be difficult to diagnose by angiography. A knowledge of the venous anatomy, however, is more important before an open rather than an endovascular aneurysm repair.

Unsuspected vascular and extravascular abnormalities are not uncommon in patients with aneurysms. The former includes additional visceral and pelvic aneurysms, aortoiliac occlusive disease, and occasional unsuspected venous occlusion involving the inferior vena cava and pelvic veins. Venous pathology can be reliably detected only on delayed CT images, thus adding to the time and cost of the examination.

The size and location of additional aneurysms may alter the surgical planning. Similarly, identification of a horseshoe kidney often dictates open rather than endovascular treatment, usually because of the caudal origin of multiple renal vessels.

Other concomitant conditions may be discovered during CTA. These include unsuspected primary and metastatic neoplasms. Silent primary tumors involving the kidney, pancreas, liver, and adrenal glands, lymphadenopathy, as well as visceral, bone, and lung metastases all are likely to be detected using a combination of spiral CTA and delayed conventional images through the abdomen. Newly diagnosed serious comorbidities often modify the surgical decision.

CT TECHNIQUES

Conventional

Conventional or incremental abdominal CT is typically obtained with contiguous 7-mm, 8-mm, or 10-mm thick slices from the diaphragm to the iliac bifurcation. Thin

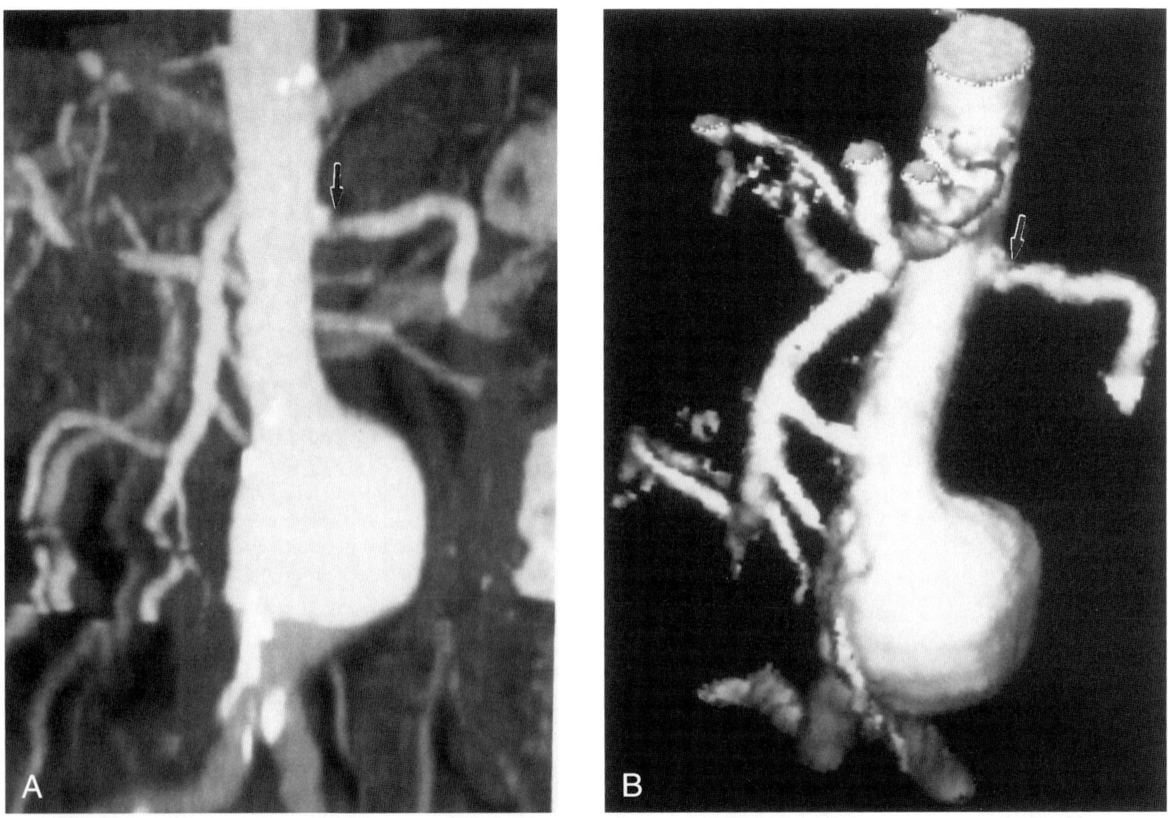

Figure 3.11. Renal artery stenosis imaged with three-dimensional rendering. A left renal artery stenosis (*arrow*) in a patient with an infrarenal AAA is better demonstrated with maximum-intensity projection (**A**) than with shaded surface display (**B**).

collimation images (4–5 mm) can be used through the region of the expected aneurysmal neck (renal hila). Both IV and oral contrast material are usually used; however, the former is not essential. From 500 to 800 mL of oral contrast are administered 45 to 60 minutes before CT, with the last 200 to 300 mL given 10 to 15 minutes before the scan. A total of 120 to 180 mL of 60% iodinated contrast material are administered intravenously by means of a power injector, using a rate of 2 mL per second and a 40- to 50-second scan delay.

Spiral CTA

The rate of injection, timing of injection-to-scan delay, and scanning parameters are critical for obtaining a high-quality diagnostic CTA study.

Before a patient undergoes spiral CT, the cranial and caudal extent of the imaging volume must be determined with a noncontrast incremental CT obtained using a 10-mm slice thickness and 10-mm or 15-mm spacing. In addition, the optimal field-of-view has to be defined for optimization of the spatial resolution.

Ideally, the smallest possible field-of-view is chosen so that only pertinent anatomic structures are included. It is also desirable to match the field-of-view and the craniocaudal length of anatomic coverage, which results in the optimal use of the entire 512 × 512 display matrix in any rendering plane and improves spatial resolution. A field-of-view for an abdominal aortic CTA varies between 18 and 26 cm.

An abdominal CTA is usually performed during a single 30-second breath-hold. A second, back-to-back, 30- to 40-second helical scan can be obtained to include the pelvis and proximal femoral arteries. Typical scanning parameters for evaluation of the abdominal aorta include 3-mm collimation with pitch two, acquired from the celiac artery to the aortic bifurcation.

A variable pitch can be used to optimize the spatial resolution and the length of coverage (9,18). Using 5-mm collimation and pitch two, we obtain a second pelvic helical scan after a 7- to 10-second breathing and tube cooling time. The images are reconstructed at every 2 mm (33% overlap) in the abdomen and 3 mm (40% overlap) in the pelvis.

Oral contrast material is not administered.

IV contrast medium, usually a 60% solution (approximately 300 mg/mL of iodine), is delivered at a rate of 3 to 5 mL per second by means of a power injector via a 20-gauge angiocatheter that is generally placed in the antecubital vein. At an injection rate of 3 mL per second, a scan delay of 25 to 35 seconds is adequate for most patients. Using the latter technique, the entire aortoiliac segment can be covered with a single IV injection of contrast material ranging from 120 to 160 mL, depending on the patient's weight.

At an injection rate higher than 3 mL per second, a test scan is necessary for calculation of the injection-to-scan delay time that would provide the maximum concentration of intra-arterial contrast material. The test scan consists of serial axial scans obtained at the same level after IV injection of 20 mL of contrast medium delivered at the same rate as the subsequent total volume of contrast.

The appropriate scan delay is determined by measuring the density in the central portion of the aorta and generating a time-density curve at the intended starting point of the spiral CT (9). The peak enhancement reflects the optimal scanning time corresponding to the individual patient's hemodynamics. During the test injection, the patient is asked to suspend respiration for a 30-second period after hyperventilation. The same breath-holding technique is used during the helical scan.

POSTOPERATIVE EVALUATION OF PATIENTS WITH AAAs TREATED WITH ENDOVASCULAR GRAFTS

The technique used for postoperative spiral CTA is identical with that for preoperative aortic CTA described above. The scan should start 1 to 2 cm above the proximal stent and continue for the maximum possible duration (30 to 32 seconds). A second, back-to-back spiral CT of the pelvis is obtained after a 7- to 10-second breathing time, without any additional IV contrast material. The pelvis is scanned to screen for potential complications at the transfemoral access site and also to assess the major pelvic vessels and the distal end of the aortofemoral graft. Additional conventional 10-mm thick sections through the device are obtained with a 3- to 5-minute delay. Oral contrast is not administered for these studies.

At our institution, CT follow-up is obtained within the first week, at 3 months after the procedure, and then at 6-month intervals.

The patency and position of an endovascular device are initially assessed by an intraoperative angiogram after the completion of the procedure. Further evaluation of the prosthesis in the early and late postoperative periods can be effectively performed using spiral CT (10).

All components of an endovascular prosthesis, including metallic stents and faintly radiopaque grafts, can be visualized on unenhanced CT. Devices of various design have a characteristic appearance. Straight (tube) and bifurcated as well as aortoiliac grafts can be recognized. A tube prosthesis can be either aortoaortic or aortofemoral. The latter is supplemented by a femorofemoral bypass and occlusion of the contralateral iliac artery (Fig. 3.12).

Some stents are completely covered with the graft fabric (covered stent), whereas others have graft material attached only to a small portion of the stent (uncovered stent). The cranial extent of an uncovered stent may overlap the origins of the renal arteries (Fig. 3.13A) with-

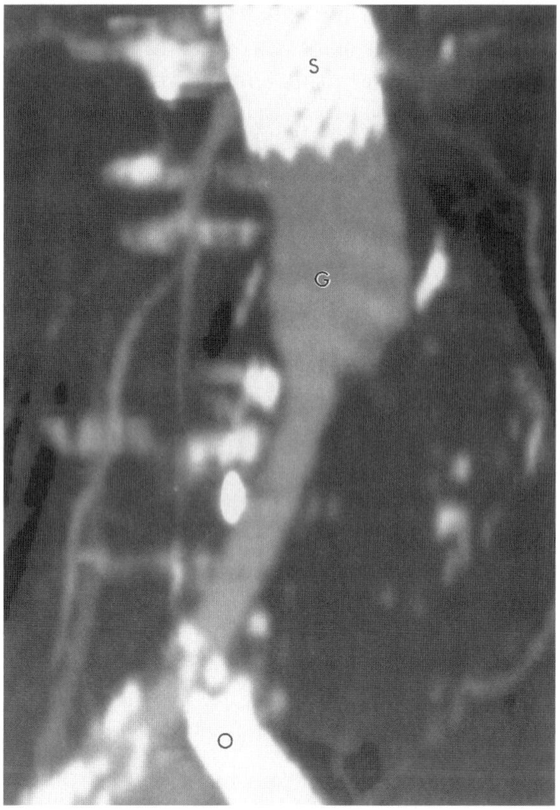

Figure 3.12. Straight aortofemoral endovascular device. Maximum-intensity projection image obtained 24 hours after endoluminal grafting shows a Palmaz stent *(S)* in the proximal aneurysm neck. A tapered graft *(G)* extends to the right common femoral artery (not shown). There is an occlusive stent *(O)* in the left common iliac artery.

out altering renal perfusion. Alternatively, a covered stent must be placed caudal to the renal arteries (see Fig. 3.13B and C).

A balloon-expandable, uncovered Palmaz stent (Johnson & Johnson Interventional Systems, Warren, NJ) is visualized as a thin metal cylinder with diamond-shape openings (see Fig. 3.12). A Wallstent (Schneider, Minneapolis, MN), also uncovered, can be recognized by the very thin braided wires composing its wall. A self-expanding, Z-type EVT stent (Endograft; Endovascular Technologies, Menlo Park, CA) is constructed of a thick, characteristically bent wire, covered with Dacron fabric. The attachment hooks of this device are well visualized within the aortic wall after successful deployment. Any other positioning of the hooks, particularly within the aortic lumen or thrombus, suggests a deficient attachment or fracture.

Fatigue fractures of these devices have been described (7). They are best seen on plain films and can be completely missed on the CT scan (Fig. 3.14), which has a lower spatial resolution than the film-screen combination. When appropriately placed, a stent abuts and supports the aortic wall and no contrast material or thrombus should be seen between the stent and the wall of the aorta. Alternatively, either contrast or thrombus separating the stent from the vascular wall is indicative of an insufficient seal.

In the early postoperative period (within 72 hours), a small amount of air can be present around the device (Fig. 3.15). If trapped within a clot, air bubbles may persist for several weeks (10). A newly formed thrombus around the graft can be recognized by the high density of the clot on noncontrast CT. However, thrombus cannot be distinguished from flowing blood. Accurate assessment of intraluminal and perigraft flow (when present) requires contrast-enhanced CT.

The technically successful placement of an intravascular device results in the exclusion of the aneurysm and provides a conduit that is opacified with contrast material extending into the distal vasculature. The aneurysmal sac surrounding the graft should be completely thrombosed and, therefore, without enhancement (see Fig. 3.15).

In some cases, a blood-tight seal between the endovascular prosthesis and the aorta is difficult to achieve. In these instances, contrast material may appear between the aortic wall and the endovascular graft, denoting perigraft leaks or channels (Fig. 3.16). Proximal or distal channels originating at the respective ends of the device can be identified. In our study, spiral CT was more sensitive than intraoperative angiography for the detection of perigraft leaks (10). Failure to detect such leaks in the operating room may be due to the technical limitations of portable, digital angiographic imaging equipment.

Alternatively, the possibility of initial sealing with the subsequent development of a perigraft channel cannot be excluded.

Persistent perigraft channels have been reported in up to 22% of patients treated with straight devices (7). The natural history of these channels is unknown. Some small leaks can completely close, whereas others may persist throughout the postoperative period (5). Certain perigraft channels are thought to persist owing to the backflow from patent branch vessels such as the inferior mesenteric or lumbar arteries (5,6,42). We observed several cases with small persistent perigraft channels communicating with the branch vessels, but no angiographic confirmation was obtained to define the direction of flow in these channels. Occasionally, the initially well-sealed aneurysmal sac develops a late perigraft channel during follow-up (10).

On CT, a Dacron graft is usually round in cross-section, similar to the ex vivo shape. The excessive length or twisting during the placement may result in deformity of the graft. EVT grafts have metallic markers along the opposite sides (see Figs. 3.14 and 3.15), which provide real-time fluoroscopic guidance and aid in avoiding twisting of the device during insertion. Graft deformity may also be observed in association with large perigraft leaks when the pressures inside the graft and in the perigraft

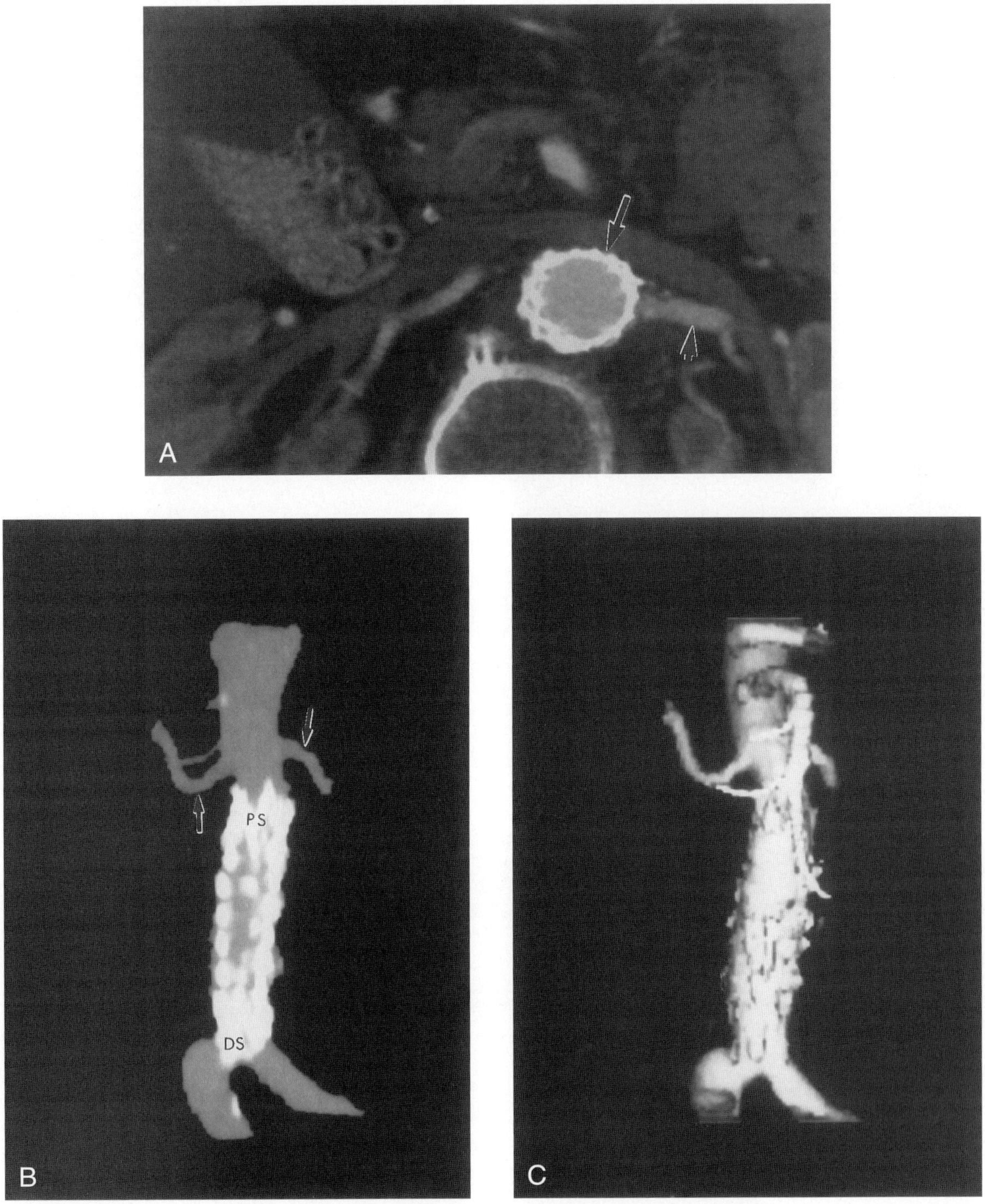

Figure 3.13. CT demonstration of an endovascular stent position. **A.** Helical CT shows a balloon-expandable Palmaz stent (*arrow*) at the origin of the left renal artery (*arrowhead*). Because the stent is uncovered, the renal artery remains patent. **B.** CTA with maximum-intensity projection rendering of a different patient shows that the proximal covered stent (*PS*) of an EVT device is appropriately positioned caudal to the renal arteries (*arrows*). The distal stent (*DS*) is at the lower neck of the aneurysm, which is completely excluded. **C.** Shaded surface display rendering in the same patient as in **B** provides very limited diagnostic information because the metallic stent cannot be distinguished from intravascular contrast material.

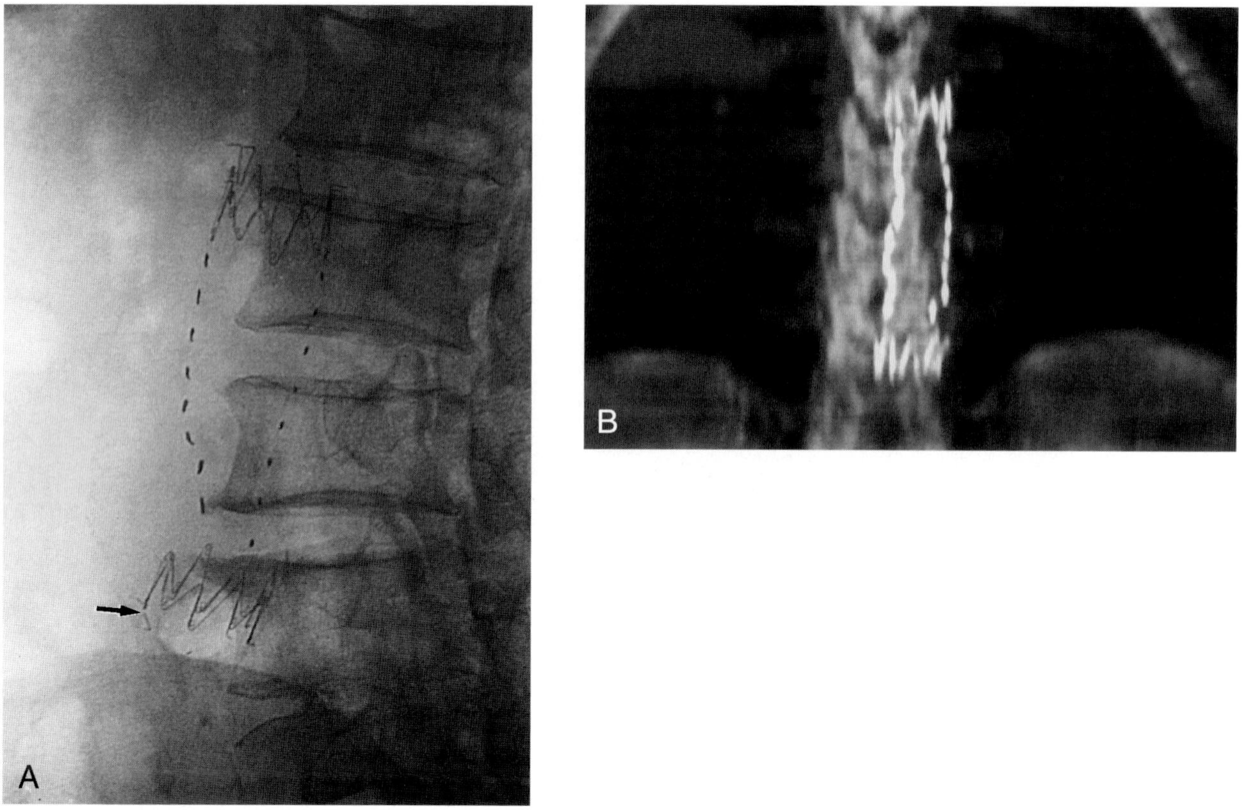

Figure 3.14. Fatigue fracture of an EVT device. A fractured hook (*arrow*) of the distal stent is apparent on plain film (**A**), but it cannot be identified on CTA with maximum-intensity projection rendering (**B**).

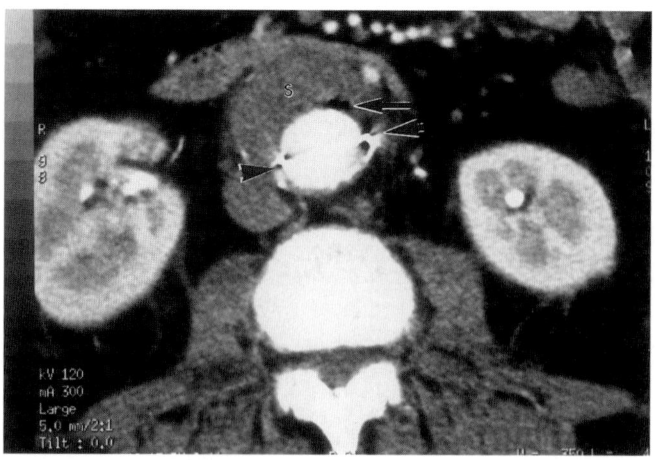

Figure 3.15. Successful placement of an endovascular graft. Spiral CT 48 hours after an endovascular procedure shows an EVT graft with metallic markers (*arrowheads*) in the proper position and without deformity. The aneurysmal sac (*S*) is thrombosed. A postprocedural air bubble can be seen adjacent to the graft (*arrow*).

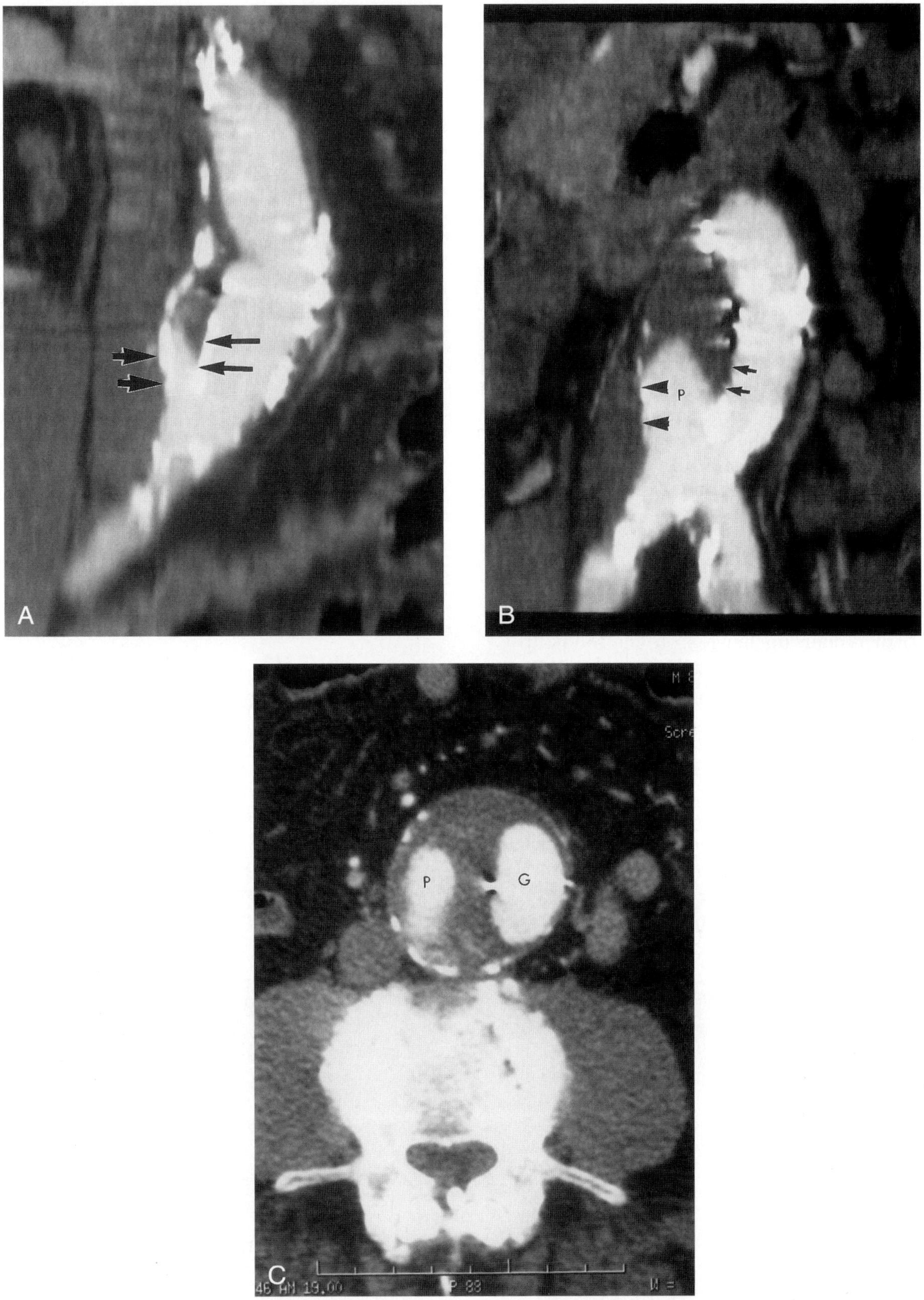

Figure 3.16. Distal perigraft channel. **A.** Multiplanar reformation shows that the contrast material extends between the distal stent (*arrows*) and the aortic wall (*arrowheads*). **B.** Multiplanar reformation 3 months later reveals enlargement of the perigraft channel (*P*) with further displacement of the stent (*arrows*) away from the aortic wall (*arrowheads*) and deformity of the graft. **C.** Axial image taken at the same time as in **B** shows a deformed graft (*G*) and a perigraft channel (*P*).

channels may be almost the same, leading to distortion of the round configuration.

A minor change in the graft contour from slightly ovoid to round and vice versa can be observed with successfully inserted EVT devices, possibly related to the differences in systolic and diastolic pressures (10).

Both axial and multiplanar reformatted images accurately show the patency of the endovascular prosthesis and the distribution of intravascular contrast material. Multiplanar reformation often clarifies the complex anatomy of the tortuous aorta and branch vessels by showing the precise position and configuration of the device in optimal planes (15,24). We found that three-dimensional reformation is of limited value in the assessment of endoluminal grafts. Shaded surface display rendering shows only the surface of the contrast column, and stents are poorly separated from intravascular contrast, appearing as irregular bulges of the surface (see Fig. 3.14).

Maximum-intensity projection can display the entire device; however, intraluminal contrast material overlaps with a high-density stent (15). Consequently, axial and maximum-intensity projection images are most valuable in postoperative evaluation and follow-up (see Fig. 3.13C).

A successfully treated aneurysm may decrease in size (Fig. 3.17) or remain unchanged on late follow-up CT (10). Parodi (5) found a 10% to 20% decrease in the size of large successfully treated AAAs. The decrease in aneurysmal size was attributed to clot retraction and depressurization of the sac (3). Some successfully treated aneurysms show no reduction in size, perhaps because of heavy calcification, perianeurysmal fibrosis, or their relatively small size.

Any aneurysm enlargement seen postoperatively indicates an inability of the device to protect the aorta from rupture and, therefore, a failure of the procedure. The most likely cause of aneurysm enlargement is the presence of either a persistent or a later-appearing perigraft channel (10).

PELVIC ANEURYSMS

Technique

Before spiral CT, a noncontrast incremental scan is obtained from L3 to the symphysis pubis with 10-mm collimation at 10-mm intervals.

The cranial extent of the spiral sequence begins 2 cm above the aortic bifurcation; caudally, the study includes the femoral arteries. We use 3-mm collimation, 1.6 to 2 pitch, with retrospective reconstruction at 2-mm intervals. An injection rate of 3 to 5 mL per second can be used with the total volume of 60% contrast medium between 100 and 150 mL. We use a 3-mL injection rate, and the injection-to-scan delay varies between 30 and 40 seconds. With higher injection rates and in patients with low cardiac output, a test scan at the level of the aortic bifurcation is necessary to determine the optimal scan delay time (see description of test scan in Principles of Spiral CT).

The entire spiral scan usually takes no more than 30 seconds, and patients are free to leave within 20 to 30 minutes after their arrival. CT is usually well tolerated by patients.

Preoperative Evaluation

Aneurysms involving the iliac as well as the common femoral arteries are well depicted by CT. Most can be identified and measured without IV contrast material. In addition to aneurysm size, the diameter of the femoral and iliac arteries must be measured to determine the feasibility of the endovascular technique. Iliac arteries smaller than 8 mm are usually inadequate for passage of a typical device delivery system (42).

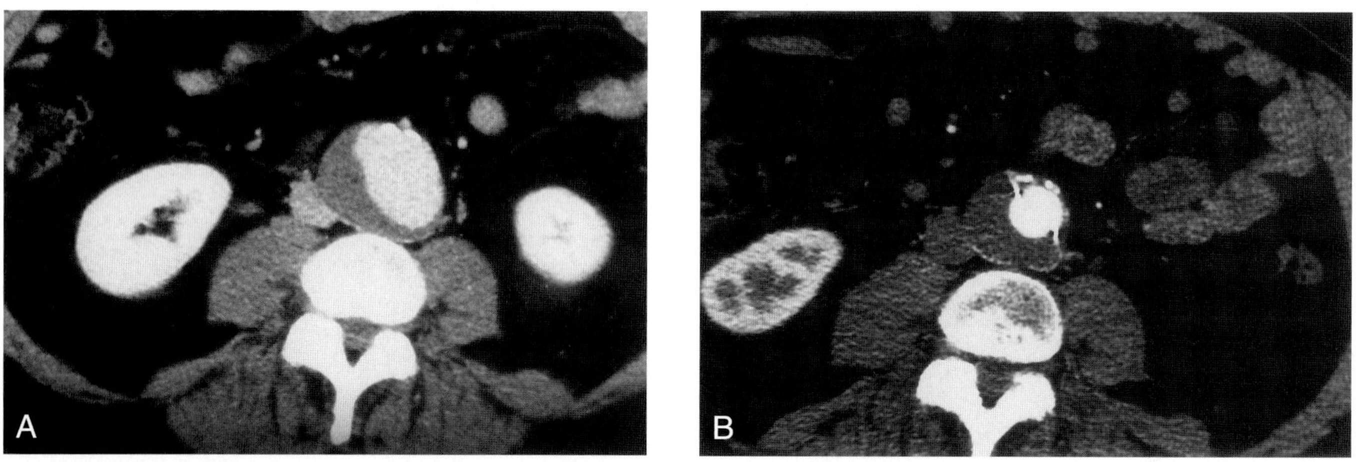

Figure 3.17. Diminished aneurysm size after successful endovascular grafting. **A.** Preoperative helical CT shows an infrarenal AAA with a moderate amount of intraluminal thrombus. (Reprinted with permission from Rozenblit A, Marin ML, Veith FJ, et al. Endovascular repair of abdominal aortic aneurysm: Value of postoperative follow-up with helical CT. AJR 1995;165:1473–1479.) **B.** Helical CT at the same level as in **A** 1 year after endovascular repair shows approximately a 30% interval decrease in the size of the aneurysm.

Contrast-enhanced conventional CT improves the visualization of pelvic aneurysms; however, owing to the tortuosity of the pelvic vessels, the exact origin of the aneurysm can be difficult to determine. Because of insufficient opacification and small size, some pelvic aneurysms may go undetected; the overall accuracy of CT assessment of the common iliac arteries is 84% (35). In addition, the status of adjacent vessels is difficult or impossible to judge on conventional CT.

With spiral CTA, including three-dimensional and multiplanar reformation, pelvic arteries are displayed in exquisite detail. The origin of the aneurysm can be directly visualized despite the presence of tortuous vessels, and the patency of the feeding artery as well as all major pelvic arteries can be evaluated. We found that the patency of the hypogastric artery is significant when planning the endovascular treatment of an aneurysm that originates from either the common or the external iliac artery, because the backflow from the patent ipsilateral hypogastric may cause a persistent perigraft channel in the postoperative period. Consequently, at our institution the ipsilateral hypogastric artery is embolized with angiographic coils before the placement of an endovascular stent, provided the patency of the opposite hypogastric artery is documented.

Although spiral CT with CTA is certainly capable of evaluating the major pelvic vessels, the role of this method in the preoperative assessment of pelvic vascular pathology is still limited. Angiography remains the "gold standard" because it demonstrates the entire regional hemodynamics including the collateral and runoff vessels, which can be evaluated at the same session. As of this writing, no large series comparing the two modalities for assessment of the pelvic vessels are available in the literature.

Postoperative Assessment

The same technique just described for preoperative pelvic spiral CT is used for postoperative assessment. As with postprocedural CTA of the abdominal aorta, axial and reconstructed images of the pelvis can confirm adequate positioning and patency of an endovascular prosthesis, completeness of thrombosis in the aneurysmal sac, and the presence of any inappropriate perigraft con-

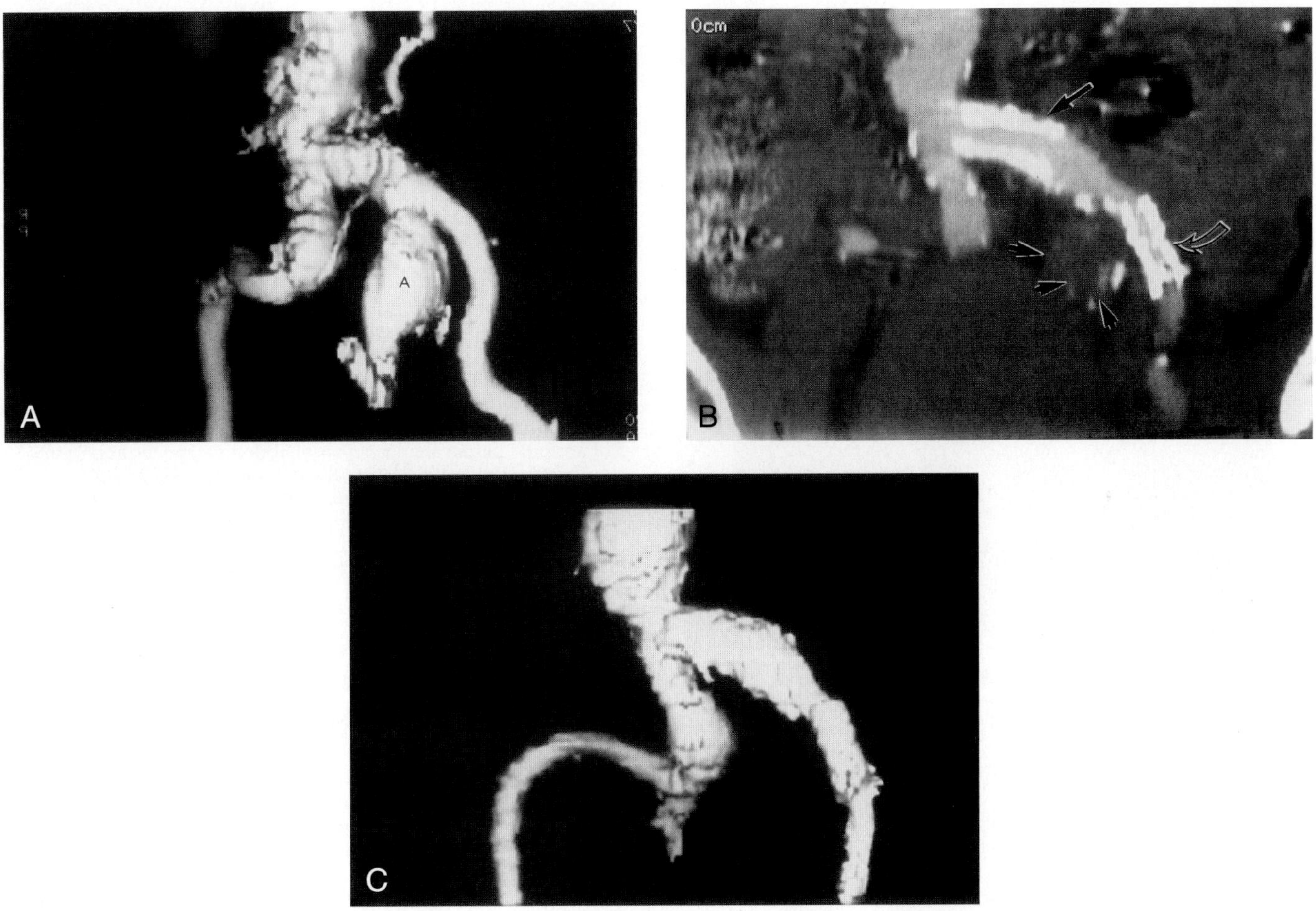

Figure 3.18. Iliac aneurysm. **A.** Preoperative CTA with shaded surface display rendering shows a large left hypogastric aneurysm (*A*). **B.** CTA with multiplanar reformation 3 months after endovascular grafting shows a thrombosed aneurysmal sac (*arrowheads*). The proximal stent (*arrow*) is in the left common iliac artery and the distal stent (*open arrow*) in the external iliac artery. **C.** Shaded surface display rendering confirms the exclusion of the hypogastric aneurysm, which is not seen.

trast material. Also, the patency of the hypogastric artery can be assessed as well as the position of angiographic coils in patients who were treated with embolization before endovascular graft placement. Early postoperative complications such as pelvic or groin hematomas can be clearly seen on CT without contrast.

In our experience, most pelvic aneurysms have been successfully treated with endovascular grafting (Fig. 3.18) with no evidence of perigraft channels. However, we were unable to document any decrease in size on follow-up CTs for up to 20 months after treatment. This may be related to the relatively small size of pelvic aneurysms and the usually large, preexisting thrombus present compared with AAAs.

OCCLUSIVE DISEASE

Aortoiliac occlusive disease is often evaluated in combination with run-off vessels. This is traditionally and appropriately assessed by angiography, which can provide unique information regarding vascular occlusion, stenosis, levels of reconstitution, and collateral circulation. Conventional or spiral CT can provide only limited diagnostic information (Fig. 3.19).

The degree and extent of wall calcification are usually of concern with open surgery. For endovascular grafting procedures, evaluation of the luminal thrombus and detection of atherosclerotic plaques, particularly with ulceration, can be important to avoid potential embolic complications. Assessment of the vascular lumen can be achieved with contrast-enhanced CT, and a spiral mode is preferred. A routine CTA technique (described in the previous text) is used. The study either can be confined to the pelvis or, more commonly, may include the entire aortoiliac segment, depending on clinical needs.

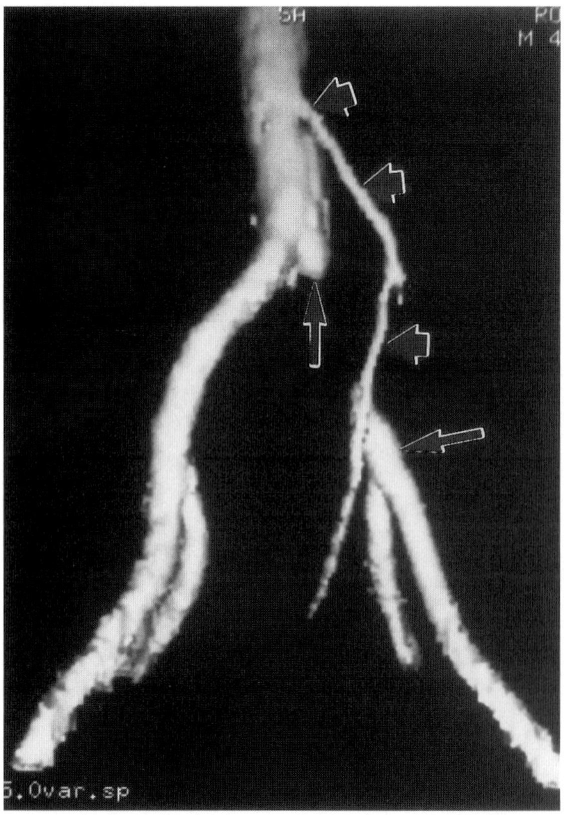

Figure 3.19. Vascular occlusive disease. CTA with shaded surface display rendering shows a long segment of occlusion in the left common iliac artery (*arrows*). The remaining major pelvic vessels as well as the inferior mesenteric artery (*arrowheads*) are patent.

CONCLUSIONS

CTA is a rapidly developing technique with many applications. Currently, it is the best single imaging modality for patients treated with endovascular grafts. The image quality continues to improve because of the progress being made in hardware and software. X-ray tubes with higher power and heat capacity increase both the signal-to-noise ratio and the length of anatomic coverage (43). This may also permit the use of a smaller volume of contrast material, with subsequent improved safety and reduced cost and preservation of a high image quality. In the near future, high-speed computers may permit storage of the raw scan data, allowing truly volumetric analysis of the entire data rather than the presently used manipulation with reconstructed axial images.

References

1. Marin ML, Veith FJ. Endovascular stents and stented grafts for the treatment of aneurysms and other arterial lesions. Adv Surg 1996;29:93–109.
2. Marin ML, Veith FJ, Cynamon J, et al. Initial experience with transluminally placed endovascular grafts for the treatment of complex vascular lesions. Ann Surg 1995;222:449–469.
3. Parodi JC, Palmaz JC, Barone HD. Transfemoral intraluminal graft implantation for abdominal aortic aneurysms. Ann Vasc Surg 1991;5:491–499.
4. Marin ML, Veith FJ. Transfemoral repair of abdominal aortic aneurysm. N Engl J Med 1994;331:1751.
5. Parodi JC. Endovascular repair of abdominal aortic aneurysms and other arterial lesions. J Vasc Surg 1995;21:549–557.
6. Moore WS, Vescera CL. Repair of abdominal aortic aneurysm by transfemoral endovascular graft placement. Ann Surg 1994; 220:331–341.
7. Yusuf SW, Hopkinson BR. Endovascular repair of aortic aneurysm. Br J Surg 1995;82:289–291.
8. Bluemke DA, Chambers TP. Spiral CT angiography: An alternative to conventional angiography. Radiology 1995;195:317–319.
9. Rubin GD, Dake MD, Semba CP. Current status of three-dimensional spiral CT scanning for imaging the vasculature. Radiol Clin North Am 1995;33:51–70.
10. Rozenblit A, Marin ML, Veith FJ, et al. Endovascular repair of abdominal aortic aneurysm: Value of postoperative follow-up with helical CT. AJR 1995;165:1473–1479.
11. Kalender WA, Seissler W, Klotz E, Vock P. Spiral volumetric CT with single-breath-hold technique, continuous transport, and continuous scanner rotation. Radiology 1990;176:181–190.
12. Kalender WA. Technical foundations of spiral CT. Semin Ultrasound CT MRI 1994;15:81–89.
13. Heiken JP, Brink JA, Vannier MW. Spiral (helical) CT. Radiology 1993;189:647–656.
14. Semba CP, Rubin GD, Dake MD. Three-dimensional spiral CT angiography of the abdomen. Semin Ultrasound CT MRI 1994;15:133–138.

15. Rubin GD. Three-Dimensional helical CT angiography. RadioGraphics 1994;14:905–912.
16. Zeman RK, Silverman PM, Vieco PT, Costello P. CT angiography. AJR 1995;165:1079–1088.
17. Polacin A, Kalender WA, Marchal G. Evaluation of section sensitivity profiles and image noise in spiral CT. Radiology 1992;185:29–35.
18. Zeman RK, Silverman PM, Berman PM, et al. Abdominal aortic aneurysms: Findings on three-dimensional display of helical CT data. AJR 1995;164:917–922.
19. Costello P, Gaa J. Spiral CT angiography of abdominal aortic aneurysms. RadioGraphics 1995;15:397–406.
20. Rubin GD, Dake MD, Napel S, et al. Spiral CT of renal artery stenosis: Comparison of three-dimensional rendering techniques. Radiology 1994;190:181–189.
21. Napel SA, Marks MP, Rubin GD, et al. CT angiography with spiral CT and maximum intensity projection. Radiology 1992;185:607–610.
22. Kuszyk BS, Heath DG, Ney DR, et al. CT angiography with volume rendering: Imaging findings. AJR 1995;165:445–448.
23. Costello P, Depuy DE, Ecker CP, Tello R. Spiral CT of the thorax with reduced volume of contrast material: A comparative study. Radiology 1992;183:663–666.
24. Rubin GD, Dake MD, Napel SA, et al. Three-dimensional spiral CT angiography of the abdomen: Initial clinical experience. Radiology 1993;186:147–152.
25. Rozenblit A, Bennett J, Kopp A, et al. Preoperative assessment of aortic branch vessels with CT-angiography in patients with abdominal aortic aneurysm. AJR 1996;166:38 (Abstract).
26. Rubin GD, Walker PJ, Dake MD, et al. Three-dimensional spiral computed topographic angiography: An alternative imaging modality for the abdominal aorta and its branches. J Vasc Surg 1993;18:656–665.
27. Rubin GD, Napel S. Increased scan pitch for vascular and thoracic spiral CT. Radiology 1995;197:316–317.
28. Brink JA, Lim JT, Wang G, et al. Technical optimization of spiral CT for depiction of renal artery stenosis: In vitro analysis. Radiology 1995;194:157–164.
29. Lasser EC. Pretreatment with corticosteroids to prevent reactions to IV contrast material: Overview and implications. AJR 1988; 150:257–259.
30. Barrett BJ, Carlisle EJ. Metaanalysis of the relative nephrotoxicity of high- and low-osmolality iodinated contrast media. Radiology 1993;188:171–178.
31. Blumke DA, Fishman EK. Spiral CT. Initial experience with vascular applications. Clin Imaging 1994;18:107–112.
32. Siegel CL, Cohan RH. CT of abdominal aortic aneurysms. AJR 1994;163:17–29.
33. Todd GJ, Nowygrod R, Benvenisty A, et al. The accuracy of CT scanning in the diagnosis of abdominal and thoracoabdominal aortic aneurysms. J Vasc Surg 1991;13:302–310.
34. Tennant WG, Hartnell GG, Baird RN, Horrocks M. Radiologic investigation of abdominal aortic aneurysm disease: Comparison of three modalities in staging and the detection of inflammatory change. J Vasc Surg 1993;17:703–709.
35. Papanicolaou N, Wittenberg J, Ferrucci JT Jr, et al. Preoperative evaluation of abdominal aortic aneurysms by computed tomography. AJR 1986;146:711–715.
36. Vowden P, Wilkinson D, Ausobsky JR, Kester RC. A comparison of three imaging techniques in the assessment of an abdominal aortic aneurysm. J Cardiovasc Surg 1989;30:891–896.
37. Zeman RK, Silverman PM, Berman PM, et al. Abdominal aortic aneurysms: Evaluation with variable-collimation helical CT and overlapping reconstruction. Radiology 1994;193:555–560.
38. Van Hoe L, Baert AL, Gryspeerds S, et al. Supra- and juxtarenal aneurysms of the abdominal aorta: Preoperative assessment with thin-section spiral CT. Radiology 1996;198:443–448.
39. Andersen PE, Lorentzen JE. Comparison of computed tomography and aortography in abdominal aortic aneurysms. J Comput Assist Tomogr 1983;7:670–673.
40. Gomes MN, Choyke PL. Pre-operative evaluation of abdominal aortic aneurysms: Ultrasound or computed tomography. J Cardiovasc Surg 1987;28:159–166.
41. Gomes MN, Choyke PL. Improved identification of renal arteries in patients with aortic aneurysms by means of high-resolution computed tomography. J Vasc Surg 1987;6:262–268.
42. Sayers RD, Thompson MM, Nasim A, Bell PRF. Endovascular repair of abdominal aortic aneurysm: Limitations of the single proximal stent technique. Br J Surg 1994;81:1107–1110.
43. Rubin GD, Napel S, Leung AN. Volumetric analysis of volumetric data: Achieving a paradigm shift. Radiology 1996;200:312–317.

Chapter 4

Application of Magnetic Resonance Angiography of an Aortic Phantom to Determine the Lumen Length of an Abdominal Aortic Aneurysm

Gerald Dorros, Yu Liu, Cynthia Sampson, George Latus, L. Samuel Wann, Michael Jaff, and Juan C. Parodi

This work was supported by an unrestricted research grant from General Electric Medical Systems, Milwaukee, Wisconsin.

The incidence of abdominal aortic aneurysms (AAAs) appears to be increasing and is estimated to be between 30,000 and 75,000 per 1 million persons. Approximately 15,000 deaths per year are the result of AAA rupture. This condition warrants particular attention because it is the 13th leading cause of death in the United States. Therefore, techniques that could significantly reduce surgical morbidity and mortality should be carefully evaluated.

Aneurysmal dilatation of the aorta presents a condition that, once initiated, continually worsens. Aneurysmal enlargement depends on the tangential stress that disrupts the spherical wall, the radius, and the transmural pressure. As a result, an increase in radius produces a multifold increase in wall tension, and rupture or leakage occurs when the wall tension exceeds the intrinsic strength of the wall. Thus, situations that portend leakage or rupture (25% to 40% within 5 to 8 years) should be identified so that appropriate therapy can be initiated. The documented presence of an enlarging aneurysm (1.0 cm or more per year), an aneurysm more than 5.0 cm in diameter, or symptoms consistent with leakage with or without confirmation may be sufficient indications to intervene (1,2). The mortality rate of patients with emergency surgical aneurysm repair is significantly increased and can approximate 50% (3), which sharply contrasts with the mortality rate associated with elective AAA repair (2% to 3% in relatively healthy patients and 8% to 10% [4] when concomitant medical problems coexist, such as coronary artery disease).

Recently, transfemoral endoluminal graft placement has been introduced as an alternative to open surgical repair (5). However, because each device is customized for the patient and, once inserted cannot be easily removed, specific information, particularly the luminal length of the aneurysm, is required to perform endovascular techniques. The graft length must be concordant with that of the aneurysm lumen—not too short and not too long. Because an aortic aneurysm can severely distort the aorta by marked tortuosity, the aorta's actual length cannot be accurately measured using a craniocaudad linear angiographic measurement.

Magnetic resonance imaging techniques enable acquisition of data in any plane. The source images of two-dimensional (2-D) time-of-flight (TOF) magnetic resonance angiography (MRA) permit the center of the aneurysm's lumen to be visually determined. This allows the aneurysm's lumen length to be measured using the Pythagorean theorem.

We examined this thesis by using a phantom model of a tortuous aorta to assess the accuracy of lumen length measurements by MRA.

METHOD

Phantoms of tortuous aortas were constructed using water-filled plastic tubing arbitrarily determined to be 7 mm in diameter. The tubes of varying lengths (80 to 230 mm) were formed into "S" or "C" shapes to simulate a tortuous aneurysm. Phantoms were imaged on a 1.5T Signa (GE Medical Systems, Milwaukee, WI) MR scanner using an axial 2-D TOF spoiled-gradient recalled echo acquisition (TR/TE = 30/6 ms, flip angle 45° (matrix 128 × 256, FOV 320.0 mm, 2 averages, flow compensation, slice thickness 2.9 mm), and a T1-weighted spin-echo acquisition (TR/TE = 800/16 ms, matrix 128 ×

256, FOV 320.0 mm, slice thickness 4 mm, and 1-mm gap). Using the Pythagorean theorem, the length of a segment of the phantom tubing between two adjacent image slices was calculated using:

$$\left[L = \sum_{n=1}^{M} \sqrt{(X_{n+1} - X_n)^2 + (Y_{n+1} - Y_n)^2 + l^2}\right]$$

where L was aneurysm length, l was the thickness of each image, M was the total number of axial images covering the aneurysm, and (X_n, Y_n) was the coordinate of the aneurysm's center point on image n.

The results of the MRA phantom study of AAA length measurements summarized in Table 4.1 indicate both measured length and accuracy percentage values (6). It is not surprising that in the 2-D TOF MRA acquisitions, the results of 10.0-mm slice thickness were less accurate than for the 2.9-mm slice thickness; however, the 10.0-mm slice thickness was used in the 230.0-mm tubing length study. Similarly, in the spin-echo acquisitions, a slice thickness/gap of 8.0/2.0 mm was not used for other tubing length studies. These data were the basis for obtaining 2.9-mm slices in the clinical study patients.

APPLICATION TO HUMAN STUDIES

The latter method was used to determine the lumen length of aneurysms. Patients with AAAs were studied according to a protocol approved by our institutional review board. Informed consent was obtained from each patient. Axial and sagittal spin-echo images (8-mm slice thickness with 2-mm gap) and 2-D TOF MRAs (2.9-mm slice thickness) (Fig. 4.1) of the abdominal aorta were acquired from the level of the renal arteries to the neck of the aneurysm. This imaging covered the entire AAA to beyond the distal limit of the aneurysm, often below the hypogastric arteries. The relationship of the aneurysm to the vertebral bodies of the spine was noted from the sagittal image.

Table 4.1. Summary of the Accuracy of the MR Phantom Study for Abdominal Aortic Aneurysm Length Measurements

	2-D TOF				Spin-echo			
	Slice thickness				Slice thickness/gap			
	10.0 mm		2.9 mm		8.0/2.0 mm		4.0/1.0 mm	
Tubing Length (mm)	Length (mm)	%	Length (mm)	%	Length (mm)	%	Length (mm)	%
230.0	216.7	94.2	228.4	99.3	221.2	96.2	229.9	99.9
160.0	N/A	N/A	153.8	96.1	N/A	N/A	154.3	96.4
140.0	N/A	N/A	127.1	90.8	N/A	N/A	130.7	93.3
120.0	N/A	N/A	111.7	93.0	N/A	N/A	111.1	92.5
100.0	N/A	N/A	94.5	94.5	N/A	N/A	93.2	93.2
80.0	N/A	N/A	71.9	89.9	N/A	N/A	74.3	92.9

2-DTOF-two-dimensional time-of-flight; N/A = not available.

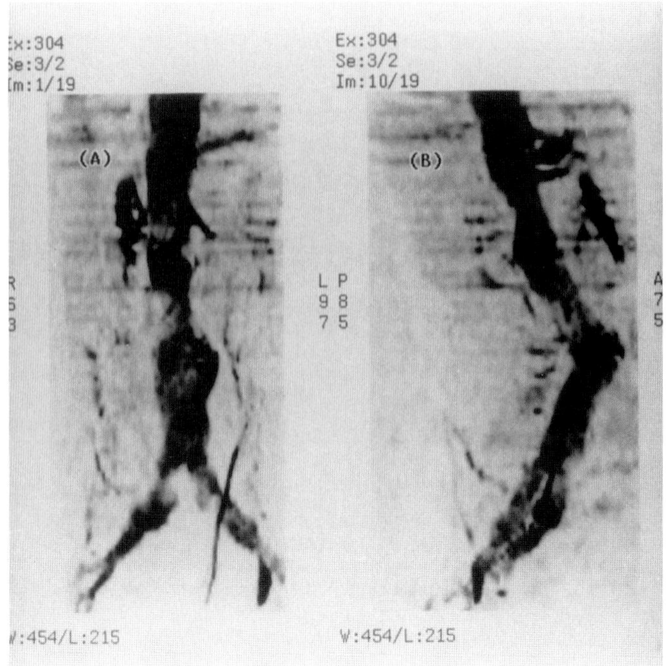

Figure 4.1. Coronal (**A**) and sagittal (**B**) views of a 2-D TOF MRA of an abdominal aortic aneurysm. The aneurysm is tortuous in the sagittal view (**B**), although it is straight in the coronal view (**A**), which is similar to the anteroposterior view in x-ray angiography. In addition, bilateral common iliac artery aneurysms and stenoses are demonstrated.

The 2-D TOF MRA of an AAA (see Fig. 4.1) appeared straight in the coronal view (i.e., anteroposterior x-ray angiography) but tortuous in the sagittal view. Therefore, measurements based on the anteroposterior view of x-ray angiograms would seriously underestimate an aneurysm's lumen length.

LIMITATIONS OF THE TECHNIQUE

The accuracy of the phantom-length measurement depends not only on the curvature of the tubing, but also on the regular phantom contour. However, with irregular geometric shapes, the accuracy of these measurements should be reduced, such as when the phantom (tubing) is lying entirely in one image slice (perhaps too large a slice). As such, the length of the tubing cannot be measured unless the direction of the imaging slices is changed.

Furthermore, the accuracy of measurements is limited by the slice thickness. Generally, thin slices are more accurate. When the signal-to-noise ratio (SNR) is too low, visualization of the aneurysm may be poor, making precise recordings difficult. The slightly more accurate measurements using the spin-echo acquisition must be balanced against the fact that the 2-D TOF axial source images save scan time (i.e., 2.9-mm thick axial images were acquired during a 2-D TOF MRA). In addition, the 2-D TOF images used to determine the relationship of the aneurysm to the renal arteries and aortoiliac bifurcation (Fig. 4.2) are supplementary information to the sagittal spin-echo images, which reference the aneurysm to the bony landmarks of the spine.

The percentage of error depends on the phantom length, because the absolute error caused by slice thickness and curvature of the phantom would have a higher value of percentage error when a shorter phantom is measured.

Furthermore, when the slice is significantly oblique, the determination of the center of the aneurysm on each image can be distorted. This problem would be magnified if large slices were obtained and the tortuosity produced significant lateral deviations, which would cause the oblique slices to magnify the error in lumen-length measurements.

In actuality, during human studies the appearance of the lumen is brighter than the surrounding wall or mural thrombus in a 2-D TOF acquisition and darker than the surrounding mural thrombus in a spin-echo acquisition. Thus, the usual estimate of the lumen center is not as difficult as might be inferred from the previous comments.

DISCUSSION

Endoluminal AAA repair requires a customized preparation that intrinsically demands an accurate assessment of the aortic wall's outside diameter (to which the stent will be expanded) and the aneurysm's lumen length. This will determine the graft's length (so as not to be too long or too short). After the customized stent-graft is inserted, it is not possible to remove it through its insertion site. Aneurysmal distortion causes the aorta to course through the anteroposterior, oblique, and lateral planes. Thus, a method that could account for this three-dimensional movement would be required to determine an accurate lumen length. Because MRA can accurately measure the length of a tortuous aortic phantom, an argument can be made for its potential applicability to measure the aneurysm's lumen length. However, some pitfalls that could make a significant impact on the measurements. Including the aneurysm's shape and convolutions, which could cause it to be severely sectioned obliquely and result in a visual misplacement of the aneurysm's central lumen point.

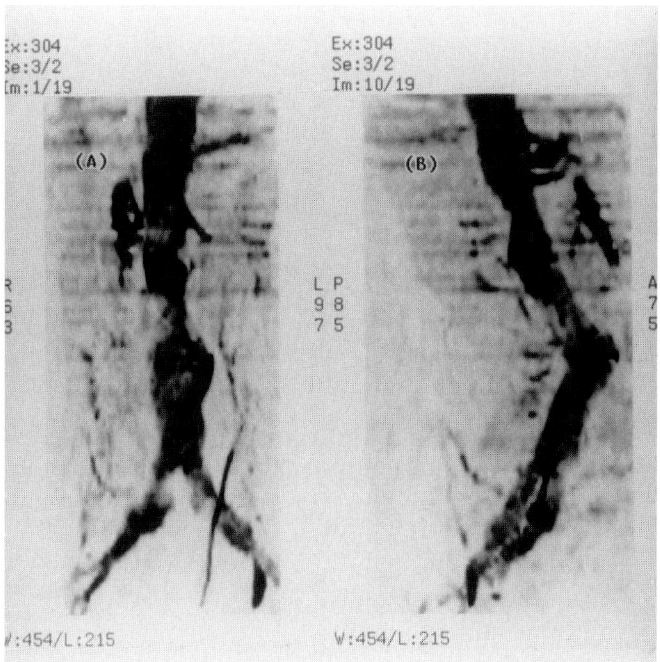

Figure 4.2. Coronal (**A**) and sagittal (**B**) views of a tortuous aneurysm, which extends to the bifurcation of the aorta.

SUMMARY

Magnetic resonance imaging can accurately assess the length of a phantom that mimics the aortic directional changes created by distortion (e.g., aneurysmal). Therefore, MRA should be able to delineate the size (lumen and diameter) and relationship of an AAA to the renal arteries and aortoiliac bifurcation and hence provide an accurate measurement of the lumen length of an aneurysm. MRA may be an important addition to the procedure of endoluminal aneurysm repair.

Acknowledgments

The authors wish to thank Tonya Hollrith, RTR, Cynthia Catella, RTR, and Gail Mueller, RN, for their assistance in collecting the patient data.

References

1. Nevitt MP, Ballard DJ, Hallett JW Jr. Prognosis of abdominal aortic aneurysm: A population based study. N Engl J Med 1989;321: 1009–1014.
2. Johansson G, Nydahl S, Olofsson P, Swedenborg J. Survival in patients with abdominal aortic aneurysms: Comparison between operative and non-operative management. Eur J Vasc Surg 1990;5: 497–502.
3. Ouriel K, Geary K, Green RM, et al. Factors determining survival after ruptured aortic aneurysm: The hospital, the surgeon and the patient. J Vasc Surg 1990;11:493–496.
4. Johnston KW. Multicenter prospective study of non-ruptured abdominal aortic aneurysm. Part II. Variables predicting morbidity and mortality. J Vasc Surg 1989;9:437–447.
5. Parodi JC, Palmaz JC, Barone HD. Transfemoral intraluminal graft implantation for abdominal aortic aneurysm. Ann Vasc Surg 1991;5:491–499.
6. Liu Y, Sampson C, Dorros G, et al. Magnetic resonance acquisition to determine the lumen length of a tortuous phantom aorta. J Endovasc Surg 1995;2:272–277.

Chapter 5

Ultrasound Imaging Techniques for Evaluation of Transluminally Placed Endovascular Grafts

Ross T. Lyon, George L. Berdejo, and Frank J. Veith

The planning, deployment, and postoperative surveillance of transluminally placed endovascular grafts (TPEGs) depend on the visualization and objective assessment of the lesion being treated, the adjacent inflow and outflow vessels, the associated hemodynamics, and the device itself. The ability to repeatedly (and precisely) image and assess these features without associated risks to the patient has resulted in the preferential use of ultrasound imaging techniques for the evaluation of endovascular stents and stented grafts (1–3). Abnormalities detected by ultrasonography can be confirmed, when necessary, by standard-contrast angiography or by spiral CT imaging. Intravascular ultrasonography (IVUS) and external-beam color duplex ultrasonography are the two principal ultrasound imaging techniques we use for vascular and endovascular device assessment. IVUS is primarily used for intraoperative evaluation, and external-beam color duplex ultrasonography for pre- and postoperative assessment and surveillance.

ULTRASOUND IMAGING TECHNIQUES

IVUS is a catheter-based (invasive) ultrasound imaging technique that uses high-frequency (20 to 40 MHz) piezo-electric ultrasound crystals. It provides high-resolution images of the surrounding vessel and adjacent structures (4–7). With this technique, a coaxial catheter with a transducer located at its distal end is inserted through an introducer sheath into the appropriate artery (or vein) and advanced to the site of interest. IVUS enables the determination of vessel and graft dimensions, luminal contours, and the location of major branch vessels (Fig. 5.1). Because of its high-definition imaging capability, IVUS can assess the adequacy of stent-to-vessel apposition (Fig. 5.2) and rapidly identify and define intraluminal graft defects and lesions without the need for nephrotoxic contrast agents or ionizing radiation.

Unfortunately, access to patent arteries is limited, because IVUS catheters cannot be passed beyond occlusive lesions. Furthermore, these devices currently do not give hemodynamic data, and longitudinal tracking is somewhat difficult. In addition, because of its high-frequency transducers, the IVUS signal attenuates over very short distances, and more remote surrounding structures are not well seen.

Despite its limitations, IVUS does permit the rapid intraoperative, three-dimensional visualization of vessels and grafts, thus complementing the use of intraoperative angiography.

Color duplex ultrasonography combines in a single diagnostic test the imaging capabilities of B-mode ultrasound, the hemodynamic assessment enabled by Doppler spectral waveform analysis, and the angiographic-like graphics of color flow images. The resulting image displays the local anatomy and characterizes flow within targeted vessels with flow direction and velocities represented by different colors, which are easily interpreted. This combination of information allows the characterization of grafts as being either normal (no apparent anatomic or flow-related abnormalities) or abnormal and enables the characterization of lesions responsible for graft-related problems. Because this test is also relatively inexpensive, noninvasive, and without known risks, it has become the primary means of surveillance for endovascular grafts (1–3,8,9).

Duplex ultrasound imaging depends not only on the skill of the examiner and the equipment, but also on the

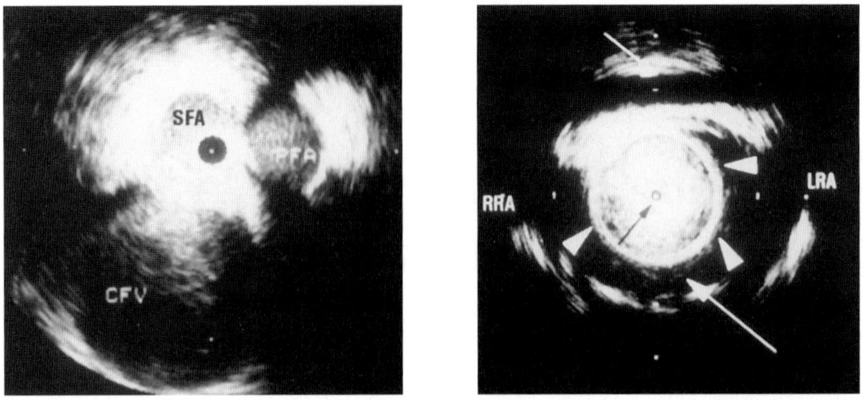

Figure 5.1. Intravascular ultrasound (IVUS) images showing landmarks and adjacent anatomy. On the left, the origin of the superficial femoral artery (*SFA*) is in the center of the image. Immediately adjacent to the SFA and to the right of the image is the profunda femoris artery (*PFA*) and just below and to the left the common femoral vein (*CFV*) is seen. The IVUS catheter resides within the lumen of the SFA. Note the blood speckle seen within the lumen of the CFV and the PFA. The picture on the right shows the aorta at the level of the renal arteries (*RRA, LRA*). The short arrow at the top of the image denotes the left renal vein crossing over the aorta in this anatomically correct image. The IVUS catheter is denoted by the black arrow in the center of the image. A large ring artifact (*arrowheads*) obscures visualization of most of the lumen of the aorta (*long, white arrow*).

patient. Examination of the abdomen and pelvis is best performed in the early morning after an overnight fast to minimize the amount of intra-abdominal gas present. This facilitates visualization of the abdominal aorta and iliac arteries (10).

APPROACH TO DUPLEX EVALUATION OF ENDOVASCULAR GRAFTS

General Considerations

Because of the variety of indications, sites, and configurations of TPEGs (11), it is important to know the original location and specific problems for which each endovascular graft was placed before assessment with color duplex ultrasound (1,3). Of special importance are the original sites of the proximal and distal stents or the distal graft termination. In addition, information regarding concomitant or adjunctive procedures such as coil embolization of branch vessels or the use of other vessel occlusion devices, stents, or supplemental standard proximal or distal arterial reconstructive procedures should be known in advance because these will also need to be evaluated. Knowledge of the indication for which the endovascular graft procedure was performed is essential because the potential problems vary based on the type of vascular lesion treated (e.g., arterial aneurysm, arterial occlusive disease, traumatic lesion, or arteriovenous fistula).

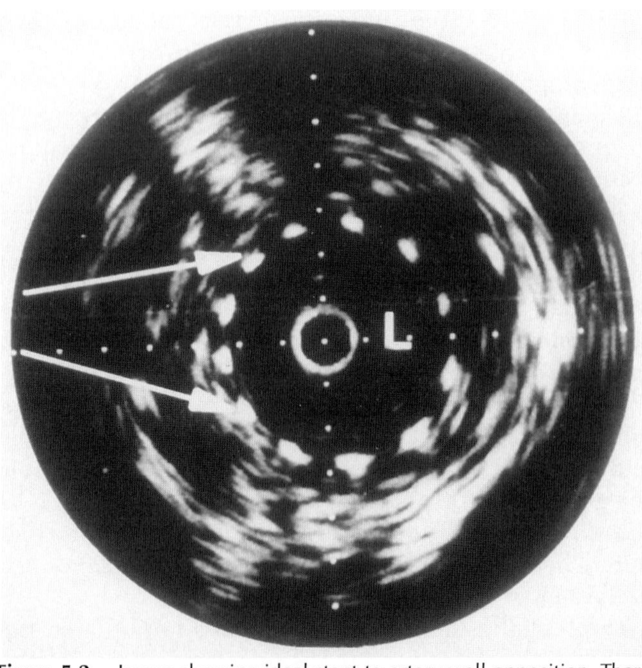

Figure 5.2. Image showing ideal stent-to-artery wall apposition. There is a circular lumen (*L*) and the echogenic struts of the stent (*arrows*) are easily seen and are in close apposition to the arterial wall.

Duplex Scan Protocol

Duplex scan assessment begins with visualization of the most proximal extent of the endovascular graft. This consists of a partially graft-covered stent that anchors the proximal end of the graft to the arterial wall (8,9,11–13). The body of the graft and adjacent inflow and outflow vessels are then scanned in both the long- and short-axis views. The B-mode scan is used to document the location of the proximal and distal stents or distal graft termination and inraluminal defects and perigraft structures.

Normally functioning endovascular grafts should have a smooth, uniform wall contour and should be closely approximated at its origin to the surrounding arterial wall. Maximum and minimum graft diameters are recorded. Spectral Doppler imaging is used to determine the velocities in the inflow vessel, the proximal stent-graft interface, the graft body, and the outflow vessel.

Color Doppler ultrasound is used to identify the graft body and native vessel flow abnormalities and spectral Doppler to grade their severity. Color flow Doppler ultra-

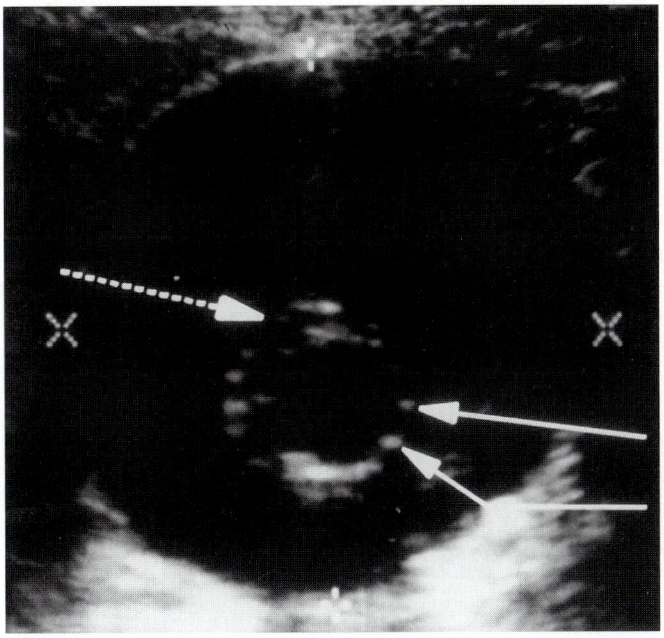

Figure 5.3. Duplex image showing crosses at the lateral and anteroposterior edges of an abdominal aortic aneurysm. These are used to obtain routine serial cross-sectional diameter measurements. The transluminally placed endovascular graft (*broken arrow*) can be seen within the aneurysm; the distal end of the proximal stent is also seen (*solid arrows*).

sound is also used to detect any perigraft flow (*endoleak*). Flow velocities and waveforms are also obtained 2 cm distal to the termination of the graft. The location of any lesion is marked on the skin, and a diagram is created of the bypass, adjacent vessels, and the site of any abnormality. A rapid longitudinal sweep (duplex cine angio) of the graft is then performed, and the examination is completed. Finally, a worksheet is generated and saved for comparison with subsequent scans.

Arterial Aneurysms

The duplex scan protocol for aneurysms treated with endovascular grafts includes longitudinal and transverse views of the graft and proximal and distal stents or the distal graft anastomosis. A maximum transverse diameter measurement (Fig. 5.3) of the arterial aneurysm is obtained and assessment of the presence of flow outside the graft is performed (i.e., continued flow within the aneurysm sac—a so-called endoleak). Finally, the native circulation distal to the graft is interrogated (3).

The goal of aneurysm repair is to exclude the aneurysmal portion of the artery from the arterial circulation, thereby preventing further enlargement, which can lead to rupture (14). Therefore, after endovascular aneurysm repair, the primary objective of the duplex examination is to determine whether or not there is persistent flow within the aneurysm sac (Fig. 5.4). Other critical objectives of the duplex examination are to determine stent location and peak flow velocity within the graft body and to detect stenotic lesions within the graft or adjacent vessels.

If flow outside the graft is detected, the next step is to

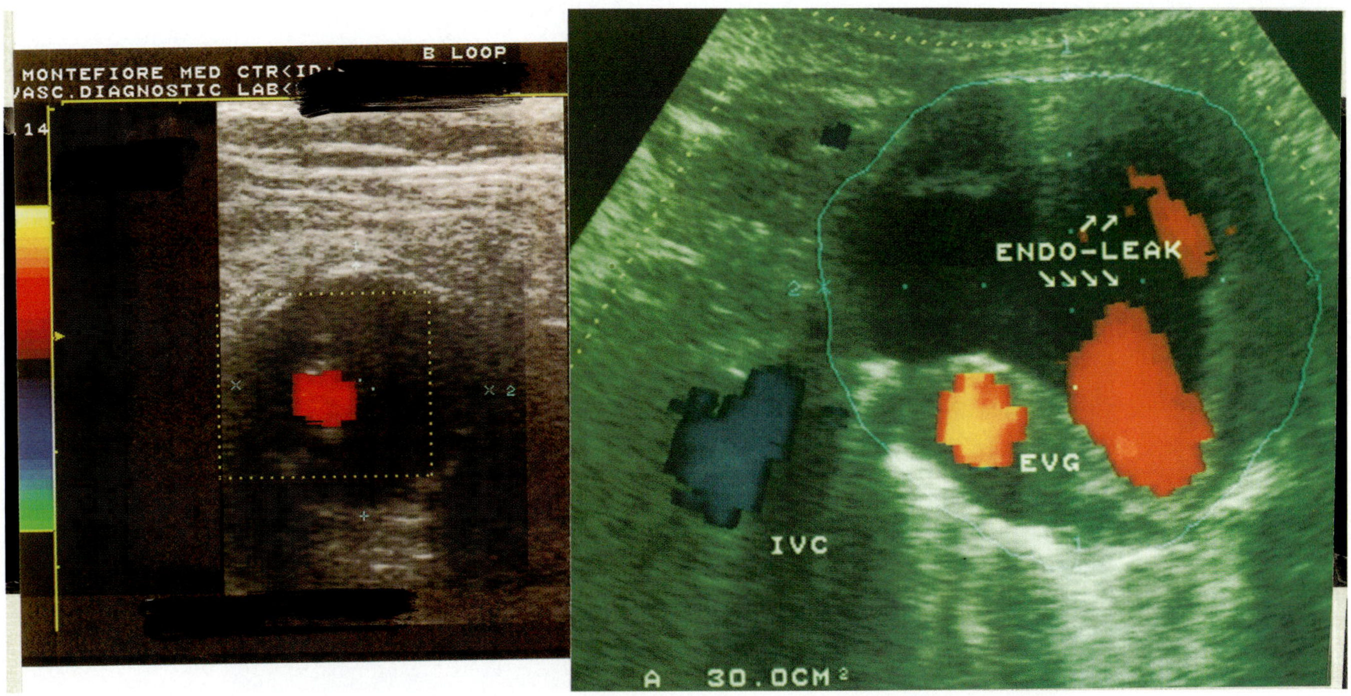

Figure 5.4. The image on the left is a color Doppler ultrasound image showing successful placement of an endovascular graft for the repair of a popliteal artery aneurysm. Flow is seen only within the lumen of the graft, and thrombus has formed in the surrounding aneurysm. The image on the right demonstrates an endoleak as persistent flow seen in the surrounding aneurysm lumen (*solid circular blue line*) outside the graft (*EVG, yellow color*). Thrombus can be seen lining part of the anterior and lateral walls. The IVC (blue color) is seen at the bottom left of the image.

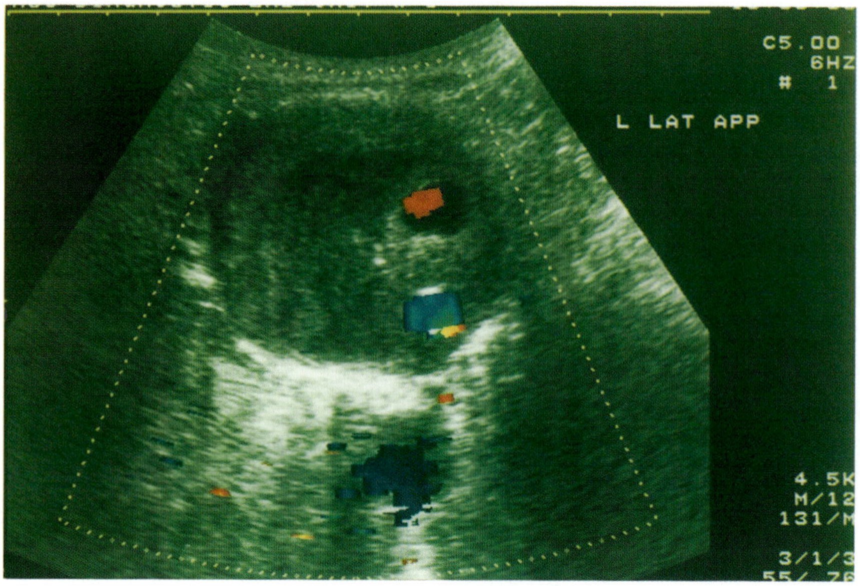

Figure 5.5. Cross-sectional color duplex image of a successfully repaired abdominal aortic aneurysm using the endovascular technique. The flow in the transluminally placed endovascular graft (bottom right of the surrounding aneurysm) has been coded as blue. The red immediately above the graft is a color flow artifact, which is the result of the movement of liquid (unclotted) blood caused by the pulsatility of the nearby graft.

identify its source. The endoleak may be the result of an anastomotic defect or may originate from backflow from patent lumbar, inferior mesenteric, or hypogastric branches.

In the early postoperative period, special care must be taken to identify an endoleak source, because graft pulsatility may result in the movement of nonclotted liquid blood (Fig. 5.5), which gives the appearance of persistent flow within the aneurysm sac—a *pseudoleak.* When an aneurysm is effectively excluded, it usually becomes stable or decreases in size over time. Any increase in size should signal the presence of persistent flow within the aneurysm sac and the continued risk of rupture. Therefore, it is important to obtain serial cross-sectional diameter measurements of the aneurysm after endovascular graft repair.

Arterial Occlusive Disease

The chief concerns with endovascular grafts performed for occlusive disease are graft patency and the presence of flow-limiting lesions within the graft, at anastomotic sites, or involving adjacent inflow and outflow vessels. Graft stenosis or occlusion can occur in TPEGs as a result of arterial dissection (secondary to stent deployment and device introduction and manipulation), extrinsic graft

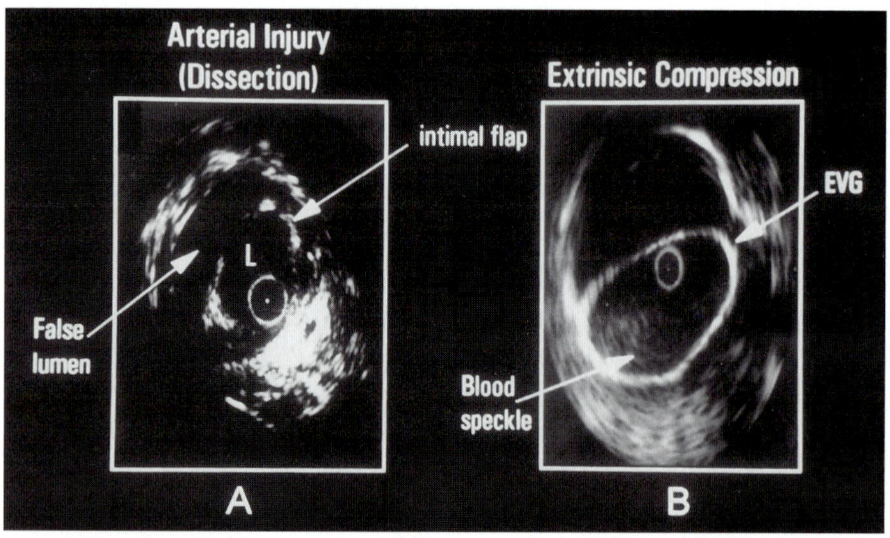

Figure 5.6. The IVUS image on the left shows dissection and an intimal flap that resulted from stent deployment proximal to a transluminally placed endovascular graft performed for occlusive disease. The IVUS catheter resides within the true lumen (*L*). This defect was not seen on intraoperative arteriography. The IVUS image on the right shows extrinsic graft compression from surrounding arterial plaque. Note the ovoid-shaped deformity of the endovascular graft.

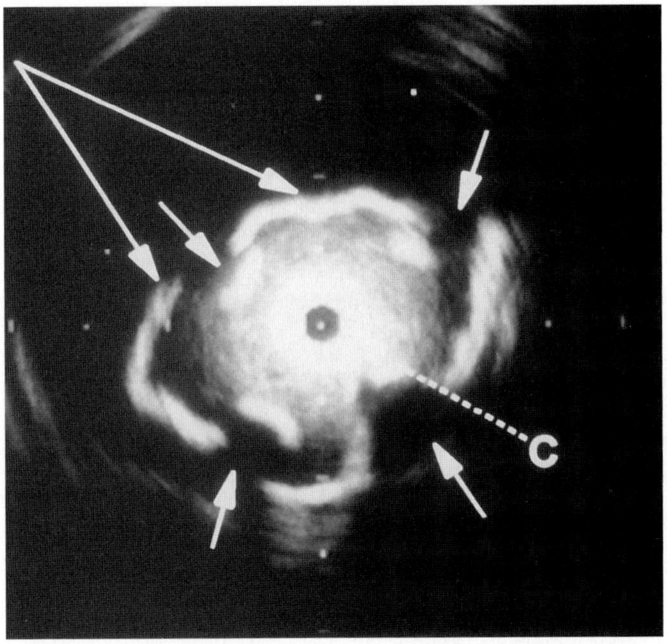

Figure 5.7. IVUS image showing longitudinal folds (*small arrows*) and luminal narrowing of an endovascular graft (*large connected arrows*) placed within an artery with a smaller diameter, resulting in graft redundancy. The IVUS catheter (*C*) is in the center of the image.

compression (Fig. 5.6), graft redundancy (Fig. 5.7), or obstructive plaque proximal or distal to the graft (Fig. 5.8). Extrinsic graft compression most often is seen as a crescent-shaped deformity of the endovascular graft. These lesions have not been found to be progressive in nature. As with standard bypass grafts, other causes of long-term graft failure, such as perianastomotic intimal hyperplasia or the development of new atherosclerotic lesions, are also possible and should be searched for (1,15–20). Accordingly, the duplex ultrasound examination includes Doppler interrogation of the adjacent inflow/outflow arteries as well as the body of the graft to determine flow velocities, obtain velocity waveforms, and document the presence of any flow abnormalities. B-mode ultrasound examination is used in conjunction with color and spectral Doppler waveform analysis to facilitate vessel and lesion identification and lesion characterization (1,9).

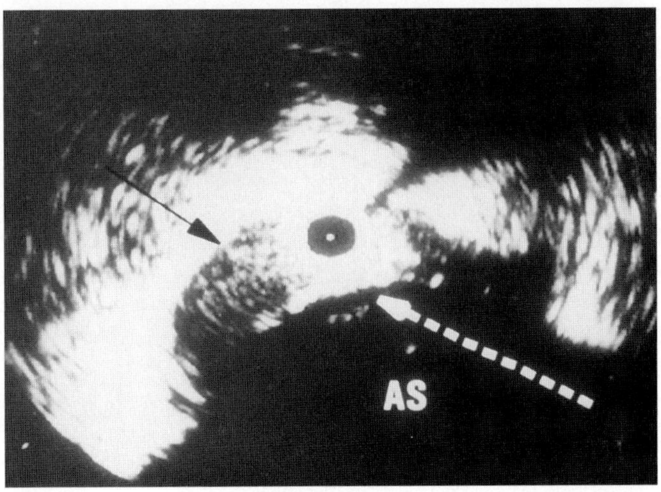

Figure 5.8. IVUS image showing severely calcified prominent arterial plaque (*broken white arrow*) that was not seen on a preoperative contrast arteriogram. The ultrasound beam is attenuated and a large acoustic shadow (*AS*) is seen. The black arrow connotes the wall of an abdominal aorta.

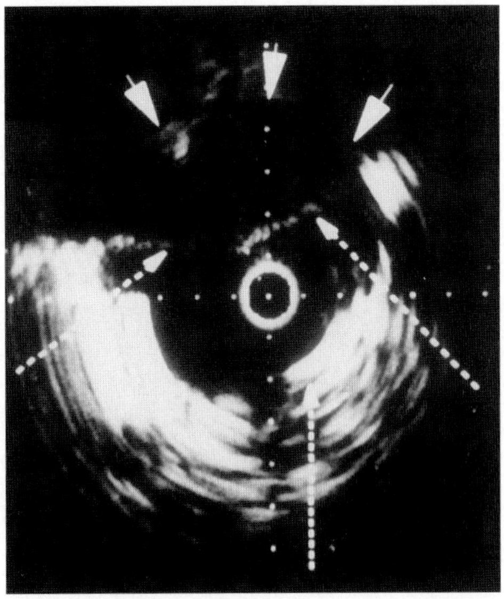

Figure 5.9. A stent deformity (fracture) is seen at the proximal end of a transluminally placed endovascular graft that was placed for traumatic injury of the subclavian artery. The stent (*broken arrows*) has an irregular configuration and is not completely apposed to the surrounding arterial wall (*arrowheads*). Compare this with the stent seen in Figure 5.2.

Traumatic Arterial Injuries

Patency, intimal hyperplasia within or adjacent to stents, and stent compression or deformity are the principal short- and long-term concerns involving TPEGs placed for the treatment of civilian traumatic lesions. Color duplex ultrasound is useful for the preoperative planning of patients undergoing endovascular grafting for trauma.

After the diagnosis of an arterial injury, diameter measurements of the involved artery proximal to the injury and the length of the arterial defect are determined. The stent, graft, and balloon angioplasty catheter size can then be selected based on these measurements. Color duplex ultrasound is used postoperatively to document exclusion of the lesion and the return of normal flow dynamics in the native circulation (2).

Arteriovenous Fistulas

Patency, persistent periprosthetic flow, and venous thrombosis are the primary concerns immediately after the treatment of arteriovenous fistulas with TPEGs. Therefore, both the artery with the endovascular graft and the adjacent vein must be thoroughly evaluated. Because a portion of the endovascular graft is exposed in the venous system and because the flow rate within the vein is suddenly decreased, these veins are prone to thrombosis, which can be detected by color duplex ultrasound.

This thrombotic phenomenon may warrant short-term systemic anticoagulation. Long-term issues include the determination of graft patency and the detection of graft lesions or stent deformity, which can result from poor or incomplete deployment and local bony compression (Fig. 5.9).

CONCLUSIONS

TPEGs have been successfully used for the treatment of a wide range of vascular problems. Although angiography and spiral CT scanning are important in the evaluation of these grafts, ultrasound imaging techniques can be the principal tools for initial, long-term, and serial postoperative assessment as well as surveillance of these devices. Detailed knowledge of the underlying pathology and individual procedures is required for adequate evaluation using ultrasound imaging techniques.

References

1. Berdejo GL, Lyon RT, Marin ML, et al. Color Doppler ultrasonography for the evaluation of endovascular PTFE stented grafts for occlusive disease. J Vasc Technol 1997;21(1):11–15.
2. Berdejo GL, Del Valle WN, Lyon RT, et al. Value of color Doppler ultrasonography in the planning and assessment of endovascular stented grafts for traumatic arterial injury. J Vasc Technol 1996;20(3):178 (Abstract).
3. Berdejo GL, Lyon RT, Sanchez LA, et al. Color duplex ultrasound evaluation of endovascular stented grafts for aneurysm repair. J Vasc Technol 1998, in press.
4. White R. Intravascular ultrasound for imaging of diseased arteries. In: Ernst CB, Stanley JC, eds. Current therapy in vascular surgery, 3rd ed. St. Louis: Mosby-Yearbook, 1995.
5. Kitney RI. Intravascular ultrasound imaging. In: Bernstein EF, ed. Vascular diagnosis, 4th ed. St. Louis: Mosby-Year Book, 1995.
6. White RA, Cavaye DM, Kopchok GE. Experimental applications of endovascular ultrasound. In: Bernstein EF, ed. Vascular diagnosis, 4th ed. St. Louis: Mosby-Year Book, 1995.
7. Yock PG, Fitzgerald PJ, Krishnankutty S, Ports TA. Clinical applications of intravascular ultrasound imaging. In: Bernstein EF, ed. Vascular diagnosis, 4th ed. St. Louis: Mosby-Year Book, 1995.
8. Marin ML, Veith FJ, Lyon RT, et al: Transfemoral endovascular repair of iliac artery aneurysms. Am J Surg 1995;170:179–182.
9. Marin ML, Veith FJ. Complications of endovascular stented grafts for the treatment of arterial lesions. In: Greenhalgh RM, Fowkes FGR, eds. Trials and tribulations of vascular surgery. London: WB Saunders, 1996.
10. Moneta GL. Duplex imaging of infrainguinal disease. In: Ernst CB, Stanley JC, eds. Current therapy in vascular surgery. 3rd ed. St. Louis: Mosby-Year Book, 1995.
11. Marin ML, Veith FJ, Cynamon J, et al. Initial experience with transluminally placed endovascular grafts for the treatment of complex vascular lesions. Ann Surg 1995;222:449–469.
12. Marin ML, Veith FJ, Cynamon J, et al: Transfemoral endovascular stented graft treatment of aorto-iliac and femoropopliteal occlusive disease for limb salvage. Am J Surg 1994;168:156–162.
13. Cragg AH, Dake MD. Percutaneous femoropopliteal graft placement. J Vasc Interv Radiol 1993;4:455–463.
14. Chuter T, Deweese JA. Treatment of abdominal aortic aneurysm by endovascular grafting. In: Ernst CB, Stanley JC, eds. Current therapy in vascular surgery, 3rd ed. St. Louis: Mosby-Year Book, 1995.
15. Bandyk DF, Schmitt DD, Seabrook GR, et al. Monitoring functional patency of in situ saphenous vein bypasses: The impact of a surveillance protocol and elective revision. J Vasc Surg 1989;9:286–296.
16. Sumner DS, Mattos MA, Hodgson KJ. Surveillance program for vascular reconstructive procedures. In: Yao J, Pearce W, eds. Long-term results in vascular surgery. Norwalk: Appleton & Lange, 1993.
17. Bandyk DF. Monitoring during and after distal arterial reconstruction. In: Bernstein EF, ed. Vascular diagnosis, 4th ed. St. Louis: Mosby-Yearbook, 1995.
18. Bandyk DF. Surveillance of lower extremity bypass grafts. In: Ernst CB, Stanley JC, eds. Current therapy in vascular surgery, 3rd ed. St. Louis: Mosby-Year Book, 1995.
19. Cato R, Kupinski AM. Graft assessment by duplex ultrasound scanning. J Vasc Technol 1994;18(5):307–310.
20. Cato R. Doppler color flow imaging of infrainguinal arterial bypass grafts. J Vasc Technol 1991;15(3):150–155.

Chapter 6

Color Duplex Ultrasound Evaluation of Endovascular Stented Graft Aortic Aneurysm Repair

Hector Lardani

It is well recognized that many imaging techniques are useful in performing endovascular grafting procedures to treat abdominal aortic aneurysms (AAAs). Spiral contrast computed tomography (SCT) and contrast arteriography are both useful in determining whether a patient is suitable to receive various endovascular graft devices. Both techniques provide important information regarding aneurysm and neck morphology and measurements. These modalities also provide information regarding branch patency and iliac size, tortuosity, and disease. Both techniques are also useful in determining the efficacy of the endovascular grafting procedure in excluding the aneurysm and successfully treating it. It is generally agreed that carefully performed SCT with delayed filming is among the most accurate and widely used methods of determining the presence or absence of endoleaks.

A smaller body of knowledge is gradually developing regarding the use and value of color duplex ultrasonography for following endovascular graft repairs of AAAs and for determining the presence and location of graft endoleaks and changes in aneurysms.

In general, all patients undergoing transfemoral endovascular graft repair of an AAA are required to have postoperative duplex evaluation according to the following schedule: A color duplex study is performed immediately after insertion of the prosthesis, 1 month later, and every 6 months thereafter or sooner, if necessary.

This chapter describes the experience with color duplex evaluation of endovascular AAA repair in one institution in which this procedure was first carried out in 1990 (1).

EQUIPMENT AND PATIENT PREPARATION

Most of our patients were studied using an Ultramark 9 HDI (Advanced Technology Laboratories, Bothell, WA) utilizing a phased array 3 to 2 MHz or a curve array C3, 4, or 3-MHz transducer. A few studies were performed with an Ultramark 7 (Advanced Technology Laboratories) using a 3-MHz phased array probe. We use a preset program for AAAs. During the procedure, we optimize the color or echo gain, the size and position of the color box, and the color velocity range. As with any abdominal ultrasound study, patients must fast for at least 8 hours to avoid artifacts produced by intestinal gas. No other preparation is required.

Patients are generally evaluated while they are in the supine position. They are instructed to relax to soften their abdominal muscles; occasionally, we ask them to bend their knees for this purpose. Sometimes it is necessary to explore the patient in the left lateral decubitus position, attempting to displace the intestinal mass forward and away from the area being insonated in the hope that the annoying air artifacts can be reduced.

TECHNIQUE

To obtain a longitudinal view of the aorta, we begin the examination at the epigastrium in the midline or slightly to the left. Important landmarks include the origin of the celiac trunk, which arises at a 90-degree angle and a few millimeters below, the origin of the superior mesenteric artery taking off at an acute angle and coursing parallel with the aorta.

Scanning distally should permit visualization of the

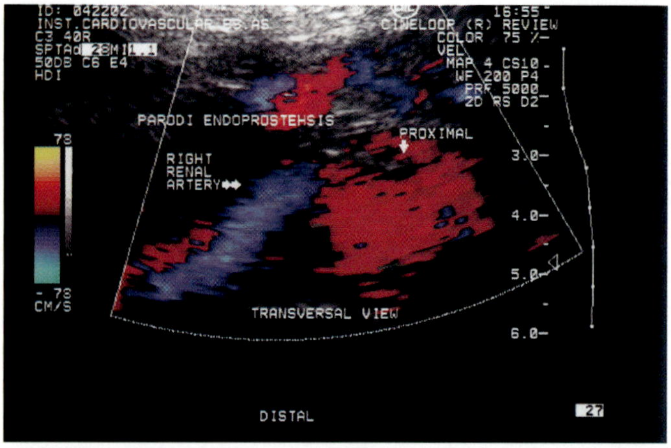

Figure 6.1. Transverse view of the proximal abdominal aorta above the insertion of an endoprosthesis showing the blue narrow flow of the intact right renal artery. The laminar flow is depicted by the homogeneous dark blue appearance of the image.

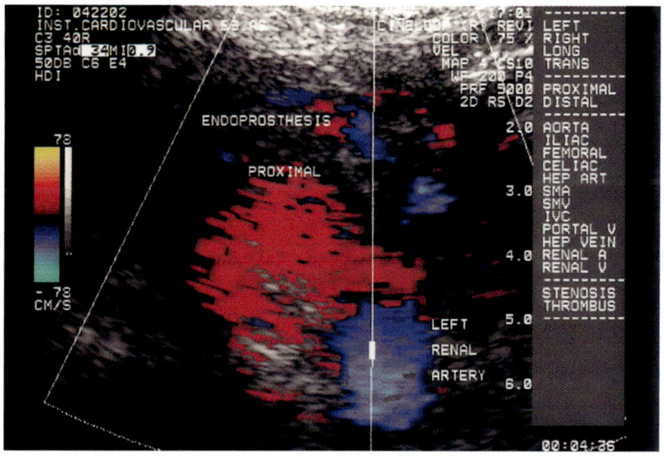

Figure 6.2. Blue image of the left renal artery in the same transverse view as in Figure 6.1, which shows an undisturbed flow. Assessment of renal artery flow is mandatory when the stent covers the renal artery ostia.

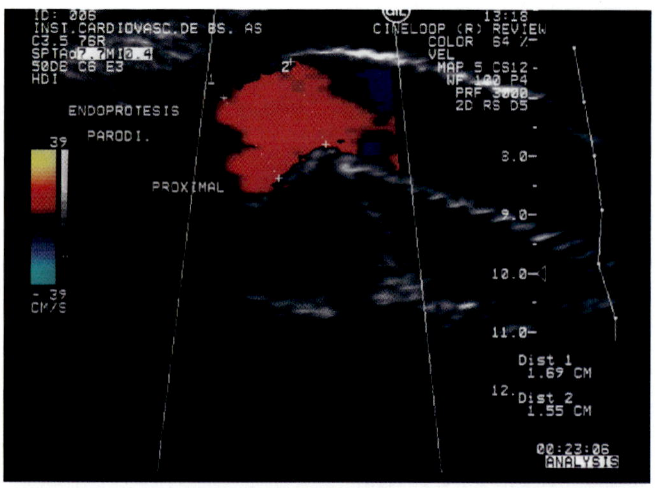

Figure 6.3. Longitudinal view of a prosthesis with the proximal segment of the stent on the left side. Initially, the lumen follows a left-to-right and inferior-to-superior direction as indicated by the red flow image, and then it bends downward. The initial diameter of the stent is 1.69 cm with no evidence of a leak.

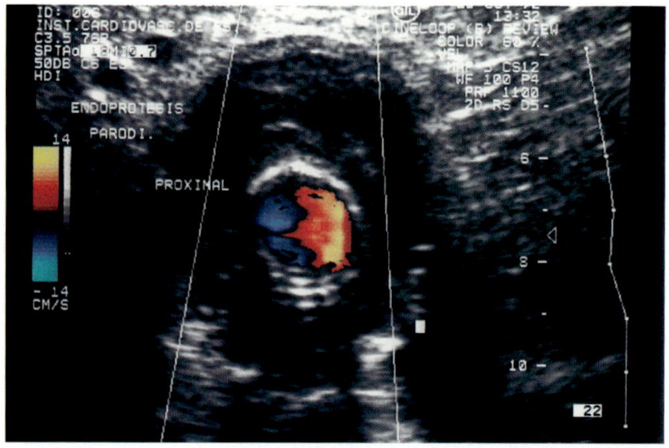

Figure 6.4. Transverse view of a graft at the site of insertion of the stent. The structure of the stent and the color flow limited to the lumen of the graft surrounded by the dark shadow of an aneurysm can be seen.

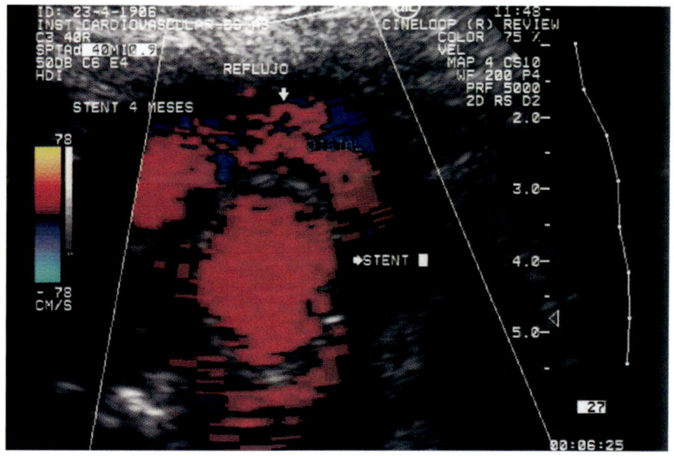

Figure 6.5. Transverse view showing a clear-cut example of a proximal leak, which is depicted in the upper part of the picture with the dark shadow of an aneurysm in the background. The round appearance of the graft structure and the color flow inside it are seen in the center of the image.

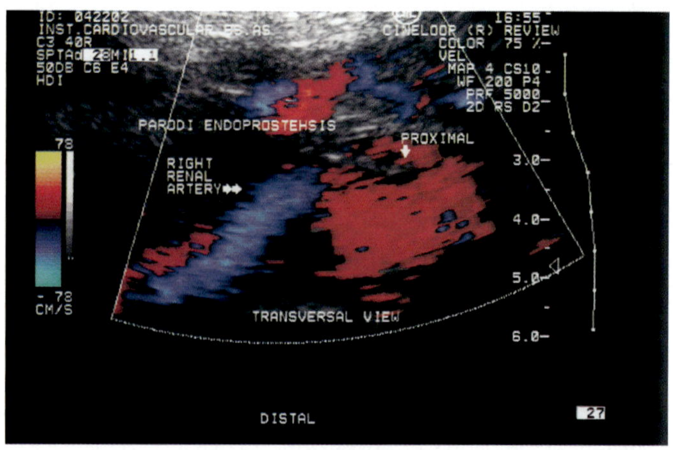

Figure 6.6. Longitudinal view of a proximal leak in the form of a narrow (0.53-cm) red color flow jet above and parallel with the central flow of the graft.

passage of the left renal vein from left to right anterior to the aorta. This important vessel is better identified in a transverse section.

Immediately deep to the left renal vein, we look for the origin of the right renal artery arising from the right anterior aspect of the round shadow of the aorta and coursing inferiorly and to the right. Typical Doppler spectral display shows an important diastolic velocity component (30% or more of the peak velocity) within this vessel and confirms the identification of the renal artery.

When scanning slightly deep to the aorta, we should be able to recognize the origin of the left renal artery arising from the left posterolateral aspect of the aorta and coursing inferiorly and to the left. The location of the renal arteries and their identification in patients with an aortic endovascular graft are important objectives of the study (Figs. 6.1 and 6.2). In six patients, the proximal uncovered portion of the stent overlapped the renal artery orifices, and no complications were noted.

The proximal segment of the prosthesis at the site of the stent fixation can be seen below the renal arteries. The diameter at this site should be measured and recorded for future follow-up, checking for possible dilatation (this area represents the proximal neck of the aneurysm). It should be insonated in both longitudinal and transverse views (Figs. 6.3 and 6.4). At this point, it is important to look for a proximal leak from the aorta into the aneurysmal sac through a hiatus caused by an imperfect seal between the aortic wall and the endovascular graft (Figs. 6.5 and 6.6). The presence of a leak may indicate the persistence of an undesirable effect exerted by the high aortic pressure on the wall of the aneurysm. If this leak is substantial, a second attempt to exclude the aneurysm may be indicated.

There is no easy way to evaluate the magnitude of a leak. We have to insonate at an adequate angle between the ultrasound beam and the direction of the jet to

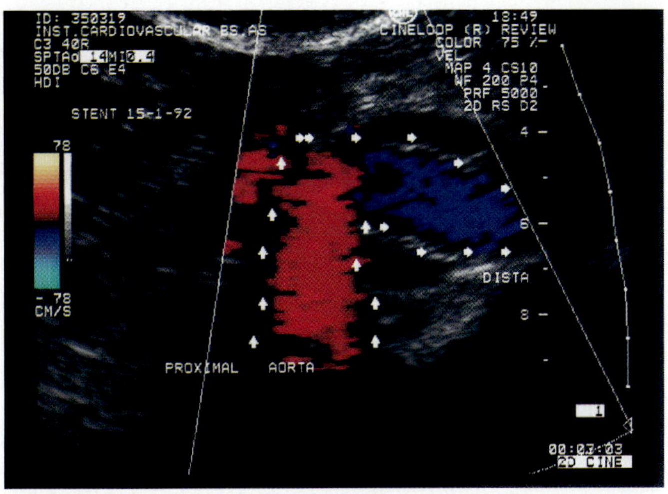

Figure 6.7. Longitudinal view showing a marked angle (more than 90 degrees) between the proximal aortic flow and the site at which the graft was inserted.

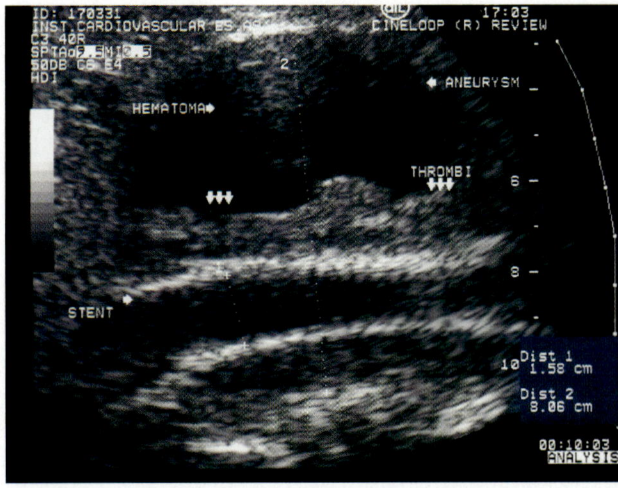

Figure 6.8. Longitudinal view showing a large aortic aneurysm (8.06 cm), with the horizontal shadow of the graft (1.56-cm). Clear evidence of clot formation in close contact with the graft walls is seen.

obtain the best possible Doppler shift and color signal. We should then estimate its magnitude by comparing the color area of the leak with the diameter of the graft. It is hoped that a more precise way of quantifying leaks will be developed in the future.

Occasionally distal to the proximal stent, the graft makes a sharp turn and follows the curves and deformations of the aorta produced by the presence of the aneurysm (Fig. 6.7). If this occurs, we record these findings and prepare for the eventuality that the aortic pulsatile flow, which predominantly pounds to one side of the graft, might generate dilatation, displacement, or a foreshortening of the graft over time. We also record the size of the aneurysm and the presence and evolution of clot formation within the sac.

We anticipate that a reduction in the size of an aneurysm and progressive clot formation might indicate a favorable outcome from the procedure (Fig. 6.8). We also evaluate the distal runoff, keeping in mind that the graft may have been inserted with the distal stent into the aortic lumen or might entirely empty into one of the iliac arteries. In the former situation, we look to see whether the aorta at the end of the graft is still dilated, that is, whether the graft is too short after the insertion procedure or retraction has occurred with the passage of time. If any of these has occurred, blood flow may be detected between the graft and the aneurysm (distal reflux), which may eventually maintain a high wall stress in the aneurysm and, in addition, prevent clot formation within the aneurysmal sac by a washout effect (Fig. 6.9).

A situation in which the reflux is significant may require the intraluminal insertion of a second graft; a severe leak may lead to a total collapse of the stent-graft, as seen in Figures 6.10 and 6.11.

Figure 6.12 shows a red image of the superior mesenteric artery above the dark shadow of the aortic aneurysm. This should not be confused with a proximal leak.

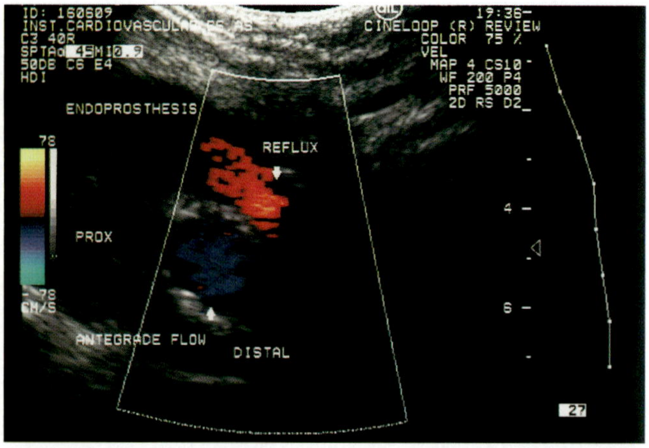

Figure 6.9. Longitudinal view of the distal end of a graft. Antegrade flow from left to right in blue (away from the transducer) and a red shadow of flow above it toward the left upper portion of the picture indicate the presence of distal reflux.

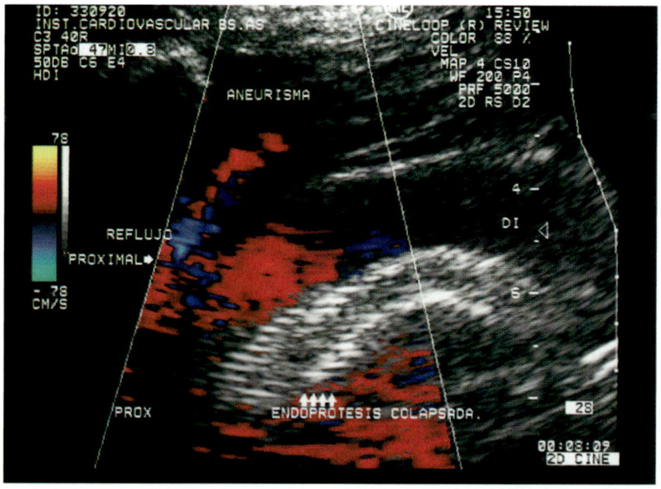

Figure 6.10. Longitudinal view of a collapsed endoprosthesis as the end stage of a severe leak. The double-striated white image of the graft is surrounded by red images above and below, indicating leaks into the aortic aneurysm.

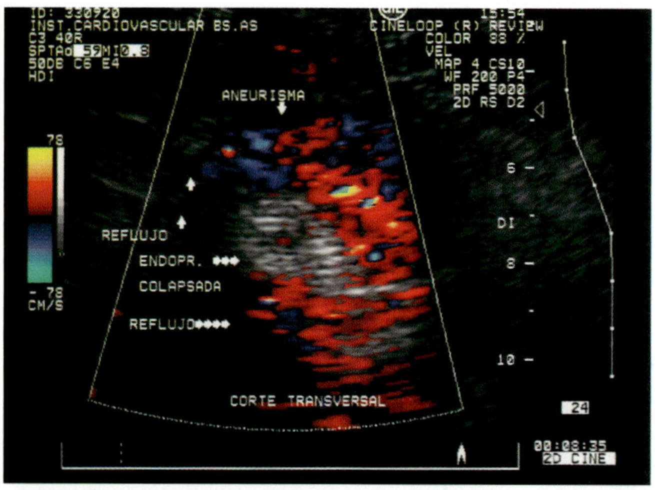

Figure 6.11. Transverse view of a nonfunctioning endoprosthesis with the red shadows of a severe leak around it.

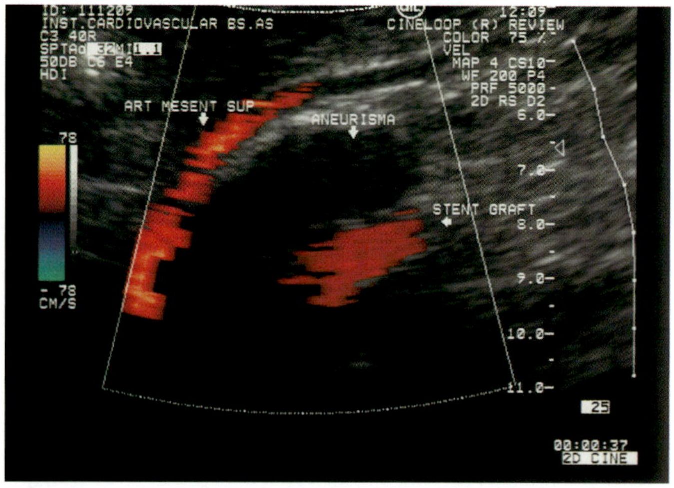

Figure 6.12. Normal homogeneous red flow of the superior mesenteric artery, which is not to be interpreted as a proximal endoprosthetic leak. The white parallel shadows of the mesenteric artery walls, the aneurysmic aortic wall below it, and the flow of a normal functioning graft inside the aneurysm can be clearly seen.

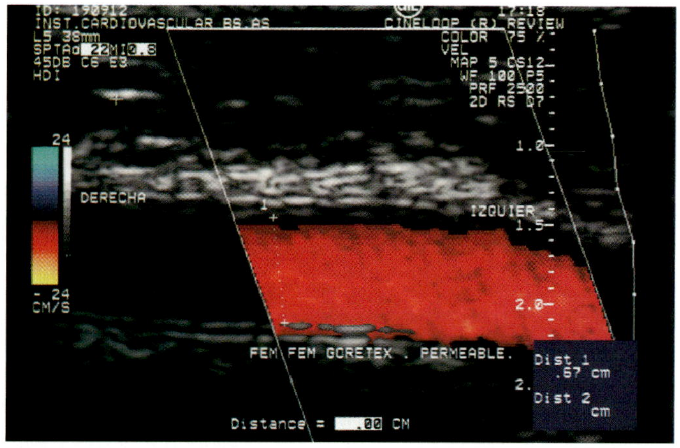

Figure 6.13. Normal color Doppler aspect of a femorofemoral graft. The double white shadow of the walls and the homogeneous dark red flow are clearly demonstrated.

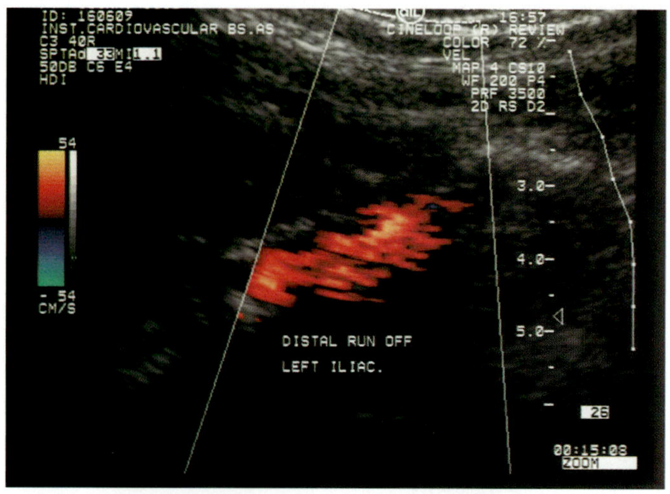

Figure 6.14. Normal distal runoff of a graft ending in the left iliac artery. The undisturbed red flow is well visualized, suggesting that there is no pressure gradient.

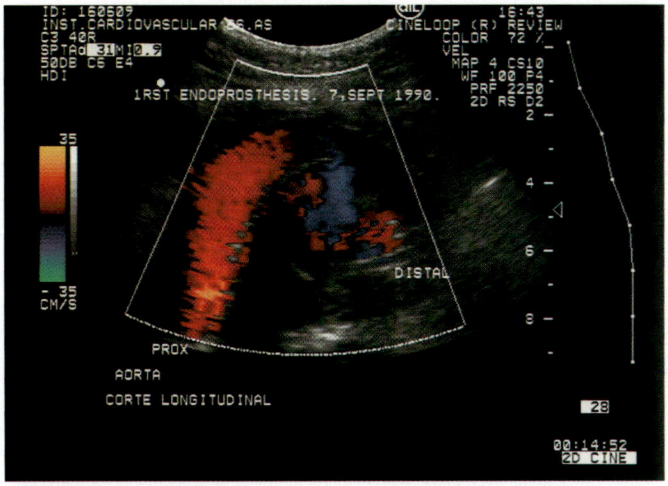

Figure 6.15. Longitudinal view of the first aortic endoprosthesis ever inserted. The marked bending of the aorta is a challenge that was successfully overcome by the graft.

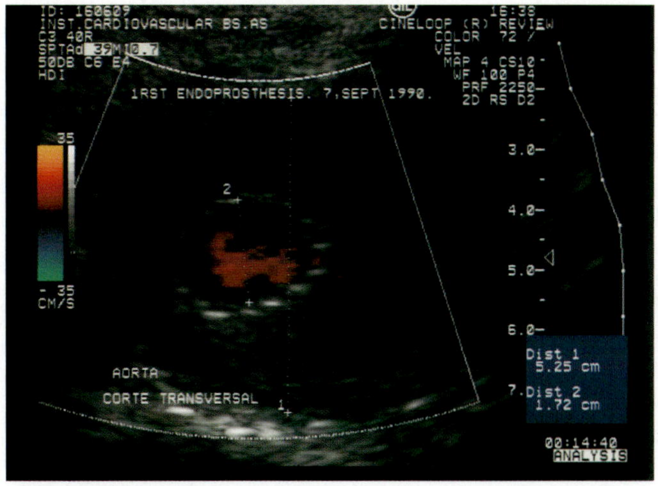

Figure 6.17. Transverse view of Dr. Parodi's first endovascular graft aneurysm repair showing an absence of leak and central prosthetic undisturbed flow.

If the distal segment of the graft is connected to one of the iliac arteries, a femorofemoral bypass is created, and its functioning is assessed as an additional important goal of the study (Fig. 6.13). Another point is to determine whether there is a pressure gradient at the site where the graft empties into the iliac arteries (Fig. 6.14).

The presence of tortuous, atherosclerotic iliac arteries is more common in high-risk patients, and the unavoidable further reduction of the lumen size by the graft may generate a pressure drop. Follow-up should determine whether these gradients will produce an unfavorable hemodynamic effect by means of high pressure and strain within the graft. Our overall impression is that proximal leaks and distal refluxes are infrequent, and, if not present during the immediate postoperative period, they are unlikely to occur during follow-up.

In a recent study, Dr. Parodi and colleagues (2) reviewed 109 procedures and found that three proximal and seven distal leaks (endoleaks) were corrected by endovascular techniques in a second procedure and that three proximal and seven distal leaks still remain untreated.

To date, we have not observed dilatation of the proximal neck or of the graft itself. After a successful procedure is achieved, the graft appears to be a stable device.

Figures 6.15, 6.16, and 17, produced 7 years after Dr. Parodi inserted his first device in an 80-year-old patient, show the excellent results obtained. In spite of the tremendous tortuosity of the aorta and after a second endoluminal procedure for a proximal leak, the prosthesis now shows perfect function with no leaks and an excellent seal at the proximal and distal necks of the aneurysm.

Recent developments include bifurcated grafts with

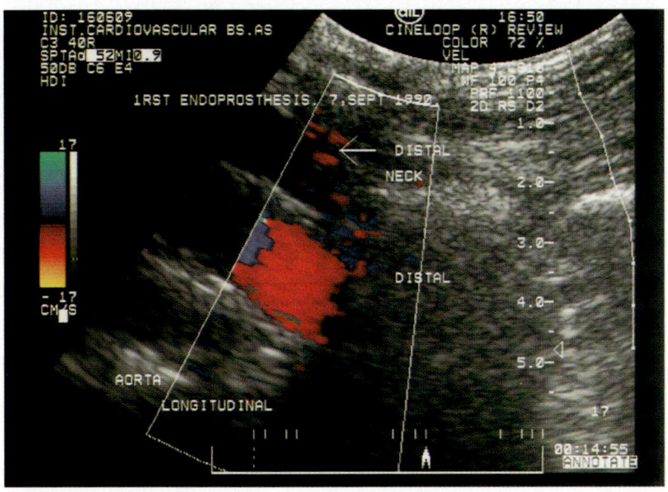

Figure 6.16. Distal end of a prosthesis with perfect sealing at the distal neck of the aneurysm. The homogeneous red shadow of the aortic flow indicates that normal flow is filling the prosthesis with no evidence of reflux.

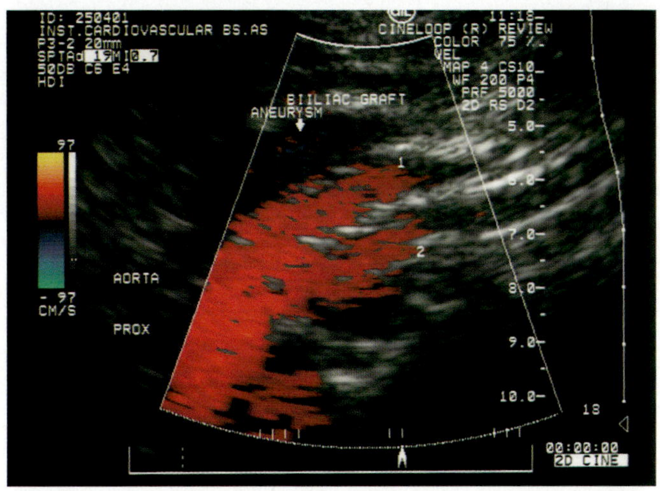

Figure 6.18. Longitudinal view of a bifurcated graft. On the left side is the proximal portion of the stent. In the center is the area of bifurcation in the superior (superficial) and inferior (deep) branches. No evidence of leak is seen.

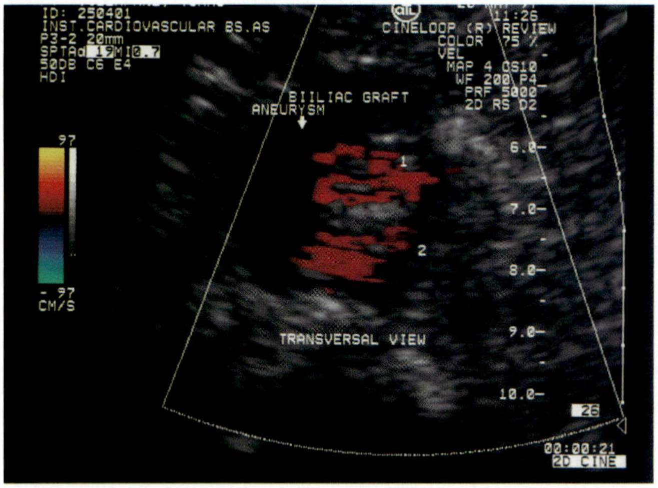

Figure 6.19. Transverse view of a bifurcated prosthesis. The superficial and deep branches are well depicted. No evidence of leak is present. Previous knowledge of the nature and characteristics of the graft should be available to the sonographer so that a bifurcated graft is not confused with a periprosthetic leak. This is particularly important when the technique and the equipment offer less than adequate color flow images.

the two distal branches attached to the iliac walls by additional stents (Figs. 6.18 and 6.19).

CONCLUSION

A noninvasive procedure with the ability to obtain anatomic and hemodynamic data with no risk or discomfort to the patient (such as a color duplex ultrasound study) may prove to be the most cost-effective technique for the assessment of endovascular graft repairs of AAAs, particularly in elderly, high-risk patients with diffuse atherosclerosis, precarious renal function, and other health problems.

References

1. Parodi JC, Palmaz JC, Barone HD. Transfemoral intraluminal graft implantation for abdominal aortic aneurysms. Ann Vasc Surg 1991;5:491–499.
2. Parodi JC, Barone A, Piraino R, Schonholz C. Endovascular treatment of abdominal aortic aneurysms: Lessons learned. J Endovasc Surg 1997;4:102–110.

Chapter 7

Anatomic Variations of Abdominal Aortic Aneurysms

Richard M. Green

One of the major advantages of a conventional open operation is the flexibility afforded the surgeon to alter the course of the procedure when conditions warrant. This is certainly true in the case of resection of an abdominal aortic aneurysm (AAA). Decisions about the size of the graft, the location of the anastomoses, and the necessity of visceral revascularization can all be made in the operating room. In addition, the graft can be tapered to accommodate variations in the diameter of the proximal and distal necks, and local endarterectomies render the use of heavily calcified vessels possible. This flexibility enables the surgeon to use straight grafts for most AAA repairs (1), thereby simplifying the procedure.

Preoperative decision-making in the conventional approach to AAA resection requires a decision to operate, which is often based on the maximum transverse diameter and a plan on where to clamp proximally. The latter choice can be made preoperatively from information obtained on a standard computed tomography (CT) scan or in the operating room. Because endovascular procedures do not allow this same flexibility, more preoperative anatomic information is required for every patient. The specific anatomic questions that must be dealt with for endovascular repair of an AAA are graft configuration and access requirements. Ultimately, the specific capabilities of each endovascular system determine whether it is appropriate for any given patient. These capabilities relate to the dimensions, flexibility, length of stents, and configuration of the graft itself, that is, straight or bifurcated. Knowledge about the anatomic variations of AAAs will also ensure that the design of endovascular devices accommodates the needs of the maximum number of patients.

DIAGNOSTIC STUDIES

The imaging techniques used to evaluate an AAA can be categorized into those used for screening and those used to plan the operation. Screening tests that provide information on the transverse diameter include ultrasonography and conventional CT scans. These tests do not completely define the aneurysm's relationship to the renal arteries, the importance of the inferior mesenteric artery, or the tortuosity or angulation of the iliac arteries. Moreover, these tests are usually sufficient in patients with AAAs undergoing conventional operation, but they do not provide the necessary information for transfemoral repair.

Specialized CT scans—either three-dimensional (3-D) or spiral—and arteriography are required when the juxtarenal segment, the visceral arteries, and the iliac arteries are being evaluated. The 3-D images provide information on the suitability of a given patient for an endovascular repair.

ANATOMIC REQUIREMENTS FOR ENDOVASCULAR GRAFTS

Regardless of the lesion, an endovascular graft must securely attach and conform to the sizes of the inflow and outflow vessels. Because current devices use stents for attachment, there must be a suitable length of normal vessel to provide a secure hemostatic connection. The length of normal aorta required depends on the requirements of the specific stent. Current models require approximately 20 mm of normal aorta for secure placement and avoidance of the renal orifices. This distance is arbitrary and may decrease as new stents are developed.

The size of the graft must conform to the diameter of the vessel at the site of attachment. Although some degree of elasticity may be tolerated, any fabric used in endovascular grafting must possess the long-term durability of current Food and Drug Administration (FDA)-approved materials. Precise measurements of the diameter at each connection are therefore required, and a straight graft must be custom-tailored to fit each patient.

Finally, access to the AAA from a remote artery, usually the femoral, requires a low-profile prosthesis with enough flexibility to accommodate the tortuosity and angulation of the iliac arteries.

VARIATIONS IN THE NORMAL AORTIC SEGMENTS (NECKS)

A series of patients undergoing elective AAA resection at the University of Rochester were evaluated with conventional and 3-D CT scans to determine the anatomic variations present. The purpose of this study was to identify the requirements of an endovascular device with stents requiring a 20-mm neck. The proximal and distal neck lengths and the transverse diameters were measured (2). The proximal necks had a mean length of 26.7 ± 4.1 mm and were longer than 20 mm in 64% of the patients and longer than 10 mm in 82%. The mean length of the distal necks or the distance from the distal end of the aneurysm to the lateral margin of the iliac arteries was 4.2 ± 1.6 mm. The distal neck was longer than 20 mm in 4.5% and longer than 10 mm in 18% of the patients. The measurements of the normal aortic segments in patients undergoing aneurysm resection at the University of Rochester are shown in Figure 7.1.

If more than 20 mm of normal aorta proximally and distally are required, only 4.5% of the patients would be candidates for a straight endovascular graft. These data have been confirmed in the phase I and phase II studies sponsored by Endovascular Technologies (Menlo Park, CA). Their screening criteria have eliminated almost 9 of 10 candidates with AAAs because of the lack of a distal neck.

The transverse diameters of the normal aorta were also measured. The mean transverse diameter was 24.9 ± 2.2 mm below the renal arteries and 27.1 ± 1.4 mm at the proximal end of the aneurysm. These values indicate that the neck itself has a cylindrical shape. The mean diameter of the distal neck was 24.6 ± 1.4 mm, which was no different from the mean diameter of the proximal neck. However, differences of more than 5 mm were found in 59% of the patients. Because the two necks are not the same size, a conventional straight prosthesis with equal diameters at each end would not be suitable unless there was considerable elasticity in the fabric.

It is not yet known whether the increased elasticity to accommodate the variations in the diameters of the attachment necks will be acceptable. If not, custom-tailored straight grafts, unique to each patient, would be required. The manufacturing and hospital supply logistics of this would be overwhelming. These anatomic variations led us to the conclusion that a bifurcated prosthesis would be necessary to treat most AAAs with endovascular grafts (3).

The 3-D CT scan shown in Figure 7.2 demonstrates the type of aneurysm that would be amenable to an endovascular straight graft repair. Both a long proximal and a distal segment of normal aorta could accommodate the stent attachments. If all AAAs were configured like this, endovascular repair with a straight prosthesis would be a

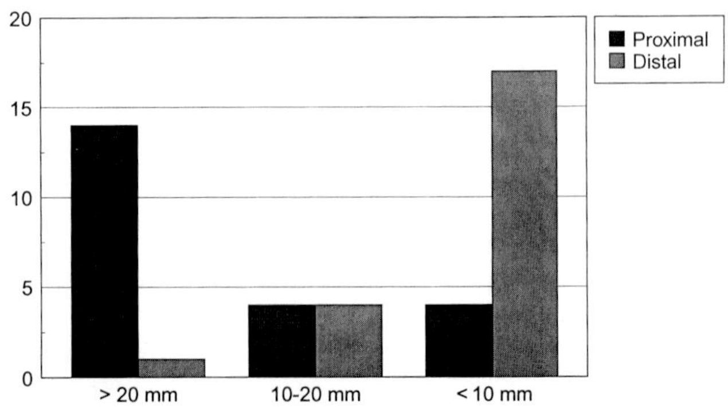

Figure 7.1. Measurements of normal aortic segments in patients undergoing aneurysm resection.

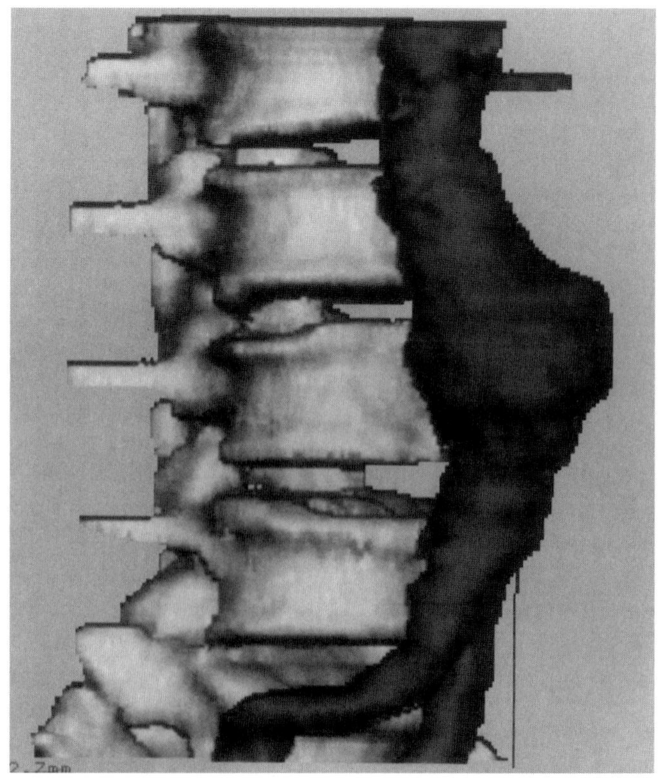

Figure 7.2. Three-dimensional CT scan of an AAA with proximal and distal necks greater than 20 mm. This is an unusual type of aneurysm occurring in less than 5% of the population referred for operation. These aneurysms are often small and not operated on but are the ideal candidates for an endovascular repair with a straight prosthesis.

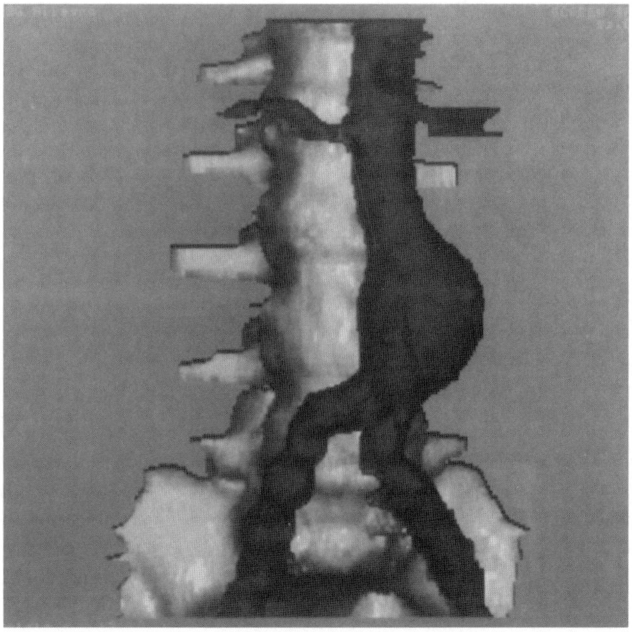

Figure 7.3. Three-dimensional CT scan of an AAA with a long proximal neck but no distal neck. This type of aneurysm could be repaired with a straight graft using conventional techniques but would require a bifurcated endovascular graft. The length of the distal neck must be measured from the lateral wall of the aorta not the iliac flow separator.

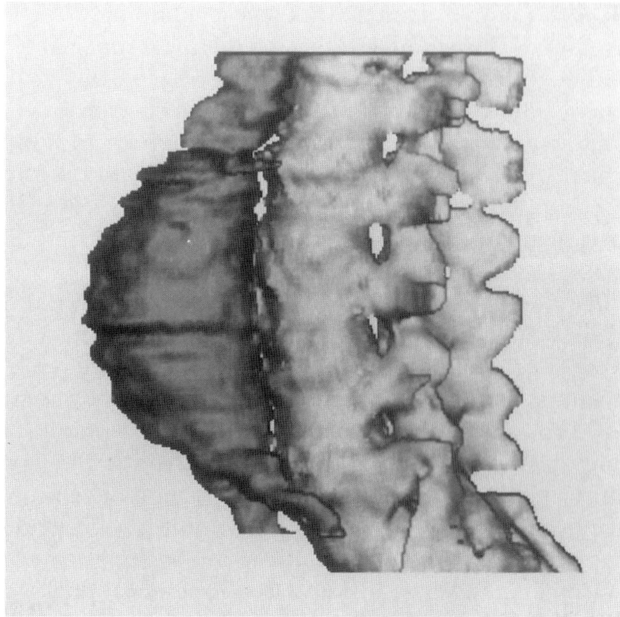

Figure 7.5. This is a large juxtarenal aneurysm. There is neither a proximal nor a distal neck. Repair will probably always require conventional techniques. This type of aneurysm comprises a small percentage of the patients currently referred for repair, but will become a significant fraction of operative volume when endovascular techniques are available for the routine cases.

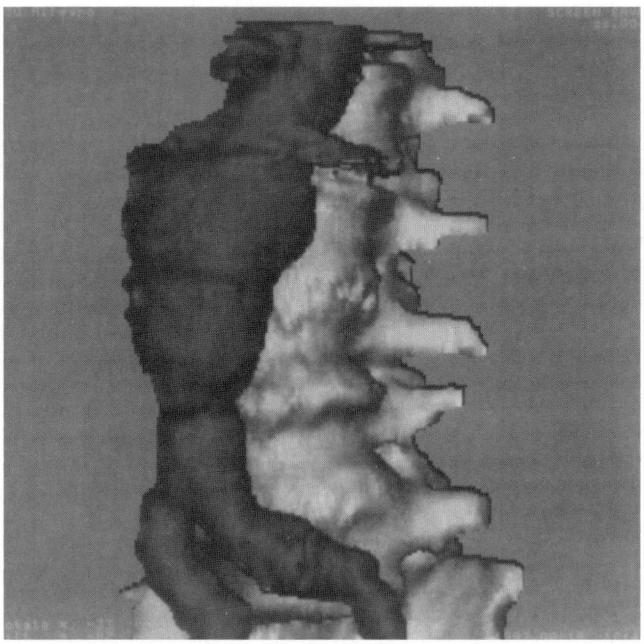

Figure 7.4. This 3-D CT scan shows a small juxtarenal aneurysm with a long distal neck. There is no satisfactory proximal neck to place the stent. Since the proximal neck is absent, repair would require a supraceliac clamp and sutured anastomosis at the level of the renal arteries.

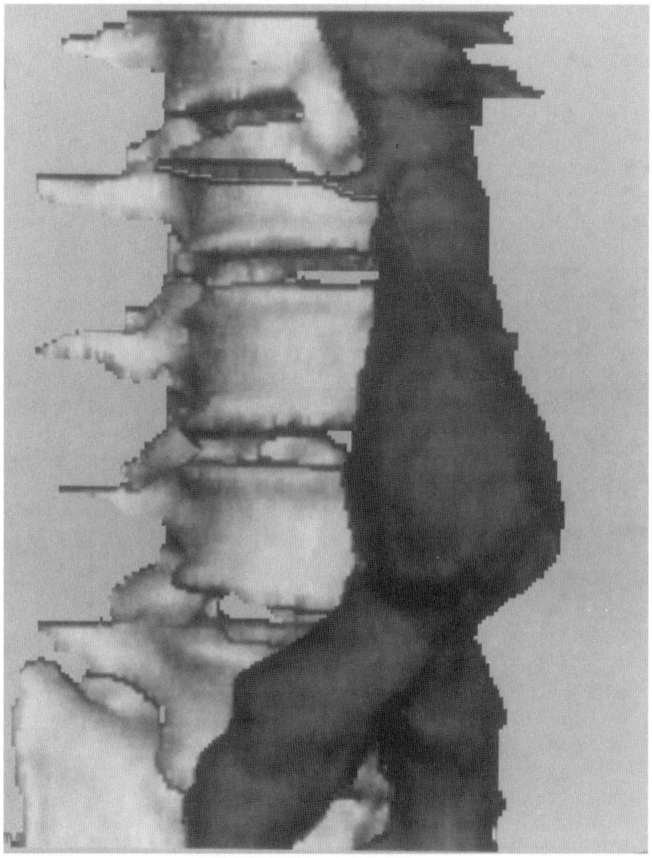

Figure 7.6. This 3-D CT scan shows iliac artery aneurysms that extend to the iliac bifurcation and require conventional repair to protect the internal iliac arteries. Iliac artery aneurysms usually end more proximally, however, and can be repaired with a bifurcated endovascular prosthesis.

relatively simple matter with the technology already developed. Unfortunately, this is a relatively unusual type of aneurysm, occurring in less than 5% to 10% of the patient population considered for aneurysm repair.

These aneurysms with both a long proximal and distal segment of normal aorta are usually small and are followed up, not resected. As the aneurysm grows in transverse diameter, it also elongates and the length of the neck, usually the distal one, shortens.

The most common type of aneurysm is shown in Figure 7.3. This aneurysm has a long proximal neck but a very short distal neck. The length of the distal neck is deceptive. It must be measured from the lateral margins of the aorta not at the iliac bifurcation flow separator; in reality, it is shorter than it appears. This type of aneurysm can be treated with a straight graft using conventional operative techniques, but would require a bifurcated prosthesis for endovascular repair.

Occasionally, an aneurysm has a long distal neck and no proximal neck (Fig. 7.4). This type of aneurysm must be corrected with conventional techniques because the proximal stents would interfere with the renal artery orifices.

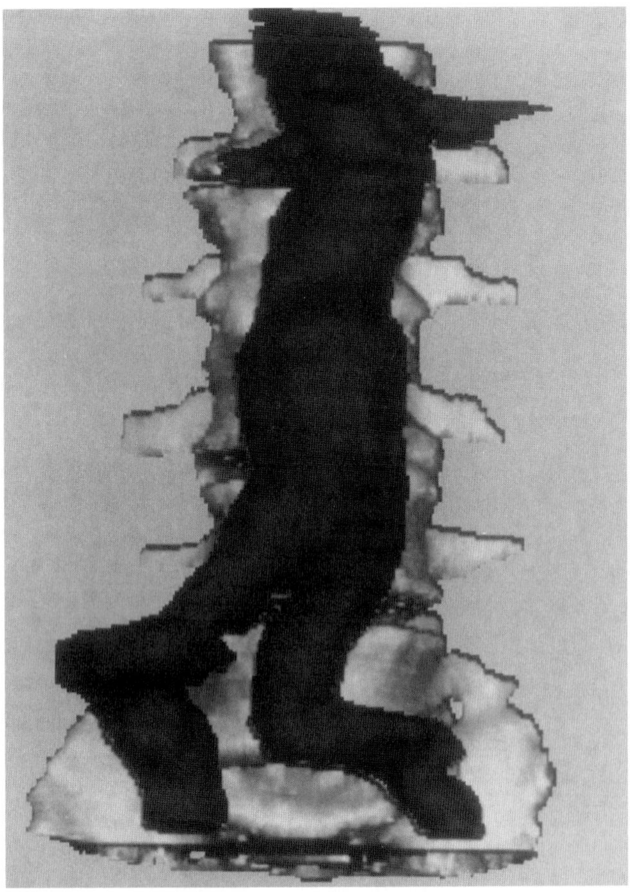

Figure 7.7. This 3-D image provides an excellent means of assessing iliac tortuosity. This bend can be crossed with a guidewire. A moveable C-arm is helpful as the device changes directions several times. Stent flexibility will be the major determining factor in crossing many of these iliac bends and kinks.

Finally, some large aneurysms have no segments of normal aorta. The 3-D CT scan shown in Figure 7.5 demonstrates a juxtarenal aortic aneurysm. These make up a small segment of the population of aneurysms seen at most centers and require conventional repair.

VARIATIONS IN THE ILIAC ARTERIES

The anatomy of the iliac arteries assumes importance in the consideration of endovascular aneurysm repair. Aneurysms, occlusive disease, tortuosity, and kinks each can influence whether an endovascular repair is feasible because of access problems and also what graft configuration is required. Iliac aneurysms were encountered in 32% of the patients and 22% of the iliac arteries studied. The aneurysms usually ended proximal to the iliac artery bifurcation, and 91% of the patients could have had bifurcated endovascular repair without interference with the internal iliac arteries. The CT scan shown in Figure 7.6 demonstrates aortic and bilateral iliac artery aneurysms, which could not be repaired with an endovascular prosthesis without compromising the internal iliac arteries.

Angulation of the iliac arteries can prevent retrograde insertion of an endovascular graft or result in plaque dislodgment and distal embolism. This is more or less of a problem depending on the flexibility and diameter of the device. Angulation can be assessed by both 3-D CT scans or arteriography or both. In the aneurysms examined, the iliac angulation exceeded 45 degrees in 16% of the patients and occurred slightly more often on the left side. The CT scan shown in Figure 7.7 illustrates an extreme degree of angulation in the right iliac artery. A flexible system over a guidewire introduced with the assistance of a moveable C-arm unit should be able to negotiate this vessel. The aneurysm seen in Figure 7.8 shows the common problem of an occlusive kink in the proximal left common iliac artery, which would clearly interfere with the insertion of an endovascular device. In these extreme cases of tortuosity or disease, development of a device that could be inserted through the subclavian artery would prevent complications of atheroembolization.

IMPLICATIONS OF ANATOMIC VARIATIONS ON THE DEVELOPMENT OF ENDOVASCULAR DEVICES

The essence of good product design is knowing at the outset what the device must accomplish. In the case of an endovascular repair of an AAA, the anatomic variations that could be dealt with during a conventional operation become critical for patient selection. Endovascular solutions cannot be expected to pattern themselves after conventional operations.

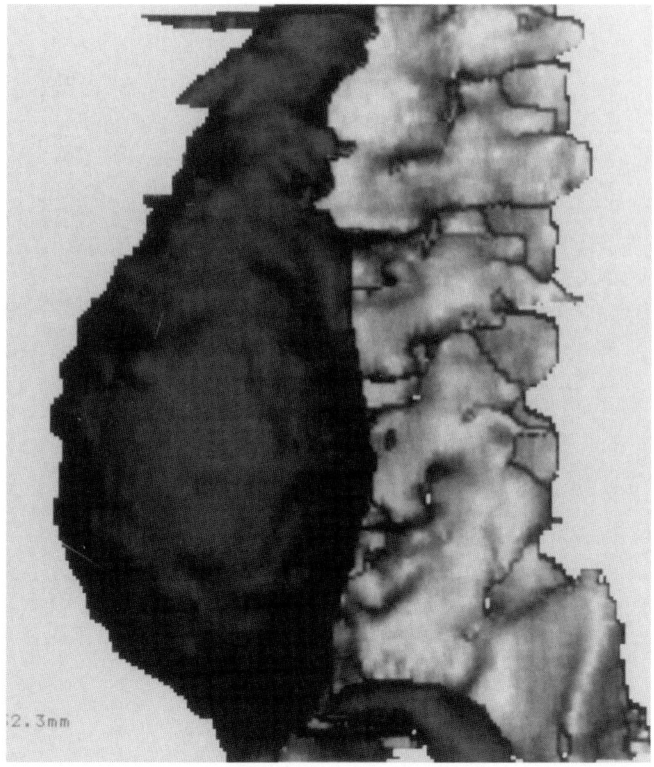

Figure 7.8. This CT scan shows a severe kink at the origin of the left common iliac artery which was associated with a significant stenosis. Access through this type of lesion will be difficult, and a device that could be inserted through the subclavian artery might facilitate the use of an endovascular graft.

For instance, in the case of AAAs, straight grafts will be used in only a small percentage of patients having an endovascular procedure compared with most patients undergoing conventional operation. Because this assumption is based on anatomic information and the properties of devices under investigation, conclusions could change with improvements in product design. If an attachment mechanism were available that required 5 mm of normal vessel, more than 90% of the patients could be treated with a straight graft.

The field of endovascular surgery is developing at an explosive rate. There is both a scientific and a commercial interest in its development. Less risky and less costly solutions to both aneurysmal and occlusive disease may be developed only if we continue to explore the basic anatomy and pathophysiology of these entities and apply that information in a scientific manner to the design of new products.

References

1. Crawford SE, Manning LG, Kelly TF. "Redo" surgery after operation for aneurysm and occlusion of the abdominal aorta. Surgery 1977;81:41–55.
2. Chuter TAM, Green RM, Ouriel K, DeWeese JA. Infrarenal aortic aneurysm morphology: Implications for transfemoral repair. J Vasc Surg 1994;20:44–50.
3. Chuter TA, Green RM, Ouriel K, et al. Transfemoral endovascular aortic graft placement. J Vasc Surg 1993;18:185–197.

Chapter 8

Stent-Graft Combination: A New Technique for the Treatment of Aneurysms

Juan C. Parodi and Claudio J. Schüonholz

The diagnosis of abdominal aortic aneurysm (AAA) has been increasingly more common during the past two decades (1). This is probably related to aging of the population as well as to the extensive use of ultrasonography and computed tomography (CT) scanning for different pathologies.

Although AAAs may occasionally cause distal embolization, rupture remains the most common and deadly complication. For the past 40 years, elective replacement with a synthetic graft has proved to be the most appropriate method to prevent AAA rupture, and it has been associated with a postoperative mortality rate of less than 5% at respected medical centers (2).

Vascular surgeons are increasingly encountering older patients with severe comorbid conditions, which can increase operative morbidity and may even elevate the mortality rate for aortic surgery to over 60% (3).

It seems inevitable that every vascular surgeon will fairly frequently encounter patients who are at a prohibitive risk for conventional graft replacement. Other alternative forms of treatment (e.g., axillofemoral bypass in conjunction with induced AAA thrombosis) have been abandoned despite preliminary reports of their initial success (4).

In 1976, we began to develop a plan for the endovascular treatment of AAAs, which was based on the fundamental principles of aortic replacement. We developed two prototypes: A self-expanding metal cage with a zigzag configuration covered by a nylon fabric and a Silastic bag with a cylindrical lumen. However, we eventually abandoned both of these prototypes because of our discouraging results in animal experiments. With the advent of stent technology in the field of endovascular treatment, we reinitiated our project in 1988 using balloon-expandable stents.

Our current approach is predicated on the concept that stents may be used in place of sutures to secure the proximal and distal ends of a fabric graft extending the length of an AAA, thoracic aneurysm, or area of intimal tearing in aortic dissections.

ANIMAL STUDIES

We used a canine aneurysm model to test the concept of intraluminal graft implantation. A 6-cm segment of the infrarenal aorta was resected and replaced with a fusiform conduit made of crimped, woven Dacron measuring 8 cm in length.

Histologic studies demonstrated endothelium graft coverage of the experimental aneurysm neck, coupled with mural thrombus formation in the body of the aneurysm. Both of these changes were commonly encountered within 1 month after implantation of the graft. Six weeks later, the aneurysms were excluded using endoluminal graft deployment techniques.

A noncrimped, knitted Dacron tube of appropriate diameter and length was attached to the stents (Palmaz stent, Johnson & Johnson Interventional Systems, Warren, NJ). Two thirds of each stent were covered by graft material, and one third was left exposed to anchor the graft to the aortic wall. The graft was folded clockwise into two folds, mounted onto a balloon catheter, placed over a guidewire, and loaded into a 14F Teflon sheath. The right femoral artery was chosen for vascular access.

Under fluoroscopic guidance, the leading stent was advanced over the guidewire into the neck of the aneurysm below the renal orifices and above the body of the aneurysm. After proper positioning of the device, the introducer sheath was withdrawn, and balloon inflation was used to secure the graft by deploying either one or two stents.

In one group of dogs, only a proximal stent was used, but in a second, larger group a distal stent was added to achieve total aneurysm exclusion. Angiography was then used to confirm correct stent-graft deployment and verify that the aneurysm was successfully excluded.

We operated on a total of 43 dogs, primarily to evaluate different models of experimental aneurysms but also to test a variety of balloons and grafts. We performed 6-month evaluations in the last group of dogs (n = 6), obtaining complete pathologic studies in all animals and color Doppler ultrasound (duplex) scans in some. Using both optical and electron microscopy, we found that both ends of the stent-grafts were covered by endothelium, that the metal stent was in direct contact with the intima of the aorta by a thin layer of myointimal hyperplasia, and that the shaft of the graft was covered by a fibrin platelet barrier.

CLINICAL EXPERIENCE

After we were satisfied that a Dacron graft could be delivered through a catheter and firmly fixed in place by balloon-expandable stents, we obtained permission from our Institutional Ethical Committee to perform a pilot clinical study in a small group of patients. Each patient was fully informed about the nature of the procedure, and each was subsequently very carefully followed up.

In September 1990, we reported the treatment of a patient with an AAA (5). Since then, we have treated 148 patients using a stent-graft combination, 109 of whom had AAAs (Table 8.1). In the past year, we gained experience with the Mintec (Mintec, Inc, Grand Bahama Island) and Vanguard (Meadox Medical, Oakland, NJ) devices, which are modular systems. Twenty-seven aortobi-iliac procedures were performed with significant success (88% initial and late success rates). These systems and their methods of application are described in by Dr. Stelter in Chapter 11.

Table 8.1. Clinical Experience With Stent-Grafts

	Number of patients
Aneurysms (n = 110)	
Aortoaortic stent-graft	
–1 stent	8
–2 stents	43
Aortoiliac stent-graft	45
Aortobi-iliac stent-graft	13
Iliac stent-graft	1 (6 combined with AAA repair)
Vascular trauma	25
Occlusive disease	13

From September 1990 to December 1996.

STENT-GRAFT DEVICE FOR THE TREATMENT OF ANEURYSMS

An 18F (internal diameter) Teflon sheath, 45 cm long, with a hemostatic valve closure at the operator end contains the balloon catheter. This catheter consists of a 9F polyethylene shaft and one or two nylon balloons that are 3.5 cm in length and either 30 mm, 25 mm, or 16 mm in diameter. The assembly contains one or two balloon-expandable stents (either two aortic stents or an aortic and an iliac stent). A thin-walled, crimped, knitted Dacron graft is sutured to the stents, overlapping 50% of the length of each stent.

The balloon-expandable stent is a cylindrical tube with longitudinal slots that adopt a diamond-shape when expanded. This design, initially described by Palmaz and colleagues (6,7), permits the stent to expand from a diameter of only 5 mm when collapsed to a diameter of more than 30 mm when expanded. The stent is 3.5 mm in length and is made of annealed stainless steel, 316L, because this alloy has been widely used in a variety of prosthetic applications.

The surface of the metal stents must be uniform, because corrosion of implanted metal pieces usually occurs at the site of cracks and crevices on the metal surface. To avoid motion between parts, the balloon-expandable stent is constructed as a single piece. Micromotion between metal surfaces disrupts the protective oxide films, allowing the area to rapidly corrode.

The knitted Dacron graft has a wall thickness of 0.2 mm and is manufactured using strong yarns with tensile and bursting strengths comparable to those of commercially available grafts. The compliance of the shaft of the graft (15%) is also similar to standard knitted Dacron grafts, but the compliance of the segment overlapping the stent is 45% to adapt to its expansion.

A transitional zone of intermediate compliance is interposed between these two segments to avoid abrupt changes in the diameter of the graft. The diameter and the length of each graft are tailored to fit each individual patient, but the grafts we most often use are 18 mm to 20 mm in diameter and 8 cm to 12 cm in length (Figs. 8.1 and 8.2).

Because of its radiopacity, a thin gold wire is sutured to both ends of the graft, permitting radiographic identification of both ends and the sides of the graft to correct for torsion during implantation.

The balloon catheter we use (Balt Company, Paris, France) is constructed of nylon material and has some degree of compliance, which allows us to use only two balloon sizes (25 mm and 30 mm). For aortoiliac grafting, either a double balloon (25 mm and 12 mm) or two independent balloons are used.

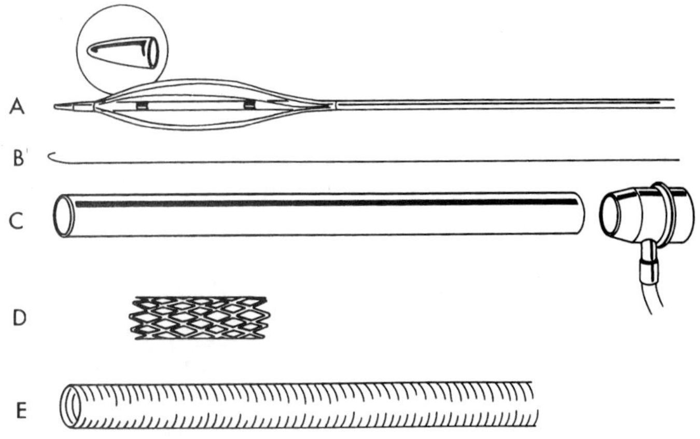

Figure 8.1. Individual components of the stent-graft device. **A.** Graft. **B.** Stent **C.** Sheath. **D.** Wide wire. **E.** Balloon.

Under fluoroscopic guidance, the device is advanced over a wire. We prefer to use a super-stiff guidewire with a diameter of 0.038 inches, because it facilitates the introduction of the device through tortuous iliac arteries, keeps the axis of the device parallel with that of the aorta, and prevents disruption of the laminated thrombus that lines atherosclerotic aneurysms by minimizing friction between the stent-graft and the aneurysm wall.

PROCEDURE

Under local or epidural anesthesia thc patient is prepared and draped as for a standard AAA resection. General anesthesia was used in three patients with thoracic aneurysms. A small incision is made over the chosen common femoral artery (the straighter and wider artery is usually selected for access). A soft-tipped guidewire is advanced in the aorta up to the level of the diaphragm, and a pigtail diagnostic catheter is then advanced over the wire and placed inside the lumen of the aorta with the tip located proximal to the renal arteries. The first injection of 30 mL of contrast medium is performed. The pigtail catheter has radiopaque marks engraved every 2 cm on its surface to facilitate length and diameter measurements using quantitative angiography.

When the infrarenal aorta or the iliac arteries or both are extremely tortuous, we prefer to insert a long wire percutaneously from the brachial artery. The wire is brought down to the femoral artery and exposed. Pulling from both ends of the wire allows us to more easily advance the device into the aorta. However, care should be taken to protect the subclavian artery with a catheter to prevent the wire from causing damage to the intima.

This maneuver is performed using a long Swan-Ganz catheter, which navigates with the arterial flow as soon as it reaches the thoracic aorta. A small amount of diluted contrast media is injected into the balloon to visualize it fluoroscopically. This small balloon catheter that navigates with flow is very useful in patients in whom the presence of friable thrombus in the infrarenal aorta increases the risk of embolization of disrupted thrombus during the procedure. As soon as the balloon catheter is exposed at the level of the femoral artery, the tip is divided using the balloon channel to advance a guidewire that will be used during the procedure.

Target areas are defined using the previously obtained images (angiogram and CT scan) and the new angiogram. These could be the proximal neck of the aneurysm and the distal cuff if the latter exists or the common iliac artery in the absence of a distal cuff. Under fluoroscopic guidance, the preloaded sheath containing the stent and graft mounted onto a balloon is placed inside the lumen of the aneurysm. Once in place, the sheath is removed, and the cranial balloon is inflated with a diluted solution of ionic contrast media and saline. The balloon is kept inflated for 1 minute and then gently deflated (Fig. 8.3).

Before proceeding with balloon inflation, the main blood pressure is dropped using a nitroglycerin solution. The pressure is kept at 70 mm Hg during balloon inflation. We have been using adenosine to stop the heart for a few seconds so that we can precisely deploy the stent in cases with short necks. Adenosine in doses ranging from 12 to 48 mg is administered intravenously, and a temporary cardiac pacemaker is inserted to improve the safety of the adenosine injection. In our experience, the use of

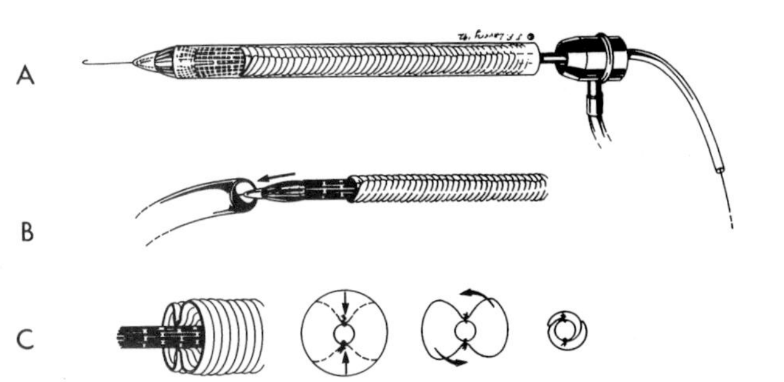

Figure 8.2. Stent-graft preparation. **A.** Fixation of the graft to the stent. **B.** Folding of the graft. **C.** Inclusion in the sheath.

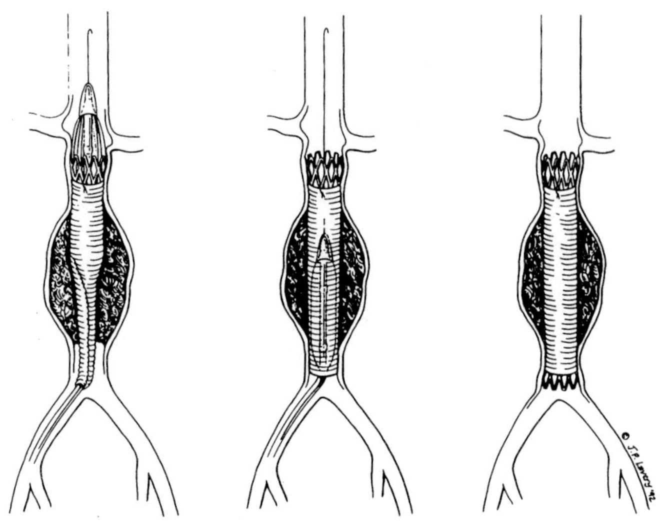

Figure 8.3. Stent-graft procedure (see text for details)

adenosine appears to be essential in patients with thoracic aortic aneurysms and dissections.

The size of the balloon is selected in advance according to the diameter of the neck of the aneurysm measured on the preoperative angiogram and CT scan. After securing the proximal stent, the second stent is placed. In some cases in which a double balloon device is used, the second balloon (either aortic or iliac) is positioned at the appropriate level and inflated, deploying the second stent. A final angiogram is performed.

Femorofemoral Bypass

In patients in whom an aortoiliac graft is placed, the procedure is completed by performing a femorofemoral bypass. The contralateral common iliac artery is occluded by means of a detachable balloon or an occluding stent (a stent covered by tubular polyester [Barone Industries, Buenos Aires, Argentina] or a polytetrafluoroethylene [PTFE] graft [W.L. Gore and Associates, Flagstaff, AZ] with one of the ends sutured, creating a blind end). Both internal iliac arteries are preserved. The internal iliac artery on the side contralateral to the access site receives its flow in a retrograde fashion from the femorofemoral bypass.

Duration of Procedure

The time of the procedure varies according to the complexity of the case. When an aortoaortic procedure is performed in a rather small aneurysm with nontortuous and wide-lumen iliac arteries, the procedure takes from 20 to 45 minutes (Fig. 8.4). When an aortoiliac graft plus a femorofemoral bypass and occlusion of the contralateral iliac artery are needed, the time in the operating room occasionally exceeds 3 hours.

Calculation of Data

We have recently used a program to elaborate the data and images obtained by spiral CT scans. The data were processed by a company (Medical Media Systems, W. Lebanon, NH) that provides an interactive CD ROM with patient-specific data. The actual diameters of the arteries and the axial lengths can be accurately calculated.

RESULTS

Between September 1990 and December 1996, 109 patients with aortic aneurysms were treated. The average age was 73 years (range 57 to 100 years); 11 patients were over 80 years old. Most of the patients had at least one associated morbid condition (Table 8.2). Fifty-one patients underwent aortic tube graft replacement, with 8 having only a proximal stent, 45 aortoiliac stent-grafts, and 13 aortobi-iliac stent-grafts. Three of the tubular grafts were implanted in the thoracic aorta—two to treat

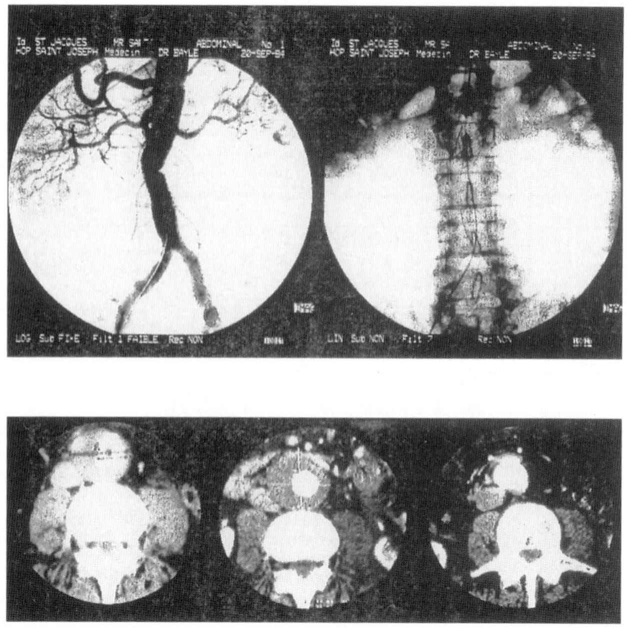

A

B

Figure 8.4. Abdominal aortic aneurysm treated with an aortoaortic stent-graft device. **A.** Angiography after treatment. **B.** One year follow-up CT scan showing complete vanishing of the aneurysmal sac.

aneurysms and one a type A dissection. Seven cases of iliac aneurysms were treated, six in combination with the treatment of an AAA and one an isolated iliac aneurysm.

Patients with successful procedures recovered very rapidly, had breakfast the next morning, and walked within 24 hours to 48 hours after the procedure. They were usually discharged from the hospital 3 to 4 days after the procedure.

All patients were followed up by clinical examination, color duplex studies every 6 months, and CT scans once per year. Angiography was performed in some patients as well as in those in whom the color duplex or CT scan suggested a sign of a leak or dilatation or any change when compared with the study performed immediately after the procedure. The average follow-up period was 34 months, with a range of 1 to 73 months.

Table 8.2. Associated Pathologic Conditions

Condition	Number of patients
Severe, chronic heart disease	53
Acute myocardial infarction	23
Severe pulmonary insufficiency	26
Renal insufficiency	6
Acute cerebral infarction	1
Two or more previous strokes	10
Hostile abdomen	2
Hepatic failure	2
Mild pulmonary insufficiency	12
Intermittent claudication	33
Disseminated intravascular coagulopathy	1
Preexisting atheroembolism	3

We define "early success" as consisting of the following:

- A graft covering the complete length of the aneurysm
- Blood-tight sealing of both ends of the graft firmly attached to the intima of the artery without thrombus interposition

We define "late success" as consisting of the following:

- Persistence of initial success criteria
- Absence of backflow from lumbar or inferior mesenteric arteries after a 6-month period
- Absence of aneurysm growth

We obtained the following results:

- Early failure in aortoaortic stent-graft procedures—23%
- Early failure in aortoiliac stent-graft procedures—24%
- Late failure in aortoaortic stent-graft procedures—30%
- Late failure in aortoiliac stent-graft procedures—5%

We have included the following in our early failures:

- Access problems that prevented proper deployment of the stent-graft
- Death and major complications (Table 8.3)
- Incomplete sealing
- Migration (one patient)

There were 10 mortalities: 4 from massive microembolization, 1 after conversion to an open procedure, 1 from pulmonary embolism, 1 from pulmonary insufficiency, and 3 after acute cardiac events (myocardial infarction and cardiac insufficiency).

All late failures were caused by perigraft leaks. All but one leak developed at the distal end owing to improper deployment of the stent-graft. The end of the graft was in direct contact with thrombus rather than with the intima of the neck of the AAA. In one patient, the proximal stent was in contact with thrombus and the patient developed a persistent leak that did not reach the aneurysmal cavity.

Besides the major complications listed in the failure group, other complications that were amenable to correction occurred in our series (Table 8.4). In addition to aortic aneurysms, three patients with aortic dissections were treated. In 1991, we successfully treated a patient with an infrarenal dissection using the "elephant trunk" technique, deploying only the proximal stent and a short polyester graft. In January 1995, a Stanford type A dissection was sealed by means of a covered Palmaz stent, and a type B dissection that was complicated by intestinal, renal, and left leg ischemia was successfully treated using the same device described for the infrarenal dissection.

Table 8.3. Major Complications

Complication	Number of patients
Massive microembolization	4
Migration of proximal stent during initial procedure	1
Hemorrhage during procedure that required iliac suture	1

Table 8.4. Minor Complications

Complication	Number of patients
Groin hematoma	2
Proximal leak (treated with a covered stent)	1
Injury to the external iliac artery (sutured)	2
Distal leak treated with a covered stent	1

DISCUSSION

Based on our experience with 110 procedures treating aneurysms (109 aortic and 1 iliac), we have made the following preliminary conclusions: The stent-graft combination procedure is feasible and, when successfully applied, has the great attraction of its simplicity. However, when treating aneurysms, we encountered more problems than we had initially predicted (8). The stent-graft procedure is simple in theory, but several details need to be attended to before one can move ahead with widespread use of this method.

Measurements of diameter and length are crucial. We learned how to obtain reliable data using enhanced CT scans, quantitative angiography, and three-dimensional reconstruction using MRI or spiral CT scanning. Scan images, intraluminal measurements, and some geometric calculations assisted us in obtaining reasonably reliable data. Understanding that elongation occurs as dilatation of the aorta develops and that elongation occurs in different planes allowed us to more accurately calculate the actual length of the artery. As stated previously, we are now using a new software that was developed by Medical Media Systems.

Access accounted for several problems we had to work out. Narrow, stenotic, and tortuous iliac arteries were responsible for these difficulties. From the beginning, the rigid stent and the large diameter of the sheath needed for the implantation were drawbacks. We overcame some of these problems by modifying the device and by using different maneuvers during the procedure. Reducing the diameter of the sheath to 18F was a remarkable advance toward defining the ideal device.

The use of one or two (parallel) extra stiff wires, the "pull-down" maneuver, and occasionally implantation of a temporary conduit on the common iliac artery were also useful in overcoming some of the problems. The *pull-down maneuver* consists of dissecting free the common femoral and external iliac arteries, lifting up the inguinal ligament, and using blunt dissection to reach the iliac bifurcation from the groin (Fig. 8.5). Small branches should be divided between suture ligatures. When the arteries are free, gentle pulling of the artery toward the feet of the patient allows the tortuous artery to become straighter, making introduction of the sheath possible. Inserting a long guidewire from the brachial artery down to the femoral artery was of great help in patients with very tortuous arteries.

Studying the morphologic changes in aneurysms over several years reveals that in the initial stage almost all aneurysms have a proximal neck and distal cuff of more than 2 cm. In the second stage, the distal cuff becomes shorter, and in the third stage, the distal cuff tends to disappear, whereas the proximal neck becomes shorter, but is still longer than 2 cm. After this stage, we found that elongation takes place, which creates tortuosity (Fig. 8.6).

Generally, the distal curve opposes the proximal one. This results in a configuration in which, if the convexity of the proximal neck is to the right, the distal cuff curves to the left, leaving the right iliac artery straight and the left with a tendency to produce a right angle. These findings indicate that in some patients a tubular graft would be applicable. As advocated by Chuter (9), when the distal aortic cuff is not present, an aortobi-iliac graft has to be used. In addition, when the angle between the iliac arteries becomes larger than 90 degrees, an aortoiliac graft should be used and a femorofemoral bypass placed.

Because of their high radial force, balloon-expandable stents appear to be ideal anchoring mechanisms. We had one case in which the stent was incompletely deployed (stent migration). Intraluminal ultrasound would be an ideal test to check for completeness of stent deployment.

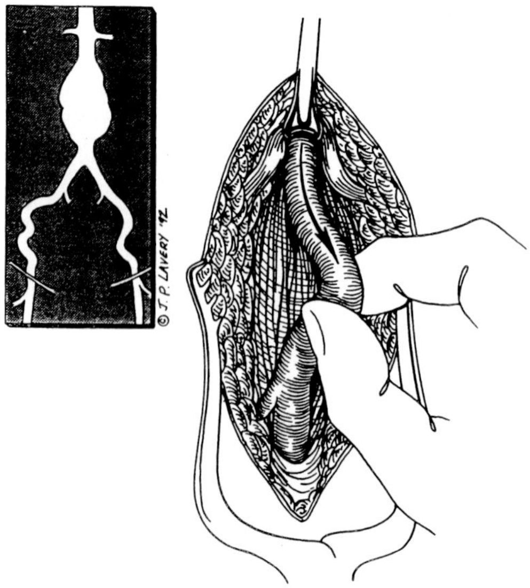

Figure 8.5. "Pull-down" maneuver (see text for details).

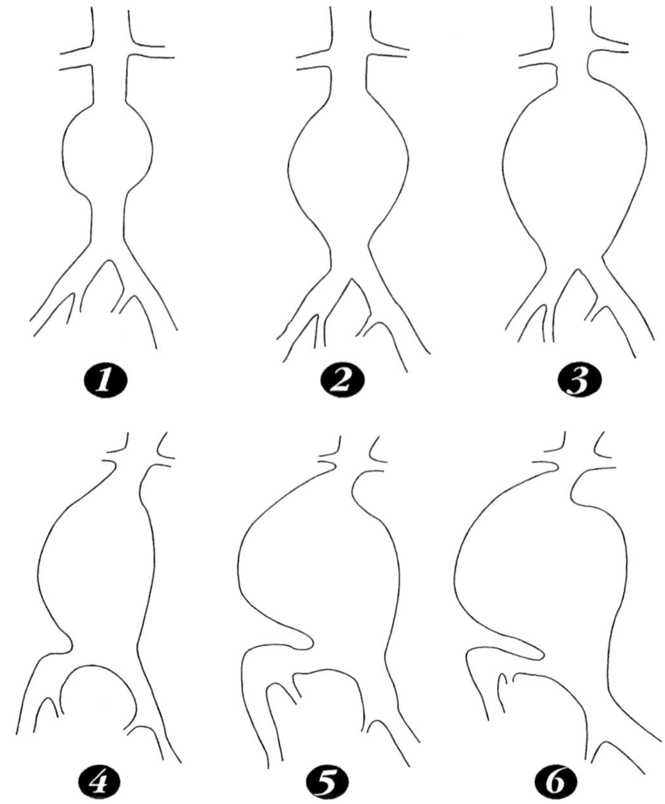

Figure 8.6. Morphology of aneurysms: Different stages (see text for details).

We consider microembolization to be the main and more dreadful problem associated with the stent-graft combination procedure. Microembolization occurred in five of our patients, four of whom died. In the remaining patient, discrete microembolization of the right foot was successfully treated with intra-arterial administration of prostaglandin E_1. In the four cases with massive microembolization, the procedures were technically difficult in patients with large aneurysms. In one patient, visceral ischemia was found on postmortem examination. In the latter patient, technical problems involving inappropriate balloon sizing occurred during the procedure. Microscopic examination, however, disclosed a nonocclusive intestinal infarction, and embolization was ruled out. A mild, reversible case of microembolization caused by incomplete deployment of the stent occurred on the side of implantation of an occluding stent after an aortoiliac graft implantation with the addition of a femorofemoral bypass.

A review of our cases of embolization shows that large and tortuous aneurysms pose an increased potential for embolization. This is probably due to the fact that when advancing the guidewire from the femoral artery into the aorta and then into the proximal neck, the operator negotiates it inside a large and tortuous chamber coated with friable material. Occasionally, it is very difficult to get the guidewire inside the proximal neck, because often within the cavity of the aneurysm the orifice of the proximal neck is small. Such maneuvers could eventually cause dislodgment of particles of the laminated thrombus. Thus, in all cases of large aneurysms with wide lumens, it is advisable to insert the guidewire percutaneously from the brachial artery.

Very often, miscalculation of the length of the aneurysm created the necessity to change the device or to use complementary procedures such as implanting a third covered stent to increase the length of the device or cover a leak.

The more intravascular manipulations we perform, the greater is the risk of dislodging particles from the aortic wall. What can we do to prevent microembolization? In cases of large aneurysms with big lumens, the soft-tipped guidewire should be introduced from the brachial artery and brought down to the femoral artery. Care should be taken to precisely measure the length and diameter of the arteries and in performing a simple, well-planned procedure.

After successfully treating patients who were admitted for spontaneous visceral and distal embolization, we believe that even in the presence of friable thrombus in the lumen of an AAA, the endoluminal procedure can be safely performed.

Although we have spent over 7 years performing clinical cases and following them up, we still have unanswered questions—some of which involve basic mechanisms. These questions are related to the potential increase of the aneurysmal neck diameter. All arteries increase in size with aging, and oversizing the stents could eventually overcome this problem. Material fatigue could cause fracture of the stents or tearing or both of the fabric graft. Backflow from the lumbar or mesenteric arteries can persist and eventually cause aneurysmal enlargement.

Using the open technique of exclusion, Resnikoff and colleagues (10) demonstrated that 2% of the patients had patent aneurysmal sacs, but only 0.28% of those in the entire series had patent sacs, which were due exclusively to an open lumbar artery. Most of the other patients had patent inferior mesenteric and hypogastric arteries.

Perhaps a combined approach should be used in some cases. Placing a polyester band around the neck of the aneurysm would create a deterrent to growth. A laparoscopic approach can be used to accomplish this banding. Using this procedure, we placed a mesh of polyester around the aorta in three pigs and were able to demonstrate its potential usefulness. Umbilical tape could cause pressure necrosis when the stent is applied inside the artery, but an open mesh could not produce necrosis because it would distribute the pressure.

Second-generation systems are being developed. These new devices will be flexible and will have a lower profile, and the graft will be totally supported by metal modular systems that comprise two or three components that would permit simple construction of a bifurcated system.

Selected cases of thoracic aneurysms and dissections are being treated. Initial experience in treating aortic dissections leads us to believe that this might be a very attractive alternative to open procedures. Stent-grafts with side branches will make it possible to treat arch and thoracoabdominal aneurysms in the future. Spinal cord ischemia will constitute a limitation to this approach. These developments clearly present a new and different challenge for the medical industry.

CONCLUSIONS

Despite the encouraging results we have obtained, we believe that controlled trials with a large number of patients who are followed up for a long period of time are necessary before endoluminal treatment is introduced into clinical practice. In the meantime, the use of these procedures should be concentrated in a few centers that have the appropriate equipment and a skillful group of surgeons and interventional radiologists or cardiologists with experience in interventional vascular procedures.

Regarding other applications of stent-grafts, the treatment of arteriovenous fistulas and false aneurysms appears to be a simple and effective application of this principle, because it will save time and prevent bleeding and peripheral nerve injuries. In addition, the treatment of vascular trauma and false aneurysms in accessible areas is a very promising application of these procedures, as described in Chapter 2 on vascular trauma. Arterial dissections, especially aortic dissections, are probably efficiently treated endoluminally, interrupting the flow through intimal tearing. Three of our cases demonstrated this promising approach. The development of an "internal bypass" after balloon dilatation is an appealing idea in view of the failure of balloon dilatation in long stenoses or occlusions.

In theory, isolating the inner surface of the treated artery will eventually prevent interaction between the damaged intima and the circulating elements and substances or elements of the blood.

Our initial experience in treating thoracic aneurysms (one resulting in a type A dissection—a thoracoabdominal aneurysm with no compromise of the visceral arteries and a descending aneurysm) indicates that the procedure is simpler than that used for treating AAAs and is very promising.

We believe that there are two potential sources of late failures when applying this procedure. The first source, related to the device itself, is a failure of the material (material fatigue). The second is the possible dilatation of the native artery caused by the vector of force created by the radial force of the stent in addition to the naturally occurring arterial dilatation that takes place with aging.

The answer to the first concern will come from the industry. Regarding the second, we know that the area of the aorta in which the renal arteries take off is stronger than the infrarenal aorta. Thus, proximal attachment should include that area of the infrarenal neck; perhaps even the suprarenal aorta could eventually be used as an attachment site to prevent dilatation. Normal iliac arteries have proved to be resistant to dilatation. In our experience as well as that of others, the aortic distal neck is the site of maximal dilatation. Thus, preferentially distal attachment should be located on the iliac arteries.

A word of caution regarding definitive conclusions: We are finding new abnormalities in patients as long as 3 years after performing what we considered to be a successful procedure. Before the stent-graft procedure can be offered to the community as an alternative treatment for AAA repair, a clear, unremarkable, long-term experience is needed.

References

1. Melton LJ III, Bickerstaff LK, Hollier LH, et al. Changing incidence of abdominal aortic aneurysms: A population-based study. Am J Epidemiol 1984;120:379–386.
2. Brown OW, Hollier LH, Pairolero C, et al. Abdominal aortic aneurysm and coronary artery disease: A reassessment. Arch Surg 1981;116:1484–1488.
3. McCombs PR, Roberts B. Acute renal failure following resection of abdominal aortic aneurysm. Surg Gynecol Obstet 1979;148: 175–178.
4. Karmody AM, Leather RP, Goldman M, et al. The current position of nonresective treatment for abdominal aortic aneurysm. Surgery 1983;94:591–597.
5. Parodi JC, Palmaz JC, Barone HD. Transfemoral intraluminal graft implantation for abdominal aortic aneurysms. Ann Vasc Surg 1991;5:491–499.
6. Palmaz JC, Richter GM, Noeldge G, et al. Intraluminal stents in atherosclerotic iliac artery stenosis: Preliminary report of multicenter study. Radiology 1988;168:727–731.
7. Rees CR, Palmaz JC, Garcia O, et al. Angioplasty and stenting of completely occluded iliac arteries. Radiology 1989;172:953–959.
8. Parodi JC. Endovascular repair of abdominal aortic aneurysms. Adv Vasc Surg 1993;1:85–106.
9. Chuter T. Bifurcated endovascular graft insertion for abdominal aortic aneurysms. In: Greenhalgh RM, ed. Vascular and endovascular surgical techniques. Philadelphia: WB Saunders, 1994.
10. Resnikoff M, Darling C III, Chang B, et al. Fate of the excluded abdominal aortic aneurysm sac: Long term follow up of 831 patients. J Vasc Surg 1996;24:851–855.

Chapter 9

Transfemoral Endovascular Repair of Abdominal Aortic Aneurysms: Feasibility Study of the EGS System

Wesley S. Moore

Transabdominal repair of aortic aneurysms is a well-established, lifesaving surgical procedure. Approximately 40,000 aneurysms are electively repaired each year in the United States. In spite of this, about 15,000 patients die annually from ruptured abdominal aortic aneurysms (AAAs), making it the 13th leading cause of death in the United States (1). The reasons for this are multifactorial, but basically include (1) a failure to identify the presence of an AAA because of the lack of surveillance programs and the inefficiency of identifying the lesion by physical examination and (2) an unwillingness to recommend the operation for relatively small aneurysms because of a disadvantageous risk/benefit ratio.

Many centers of excellence have reported mortality rates for elective repair to be less than 5%; some report less than 3% (2). However, community-based surveys suggest that elective repair of AAAs may carry a mortality rate of 10% to 14% (3).

The introduction of a simple and safer method for repair of AAAs will have great appeal. First, the morbidity, mortality, discomfort, and disability are likely to be markedly reduced. In addition, a major benefit in terms of overall cost reduction will result.

If the latter two objectives can be achieved, surveillance programs to identify the unsuspected presence of AAAs are very likely to be increased, and the willingness to offer repair to patients with smaller aneurysms will also be increased. The net result from this activity should have a measurable benefit in further reducing the mortality associated with aneurysm rupture, the morbidity and mortality related to elective repair, and, finally, an overall unit cost reduction in the care of patients with this lethal condition.

This chapter reviews the clinical experience with the Endovascular Technologies System (EGS) (Endovascular Technologies, Inc, Menlo Park, CA) as part of a trial approved by the Food and Drug Administration (FDA) for clinical investigation (4–13).

THE EGS SYSTEM

The EGS system consists of two principal components—an introducer sheath and a catheter delivery system.

Introducer Sheath

The purpose of the introducer sheath is to provide vascular access to the aneurysm, principally for placement of the catheter delivery system, which contains the graft, but also for other manipulations, such as angiography and intravascular ultrasonography.

The current model of the introducer sheath measures 23F in outside diameter. The leading edge of the sheath comes in a collapsed configuration. The sheath holds an obturator, which, when advanced, fully expands the entire length of the sheath to a specified diameter. The proximal, or working, end of the sheath contains a double-valve system for hemostatic control during manipulation. Finally, the entire system is designed to be advanced over a guidewire into its ultimate position within the aneurysm (Fig. 9.1).

The Catheter Delivery System

The dual purpose of the catheter delivery system is to bring the graft into proper anatomic location for bridging the aneurysm and to provide a balloon catheter for

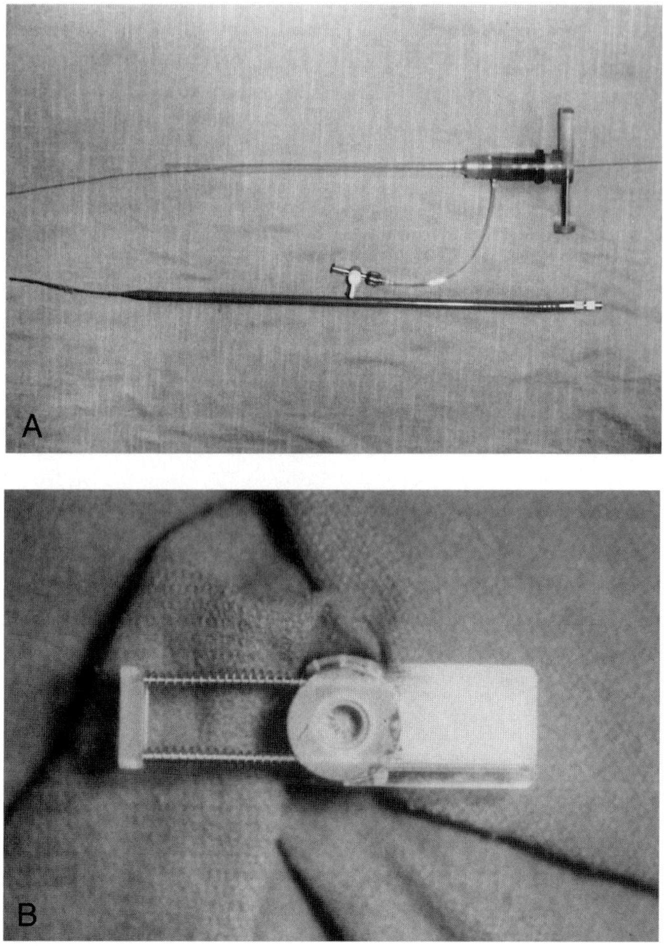

Figure 9.1. **A.** Photograph of the introducer sheath showing the obturator removed and separate from the sheath itself. The sheath is passed over a guidewire. The proximal portion of the sheath has a double-valve system. **B.** End-on view of the gate valve of the sheath. An iris diaphragm provides a hemostatic seal.

seating the proximal and distal attachment systems that are intrinsic to the prosthetic graft. The newest version of the catheter delivery system has been nicknamed the "bullet" because the leading edge is silver and it has a bullet shape when viewed under fluoroscopy. The catheter delivery system is also designed to be advanced over a guidewire within the introducer sheath on its way to positioning within the aneurysm.

The catheter delivery system contains a capsule that holds the graft together with the proximal and distal self-expanding attachment systems in a compressed configuration. The graft within the capsule is covered by a retractable jacket. Intrinsic to the catheter delivery system is a movable and expandable balloon catheter. In its collapsed configuration, the balloon initially lies within the graft.

The catheter delivery system contains two ports that can be used for heparin/saline irrigation as well as the introduction of contrast for fluoroscopic localization. A third port is used to inflate the angioplasty balloon (Fig. 9.2).

The EGS graft system now comes in three configurations: A tube graft, a bifurcated graft, and an aortoiliac system in combination with a femorofemoral graft for patients with combined aortic and iliac aneurysms.

INDICATIONS FOR USE OF THE EGS SYSTEM

Tube Graft

The tube graft configuration of the EGS system was the first prototype available. It was designed to be used in patients who have both a proximal neck and a distal aneurysmal neck proximal to the bifurcation of the iliac arteries. The proximal and distal aneurysm necks must measure at least 1.5 cm in length and have a diameter no greater than 26 mm. The patient must have an iliac and femoral arterial pathway sufficiently large in diameter to accommodate the passage of a 23F introducer sheath. The aorta must be free of aberrant vessels that might be occluded with placement of the graft, such as an aberrant renal artery. In addition, there must be no evidence of mesenteric occlusive disease, which would make the patient dependent on the integrity of the inferior mesenteric artery for intestinal circulation.

At the present time, patient characteristics for endovascular repair have been limited to good-risk patients in view of the fact that conversion to open repair is always a possibility.

Bifurcated Graft

The EGS bifurcated graft system is used for patients who have an AAA that extends to the bifurcation of the common iliac arteries. The proximal neck must be at least 1.5 cm in length and 26 mm in diameter. The ratio between body diameter and limb diameter is 2:1. Therefore, the diameter of the recipient iliac arteries cannot exceed the proposed graft limb diameter. Finally, there must be no

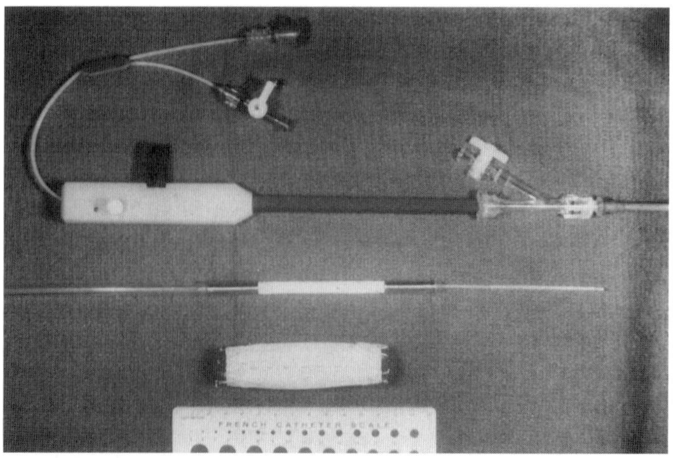

Figure 9.2. Photograph of the newer prototype, the "bullet." The photograph indicates the handle portion of the catheter delivery system, which permits capsule manipulation for graft deployment. The capsule itself is also imaged, demonstrating the graft within the capsule and covered by the outer jacket.

evidence of true aneurysmal dilatation of the recipient iliac arteries.

Aortoiliac/Femorofemoral System

To be able to accommodate patients who have a combination of an AAA and an iliac aneurysm, a tapered tube graft was designed that goes from the infrarenal abdominal aorta to the external iliac artery on one side. The contralateral common iliac artery is occluded with a blind endosoc, and circulation to the contralateral extremity is restored with the placement of a crossover femorofemoral bypass. For a patient to be a candidate for the aortoiliac/femorofemoral device, the aneurysm must have a proximal neck at least 1.5 cm in length and 26 mm in diameter. The contralateral iliac artery must not be larger than 14 mm to be potentially occluded by a blind endosoc unless a surgical ligation is anticipated.

PREOPERATIVE EVALUATION

After the patient receives a general medical evaluation and is deemed to be a suitable candidate for surgical repair, several imaging procedures are necessary to determine whether the patient is a candidate for endovascular grafting and to establish the proper measurements for graft selection. The initial imaging study is a contrast-enhanced, spiral computed tomography (CT) scan. It is recommended that cuts be made at 3.0-mm intervals. After CT criteria have determined that the patient is a potential candidate for one of the three EGS systems, the final step before offering the procedure to the patient is a contrast angiogram.

The contrast angiogram is done with a marker catheter, which is a catheter with radiopaque marks set at 1.0-cm intervals. This permits the accurate internal measurement of the length and confirmation of neck diameter at the time of angiography to compensate for a magnification factor and parallax. The angiogram also provides information concerning the status of the mesenteric circulation, the presence of any aberrant vessels that might be compromised with endovascular grafting, and any other anatomic variations, such as tortuosity or kinking of the iliac arteries, which might preclude the safe placement of the introducer sheath.

CONTRAINDICATIONS TO ENDOVASCULAR GRAFTING

The following are contraindications to endovascular grafting:

- Absence of a satisfactory proximal aneurysm neck.
- Aneurysm neck that exceeds the largest graft diameter.
- Evidence of a collateral mesenteric pattern, suggesting dependence on flow through the inferior mesenteric artery to maintain intestinal circulation.
- An aberrant renal artery that would be compromised by graft placement.
- Iliac arteries that are too small or too tortuous to permit satisfactory access by the sheath.
- Occlusive disease within the iliac artery that would compromise sheath introduction.

Patient characteristics that would limit or preclude the use of this device include associated morbid conditions that would limit the patient's life expectancy to less than 2 years and morbid obesity with weight over 350 pounds.

TECHNICAL CONSIDERATIONS

My position is that endovascular graft implants should be carried out in an operating room. My reasons include the need for the best conditions of sterility. This is because a prosthetic graft (a multifibered foreign body) is going to be surgically implanted, and the potential consequences of infection would carry the same dire outcome of prosthetic graft infection implanted in a conventional manner.

Perhaps a more compelling reason for the use of an operating room is the potential need to convert the procedure from an endovascular graft implantation to a direct surgical repair if a complication arises with the endovascular procedure. The potential exists not only for graft misdeployment but also for aneurysm rupture. Therefore, the procedure should be carried out with the patient under general anesthesia and fully prepared and draped for a conventional as well as an endovascular procedure.

The abdomen and both femoral regions should be included in the surgical preparation.

The decision as to whether to use the right or the left femoral artery as the ipsilateral artery is based on the arterial size and configuration as seen on angiography. The right femoral artery presents some advantages to the right-handed surgeon, but either side may be used. A vertical incision is made over the common femoral artery, and that vessel is fully mobilized. Circumflex branches are divided between ligature, and the leading edge of the inguinal ligament is incised to permit mobilization of the terminal portion of the external iliac artery. If the angiogram demonstrates redundancy of the iliac system, it is possible to mobilize the iliac artery in the retroperitoneum through a femoral incision. This permits the opportunity to straighten the vessel before insertion of the device.

In the case of a bifurcated or aortoiliac graft, the contralateral femoral artery is also exposed. After the vessel is fully mobilized, the next step is to obtain an angiogram with road, map imaging.

A dedicated or specialized operating room is not necessary for this procedure. We currently use a Skytron 3100 operating room table (Fig. 9.3). This table can be used for a variety of conventional operations as well as

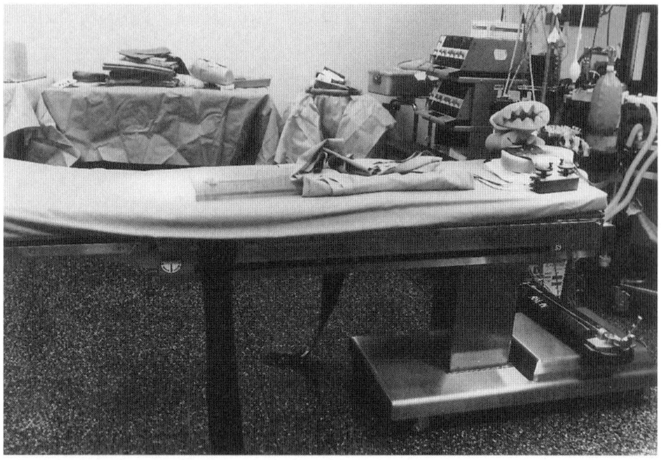

Figure 9.3. Photograph of the Skytron 3100 table demonstrating the positioning of the pedestal at the head of the table, leaving the middle and end of the table free for fluoroscopy.

for endovascular procedures. The advantage is that the table top can be moved in the horizontal plane to place the pedestal in an eccentric position. This permits a long radiolucent segment of table for imaging of the abdominal aorta with a C-arm.

When the patient is positioned supine on the operating table, a specially designed marker board is incorporated on the table top. The marker board contains horizontal cursors that can be moved by remote control (Fig. 9.4). These cursors are radiopaque and are used to designate the proximal and distal points of graft deployment.

The fluoroscopy equipment should be a high-quality, portable digital C-arm. Road-mapping capability is desirable as well as digital recording. A unit that we previously used is the OEC Diasonics C-arm (Fig. 9.5), but a variety of high-quality C-arm units are currently available. We are now using the Phillips 3000 unit (Phillips Medical Systems, Shelton, CT).

Finally, it is necessary to have a programmable pressure injector for use in obtaining the preimplant angiogram (Fig. 9.6).

Figure 9.4. Photograph of a marker board demonstrating a Lucite surface with controls of horizontal cursors that can be moved and used to mark the specific positions on the road-map image.

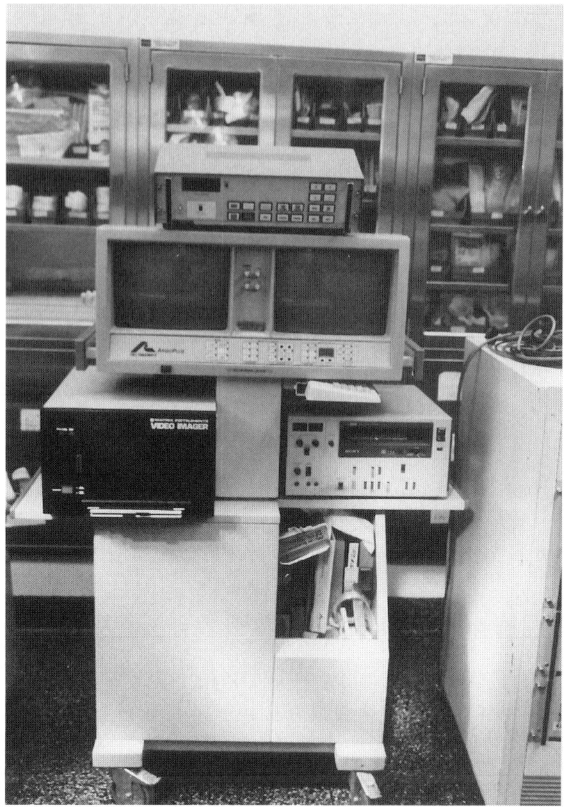

Figure 9.5. Photograph of a TV monitor for the OEC Diasonics digital fluoroscopy unit.

A 7.0F angiogram sheath is then inserted into the femoral artery over a guidewire. A pigtail angiogram catheter with 1.0-cm radiopaque marks is then passed over the guidewire and advanced under fluoroscopic control into the suprarenal aorta. The guidewire is removed, and a pressure injected aortogram is obtained.

The aortogram should include the renal arteries and the aorta down to and including the iliac bifurcations. Images preserved on the road map serve as an important overlay when carrying out subsequent manipulations (Fig. 9.7). This is the optimum time for positioning the cursors within the marker board. The proximal cursor is positioned immediately below the renal arteries in a position selected for the proximal deployment of the attachment system. The distal cursor is then placed in the distal neck of the aneurysm just proximal to the aortic bifurcation. The marker catheter is then used as a ruler to determine the graft length between the two cursors for tube grafts. The distance from the bifurcation to a drop zone is measured for the bifurcation and aortoiliac systems. The graft diameter is selected preoperatively based on interpretation of the CT and angiographic studies.

Next, the patient is systemically anticoagulated with 5000 units of heparin administered intravenously. A stiff 0.035-inch guidewire is inserted through the catheter

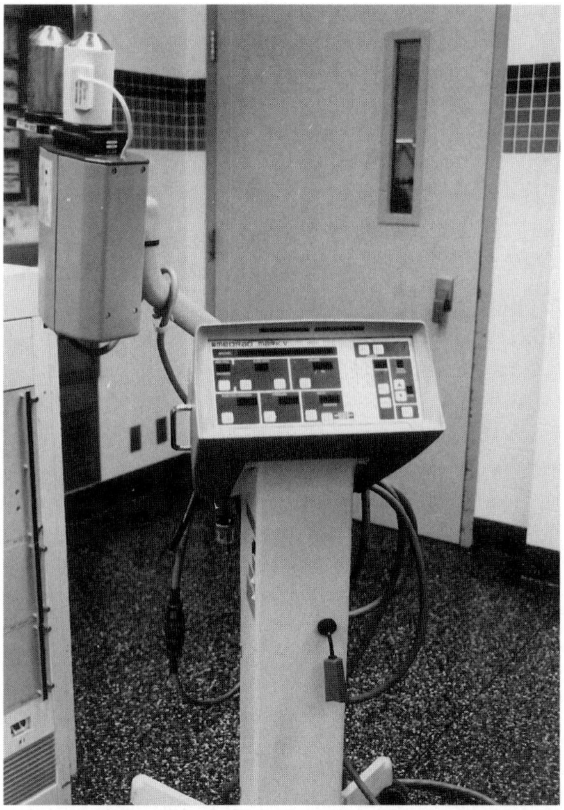

Figure 9.6. Photograph of a conventional pressure injector used to obtain optimum contrast concentration for the road map image.

and advanced to the suprarenal aorta. The catheter is then removed, leaving the guidewire in place. The distal femoral artery is clamped, and the proximal femoral artery is controlled while the angiogram sheath is removed over the guidewire.

A transverse, somewhat oblique incision is then made in the femoral artery, incorporating the sheath puncture site. The introducer sheath component of the EGS system is prepared and loaded over the guidewire. Using digital control of the femoral artery, the proximal clamp is removed, and the sheath is inserted into the femoral artery over the guidewire and fluoroscopically advanced up the iliac artery into the aneurysm sac and toward the proximal neck of the aneurysm. When the sheath is fully in place, the obturator within the sheath is advanced so as to dilate the sheath to its full capacity. The obturator is removed, leaving the sheath in place with its double hemostatic valve system competent.

The EGS catheter/delivery system is then prepared. It is loaded over the guidewire and advanced into the sheath using the double valve system for hemostatic control while the catheter/graft capsule is advanced into the sheath (Fig. 9.8). Using fluoroscopic control, the catheter system is then advanced up the sheath and beyond into the proximal neck of the aneurysm. The steps up to this point are common to all three systems.

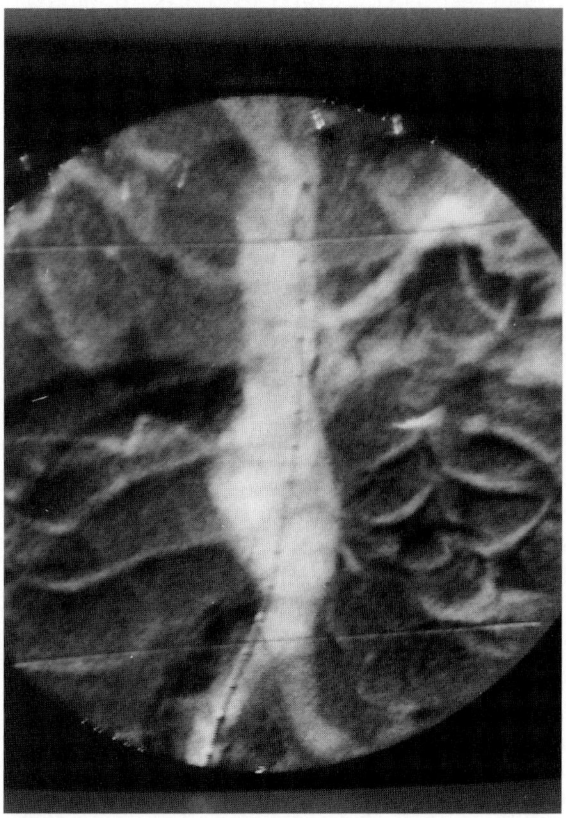

Figure 9.7. An image photographed off a TV screen demonstrating a road map image with a pigtail marker catheter in position.

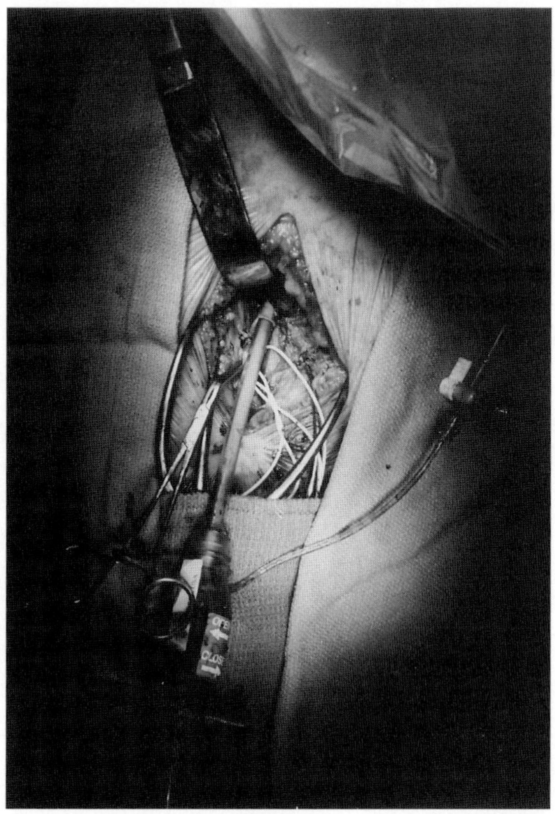

Figure 9.8. Photograph demonstrating the insertion of the sheath into the femoral artery before advancing the obturator. This is passed over a guidewire under fluoroscopic control.

TECHNIQUES

The Tube Graft

The following steps are specific for the tube graft.

- The proximal attachment system is aligned with the superior horizontal cursor. Tentative alignment of the inferior attachment system with the distal cursor is also judged for its accuracy at this point.
- The introducer sheath is then withdrawn back through the aneurysm and into the iliac artery, because its presence within the abdominal aorta is no longer necessary once the jacketed capsule is fully in position.
- The jacket, which serves as a cover to the graft, is retracted a sufficient distance to permit proximal graft deployment.
- The proximal capsule is advanced while the position is continually monitored fluoroscopically until the proximal attachment system springs open.
- When the systems open, the lock on the balloon angioplasty catheter is loosened, and the balloon is advanced so that it is centered across the proximal attachment system.
- The balloon is then inflated with radiopaque contrast to a pressure of 2.0 ATM and held in place for 1 minute (Fig. 9.9).
- The balloon is deflated, the catheter rotated, and the balloon reinflated.

With the balloon fully inflated, preparation is made for deployment of the distal attachment system.

- The jacket is fully retracted, and, by withdrawing the catheter system over the balloon shaft, the distal capsule is now withdrawn, permitting deployment of the inferior attachment system.
- The balloon is deflated and pulled back so as to be appropriately positioned to expand and fully seat the inferior attachment system.

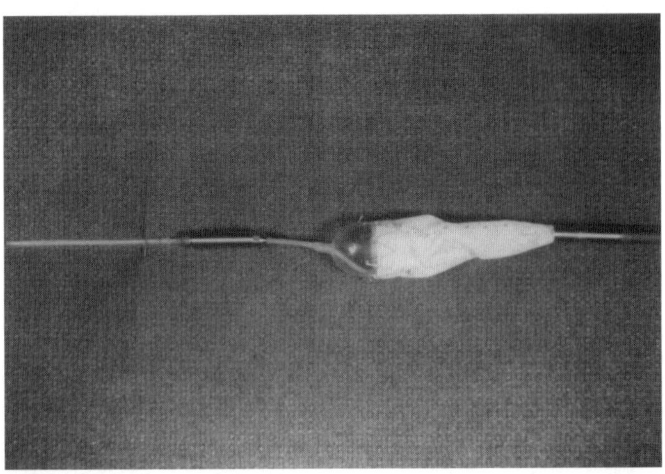

Figure 9.9. Ex vivo demonstration of the proximal deployment of the graft with the balloon catheter in position and inflated.

At this point, the balloon can be expanded or a second balloon can be advanced up the opposite femoral artery by percutaneous access to seat the inferior attachment system with a "kissing" balloon technique. The balloon catheter system is then reassembled and withdrawn through the introducer sheath.

A J wire is then inserted into a pigtail angiogram catheter and advanced into the introducer sheath as a unit. It is passed up the graft and positioned in the suprarenal location. A completion angiogram is then obtained to document the normal functioning of the graft. Particular care is taken to observe whether there is any evidence of perigraft leak, either at the proximal or distal points of graft attachment. If none is seen, the procedure is concluded.

Any evidence of anastomotic communication with the aneurysm sac may necessitate carrying out another course of balloon expansion. If so, a "kissing" balloon technique with catheters advanced from both sides is the best technique.

The Bifurcated Graft

When the sheath is in place and the angiogram has been completed, the pullwire at the end of the bifurcated graft catheter delivery system is inserted through the double-lock mechanism of the introducer sheath and the following procedure is carried out.

- The pullwire is connected to the contralateral limb, and the objective is to get the pullwire from the ipsilateral femoral artery around the bifurcation and out the contralateral femoral artery to achieve bilateral control of the bifurcated graft.
- The pullwire is advanced up the sheath into the AAA.
- The cosurgeon, working on the contralateral femoral artery, inserts a 12F sheath over a guidewire.
- An Amplatz snare catheter system (Medi-tech, Inc, Watertown, MA) is inserted through the 12F sheath with the snare positioned and opened in the AAA just proximal to the bifurcation of the iliac arteries.
- With the snare opened, the pullwire is passed through the loop of the snare under fluoroscopic control.
- The snare is tightened, capturing the pullwire.
- After the snare is tightened, together with the captured pullwire it is gradually drawn into the contralateral iliac and femoral arteries while the pullwire is gradually advanced up the iliac artery through the sheath.
- This is a good time to load the graft catheter delivery system over the Amplatz super-stiff wire (Medi-tech, Inc).
- The graft catheter delivery system is then passed through the double-lock mechanism of the EGS sheath and advanced up the sheath as the pullwire is drawn through the contralateral side.
- When the leading edge of the catheter delivery system reaches the bifurcation, the cosurgeon working the contralateral femoral artery begins to return the pull-

wire so that the graft catheter delivery system can be advanced up through the AAA into a position well above the renal arteries.

At this point, it is necessary to rotate the catheter delivery system to optimize and correct any malrotation of the bifurcated graft system as judged by the radiopaque marks seen fluoroscopically on the packaged graft.

When the catheter rotation is optimal, the jacket covering the graft is retracted while the pullwire is drawn back down the contralateral femoral sheath. After the jacket is fully retracted, the two graft limbs separate. By drawing the catheter delivery system distally, we can now position the proximal attachment system immediately below the renal arteries and the ipsilateral and contralateral graft limbs in the respective iliac arteries.

The proximal attachment system is deployed in the same manner as with the tube graft. It is seated with three successive inflations of the intrinsic balloon catheter. The contralateral distal iliac limb attachment system is deployed next and seated with an EGS balloon. Finally, the ipsilateral distal iliac graft limb attachment system is deployed, completing the deployment of the bifurcated graft system. A completion angiogram is then obtained as in the case of the tube graft.

Aortoiliac/Femorofemoral System

If both hypogastric arteries are patent and the graft limb is to be deployed in the external iliac artery, it is necessary to occlude the ipsilateral hypogastric artery. This can be done at the time of operation with placement of coils in the hypogastric artery. However, it is time-saving to do this preoperatively in the Radiology Suite in the following manner:

- The graft catheter delivery system is advanced up the sheath in a manner similar to that which is done with the tube graft.
- The proximal attachment system is deployed.
- The graft is rotated to correct any twist, and the distal attachment system is deployed in the external iliac artery. This is then seated with a balloon catheter. The contralateral femoral artery has been exposed and accessed with a guidewire.
- A transverse, oblique arteriotomy is made in the contralateral femoral artery, and another EGS sheath is inserted on that side. Usually, we use a short sheath for this purpose.
- The blind endosoc delivery system is then passed up the sheath over the guidewire and positioned in the common iliac artery.
- The endosoc is deployed and seated with a balloon catheter. This serves to prevent reflux up the common iliac artery into the aneurysm sac.
- The sheaths are removed from both sides, and the femoral arteriotomies are connected with a crossover femorofemoral bypass in a standard fashion, usually through a subcutaneous tunnel. After completion of the latter step, a completion angiogram is obtained to confirm the presence of good flow through the aortoiliac graft as well as the femorofemoral graft and make sure there is no periprosthetic leak.
- The femoral incisions are closed, and the patient is returned to the recovery room.

As soon as recovery from anesthesia has occurred, the patient is transported to a regular hospital room. Intensive care unit monitoring is not necessary. The patient may ambulate that evening and may have a progressive diet as tolerated. Discharge the following day is anticipated unless any medical problems arise that would delay a routine discharge.

Before discharge, plain posteroanterior and lateral films of the abdomen are obtained to document the position of the graft as evidenced by radiopaque markers. A color flow duplex scan of the graft is then obtained to confirm that there is good graft function without evidence of flow reflux into the aneurysm.

Finally, a contrast-enhanced CT scan or spiral CT scan is obtained, which serves as a baseline for the functioning graft within the defunctionalized aneurysm.

CLINICAL TRIAL

Endovascular graft repair of AAAs represents a totally new concept. Although initial results are beginning to accumulate and look promising, long-term results in regard to safety, efficacy, and durability are completely unknown. For this reason, endovascular graft repair of AAAs should be confined to a regulated clinical trial.

The FDA reviews and regulates new devices or new applications for existing devices within the United States. The agency's responsibility is to evaluate new proposals and to grant or deny permission for their clinical investigation. The objective is to ensure public safety and at the same time provide an environment that will be conducive to new advances in health care.

Endovascular Technologies applied for and received approval for evaluation of their device for endovascular repair of AAAs. Phases I and II trials for the tube and bifurcated grafts are complete. Phase II trial of the aortoiliac system is in progress.

RESULTS

Tube Graft

Results of the Phase I trial of the tube graft have been reported (10). Since the first implant at UCLA Medical Center on February 10, 1993 through December 6, 1994, 46 patients at 13 centers have undergone tube graft implantation. Their ages have ranged from 54 to 84 years, with a mean age of 72. A review of the results and complications within the first 30 days demonstrates no postoperative deaths. One patient suffered a mild myocardial infarction, and seven patients required conversion to open repair. No patient had major thromboembolism, but two patients showed minor petechiae with spontaneous clearing and no tissue loss.

The combined European, United Kingdom, and Australian experience has been reported (11). Thirty-one consecutive patients undergoing tube graft repair were described. All graft placements were successful, but one graft was subsequently replaced with a transabdominal repair because of a persistent perigraft leak. There was one postoperative death.

The long-term behavior of the aneurysm sac after endovascular repair has been reported (12). Thirty-four patients with tube grafts in place for 1 year have been followed up with CT scanning. The authors divided the patients into three groups. Group 1 had no evidence of a perigraft leak after implantation. Patients in group 2 were noted to have early perigraft leaks, but on follow-up CT scan were seen to have sealed within the first year. Group 3 consisted of patients who had persistent perigraft leaks.

Group 1, consisting of 20 patients representing 58% of all the groups, were shown to have a reduction in mean aneurysm sac diameter by 0.63 cm. Group 2, consisting of 7 patients (21%), showed a mean aneurysm diameter reduction of 0.34 cm. However, group 3, consisting of 7 patients (21%), showed an aneurysm sac diameter increase of 0.19 cm.

Phase II of the tube graft experience was designed to compare endovascular repair with conventional open repair. Although the results are complete, the data have not yet been published. Eighty-eight patients underwent endovascular repair of AAAs, and 90 patients who underwent conventional transabdominal repair served as controls. Eighty-three of the 88 endovascular repairs (94%) were considered technical successes. The 30-day mortality rate for the endovascular repair was 0% in contrast to a 3.3% mortality rate in those undergoing conventional repair. No serious morbidity occurred in the endovascular group compared with a 3.3% serious morbidity rate among those who underwent open aneurysm repair.

The mean hospital stay of the endovascular group was 2.4 days compared with 8.2 days for the surgical control group.

Bifurcation Graft

Phase II of the bifurcation graft experience has just been completed but the results have not yet been published. Endovascular repair using the bifurcated device was carried out in 106 patients, and 103 patients served as conventional surgical controls. This study was carried out at a total of 20 participating centers. Ninety-six of the 106 (90.6%) endovascular repairs were considered technically successful.

The mortality rate in the endovascular graft group was 1.9% compared with a mortality rate of 3.9% in the conventional surgery group. Serious morbidity occurred in 1.9% of the endovascular group compared with 2.9% in the control group. The mean hospital stay for the endovascular patients was 3.9 days compared with 7.6 days for the control group.

Aortoiliac/Femorofemoral Graft

Phase II of the aortoiliac/femorofemoral graft series is still in an early stage, and no results are available.

SUMMARY

The EGS system for transfemoral repair of AAAs has been described. Methods of patient evaluation and the technique of implantation have been outlined. A two-phase, FDA-approved clinical trial for investigation of this new device has been reviewed and the clinical results of Phases I and II of the tube and bifurcated grafts summarized.

References

1. Goldstone J. Aneurysms of the aorta and iliac arteries. In: Moore WS, ed. Vascular surgery: A comprehensive review. Philadelphia: WB Saunders, 1993.
2. Ernst CB. Abdominal aortic aneurysm. N Engl J Med 1993;328:1167–1172.
3. Veith FJ, Goldsmith J, Leather RP, Hannan EL. The need for quality assurance in vascular surgery. J Vasc Surg 1991;13:523–526.
4. Moore WS. Endovascular grafting technique: A feasibility study. In: Yao JST, Pearce WH, eds. Aneurysms: New findings and treatments. Norwalk, CT: Appleton & Lange, 1993.
5. Moore WS. Transfemoral endovascular repair of abdominal aortic aneurysm using the endovascular graft system device. In: Greenhalgh RM, ed. Vascular and endovascular surgical techniques: An atlas. 3rd ed. London: WB Saunders, 1994.
6. Moore WS, Vescera CL. Repair of abdominal aortic aneurysm by transfemoral endovascular graft placement. Ann Surg 1994;220:331–341.
7. Moore WS. The role of endovascular grafting technique in the treatment of infrarenal abdominal aortic aneurysm. Cardiovasc Surg 1995;3:109–114.
8. Colburn MD, Moore WS. Endovascular repair of abdominal aortic aneurysms using the EGS tube and bifurcated graft systems. World J Surg 1996;20:664–672.
9. Moore WS. Transfemoral endovascular repair of abdominal aortic aneurysm. J Irish Coll Phys Surg 1996;25:104–109.
10. Moore WS, Rutherford RB, for the EVT Investigators. Transfemoral endovascular repair of abdominal aortic aneurysm: Results of the North American EVT phase 1 trial. J Vasc Surg 1996;23:543–553.
11. Balm R, Eikelboom BC, May J, et al. Early experience with transfemoral endovascular aneurysm management (TEAM) in the treatment of aortic aneurysms. Eur J Vasc Endovasc Surg 1996;11:214–220.
12. Matsumura JS, Pierce WH, McCarthy WJ III, Yao JST, for the EVT Investigators. Reduction in aortic aneurysm size: Early results after endovascular graft placement. J Vasc Surg 1997;25:113–123.
13. Matsumura JS, Moore WS, for the EVT Investigators. Clinical consequences of periprosthetic leak following endovascular repair of abdominal aortic aneurysm. J Vasc Surg 1998;27:606–613.

Chapter 10

The Nottingham Experience With Endovascular Repair of Abdominal Aortic Aneurysms

S.W. Yusuf and B.R. Hopkinson

For the past 4 years, our unit has been interested in the development of endovascular surgery for aortic aneurysms with particular emphasis on the possibility of treating most aneurysms and ruptured aneurysms with an endovascular technique. The technique of endovascular repair used and some of the issues addressed in our investigations during this period are presented in this chapter.

IS ENDOVASCULAR REPAIR LESS TRAUMATIC?

Endovascular repair is considered to be an advance because it is perceived to be less traumatic than open surgery. Although this may seem obvious to the proponents of endovascular surgery, the notion needs verification by scientific methods. We therefore investigated the hemodynamic changes, colonic perfusion, and renal damage during endovascular and open aneurysm surgery.

Previous studies have shown that aortic clamping during open aneurysm surgery results in approximately a 40% increase in the systemic vascular resistance and a 15% to 35% decrease in the stroke volume and cardiac output (1). Endovascular repair prevents the need for a laparotomy, aortic sac dissection, and aortoiliac clamping, but it does involve bilateral femoral and, in some techniques, bilateral iliac clamping. Moreover, there is a short phase of interruption of flow through the aorta while the prosthesis is being deployed.

Hemodynamic Changes

The changes in blood pressure, systemic vascular resistance, cardiac output, and serum lactate levels were compared in two groups of 10 patients each undergoing open or endovascular aortic aneurysm repair (2). Our results demonstrated that cross-clamping the aorta in the open group resulted in significant increases ($P < .05$, paired t test) in the mean arterial pressure from a mean value of 98 ± 6 to 115 ± 9 mm Hg and in the systemic vascular resistance from 2413 ± 189 to 2660 ± 204 dynes/cm, with the cardiac output decreasing from 3.8 ± 0.4 to 3.0 ± 0.5 L/min.

No significant changes were seen in the endovascular group when the proximal end of the stent was deployed within the aorta. However, the endovascular group demonstrated a significant increase in the systemic vascular resistance from 2280 ± 196 to 2485 ± 240 dynes/cm during bilateral femoral clamping.

In the open group, reperfusion of the first limb resulted in a significant decrease in the mean arterial pressure from 89 ± 4 to 61 ± 3 mm Hg and in the systemic vascular resistance from 2327 ± 146 to 1686 ± 153 dynes/cm without significant changes in the cardiac output.

In the endovascular group, the only significant change during reperfusion was in the systemic vascular resistance, which decreased from 2039 ± 227 to 1673 ± 209 dynes/cm on removal of the second femoral clamp.

Renal Perfusion

Renal perfusion is also known to be compromised during open aortic aneurysm surgery (1). Although renal failure is relatively uncommon, it remains an important cause of postoperative morbidity and mortality after open aneurysm surgery. Renal blood flow falls significantly on aortic clamping, and use of the endovascular method could potentially avoid such changes. A study of subclinical renal damage during open and endovascular surgery using N-acetylglucosamine (NAG) and albumin excretion in urine, which are markers of tubular and glomerular damage, respectively, was therefore carried out (3). This study showed a significant increase in the microalbumin excretion starting at 1 hour, lasting until the third postoperative day after open aneurysm repair. No significant changes were observed after endovascular repair.

The levels of NAG in the urine also significantly increased at the same time and remained elevated until 7 days after open repair, whereas no significant changes were observed after endovascular repair.

Colonic Ischemia

Almost 50% of all patients undergoing conventional aneurysm surgery are reported to experience a fall in colonic intramucosal pH (Phm) below 7.0 (4). Colonic ischemia plays a role in the pathogenesis of multiple organ dysfunction and failure. A study of the changes in the Phm using a sigmoid balloon tonometer showed that after open surgery the Phm significantly decreased, reaching a trough at 6 hours postoperatively and gradually recovering to the preoperative level at 24 hours. On the other hand, no significant fall in the Phm was observed in the endovascular group (5).

We believe that these studies provide objective evidence to support the hypothesis that the detrimental physiologic changes associated with open surgery are reduced with endovascular surgery.

IS ENDOVASCULAR REPAIR POSSIBLE FOR MOST ANEURYSMS?

Endovascular repair of abdominal aortic aneurysms commenced in Nottingham in 1994 with the use of the Chuter bifurcated endoprosthesis (6). However, it soon became clear that only a minority of patients were anatomically suitable for a bifurcated endoprosthesis (7). To make an impact on the clinical practice and outcome of patients with aortic aneurysms, the technique must be used on the vast majority of patients.

A detailed analysis of the data on aneurysm morphology showed that only 4% of patients had a distal aortic neck for implantation of an aortoaortic graft, and only 10% were suitable for a bifurcated graft, whereas 55% were suitable for an aortouni-iliac graft (8). This encouraged us to concentrate on the development of an aortouni-iliac device (9,10) in collaboration with the Malmo group (Sweden) and Cook Europe (Denmark), which was similar to that described by Parodi (11), May and associates (12), and Marin and associates (13).

Use of the aortouni-iliac graft has increased the suitability of aneurysms for endovascular repair, and almost 70% of all aneurysms referred to our unit for endovascular repair are treated with an aortouni-iliac endoprosthesis.

TECHNIQUE OF AORTOUNI-ILIAC ENDOVASCULAR REPAIR

The endoprosthesis consists of an endoskeleton made of linked Gianturco stents (Cook Europe) covered with a thin-walled, uncrimped Dacron graft (Vascutek, Renfrewshire, Scotland). The proximal-most stent for attachment to the aneurysm neck has four caudally pointing and four cranially pointing barbs. The height of the proximal stent is 25 mm and the resting or unrestrained diameter is 45 mm. The stents are sutured to the graft with 5/0 Gore-tex suture (W.L. Gore and Associates, Flagstaff, Arizona). The prosthesis is tailored for each case according to the measurements obtained from spiral CT scan reconstructions.

Calibrated angiography using a marker catheter (Cook Europe) is performed at the time of the procedure to confirm the length of the prosthesis. The diameter of the prosthesis is approximately 10% greater than the diameters of the aortic neck and the common iliac artery. As previously described (10), the autoclaved endoprosthesis is mounted on a graft carrier and inserted into a 20F loading sheath at the time of the procedure.

The procedures are performed in the operating room usually under general anesthesia with preparation for conversion to an open procedure should that become necessary, although we have performed several procedures under epidural or local anesthesia. A mobile image intensifier (Siremobil 2000, Seimens, Erlangen, Germany) with facilities for digital subtraction imaging, playback, and hard copy output is used for intraoperative imaging. The stages of the procedure are as follows:

1. Both common femoral arteries are exposed.
2. A 3 J guidewire is introduced via an anterior wall femoral puncture on the side chosen for deployment of the iliac end of the endoprosthesis. If the arteries are particularly tortuous, then a Terumo Glidewire (Terumo, Tokyo, Japan) and a multipurpose catheter are used for negotiating the aortoiliac segment.
3. A 7F sheath is introduced and a 7F pigtail angiographic catheter with radiopaque marks at 1-cm intervals over 15 cm is introduced over the guidewire and positioned at the level of the renal artery.
4. A hand injection is usually adequate for obtaining an aortogram to identify the origins of the renal arteries and to measure the distance from the renal artery to the aortic and iliac bifurcations.
5. The angiography catheter is removed and reintro-

duced over a guidewire via a 7F sheath placed in the contralateral femoral artery.

6. A 20F sheath is introduced over the guidewire through a longitudinal arteriotomy, and the tip is positioned in the lower thoracic aorta.
7. The dilator is slowly withdrawn, and the sheath is clamped with a pair of rubber covered clamps to achieve hemostasis.
8. The preloading sheath containing the endoprosthesis is locked to the introducing sheath, and the endoprosthesis is advanced to the level of the renal artery.
9. The position of the renal artery is again confirmed with an angiogram, and the process of deployment commences. The position of the endoprosthesis and its relationship to the origin of the renal artery are constantly checked with frequent injections of small volumes of contrast through the angiography catheter.
10. The endoprosthesis is deployed as close to the renal artery as possible without compromising its origin.
11. Once the entire endoprosthesis has been released by completely withdrawing the sheath, the suture line used for mounting the endoprosthesis on the carrier is removed, allowing the proximal stent to fully expand.
12. A guidewire is introduced through the central lumen of the carrier to maintain access through the endoprosthesis before removal of the carrier.
13. The angiography catheter, which at this stage is lying between the endoprosthesis and the aortic wall, is gently withdrawn over a guidewire after unfurling the coil in the pigtail.
14. A 20F sheath is then introduced via a longitudinal arteriotomy in the contralateral femoral artery over a guidewire into the aneurysm sac.
15. The guidewire is removed and an aneurysmogram is performed by injecting through the lumen of the dilator. This is done to detect any patent lumbar or inferior mesenteric arteries.
16. If any patent branch is found arising from the aneurysm sac, then large embolization particles made from sheets of gelatin sponge are introduced into the sac.
17. A Gianturco stent covered with a Dacron graft is deployed inside the common iliac artery to act as an occlusion device.
18. The procedure is completed with a traditional femorofemoral bypass and a final angiogram to ensure that:

- The renal arteries are patent.
- There is no paragraft perfusion on the aneurysm sac— posteroanterior and lateral.
- There is no perfusion of the sac along the occluding device in the iliac artery.
- The internal iliac arteries are patent.
- The femorofemoral crossover graft and the distal superficial femoral and profunda arteries are patent.

Postoperative surveillance is carried out using a contrast-enhanced CT scan that is performed within the first week postoperatively and subsequently at 3 months, 6 months, and then annually.

IS ENDOVASCULAR REPAIR POSSIBLE FOR RUPTURED ANEURYSMS?

Almost 50% of all surgical repairs are carried out for ruptured aneurysms (14), and the perioperative mortality rate at most units remains 40% to 60% (15), leaving considerable room for improvement. Based on the information we obtained from previous studies on aneurysm morphology (16), we have been keeping a stock of sterile, tapered endoprostheses for aortouni-iliac repairs that are suitable for most aneurysms, making it possible to treat urgent and emergency cases.

Since our first endovascular repair of a contained aneurysm rupture in 1994 (17), we have attempted endovascular repairs on another nine patients with contained aneurysm ruptures. Of these nine, two became unstable during transfer to the operating room and died before the stent-graft could be deployed. The other seven cases remained hemodynamically stable, and it was possible to implant a device and achieve effective exclusion of their aneurysms. All seven survived the procedure, but one 84-year-old patient with an excluded aneurysm died 2 weeks postoperatively from renal failure without any evidence of renal artery occlusion.

On an intention-to-treat basis, successful repair was achieved in 77.7% (seven of nine) of our cases, with an overall perioperative mortality rate of 33.3% (three of nine). The anatomy was particularly complex and challenging in five of these seven patients, and six of the seven had additional risk factors.

This limited experience confirms the technical feasibility of endovascular repair of ruptured aneurysms, but further controlled studies are needed to evaluate the role of endovascular repair of ruptured aneurysms.

SHOULD ENDOVASCULAR REPAIR BE OFFERED TO HIGH-SURGICAL-RISK PATIENTS?

Endovascular repair is perceived to be less invasive than open repair. This has led to an increased referral of patients who are deemed to be high surgical risks. Of the 196 aortic aneurysm repairs performed at our unit, 103 (53.6%) were referred from other vascular units. The outcome of high-surgical-risk patients is shown in Table 10.1. In the acceptable-risk category the perioperative mortality rate was 2.3%, and 90% of the patients were alive at a median follow-up of 17 months.

Although some deaths in the high-risk group can be

Table 10.1 Outcome of High-Surgical-Risk Patients Undergoing Endovascular Repair

Risk factor	Number of patients	Perioperative mortality (%)	Survival at 17 months median follow-up (%)
Cardiac risk	56	7 (12%)	45 (80%)
Pulmonary	17	1 (6%)	14 (82%)
Renal	11	2 (18%)	5 (45%)
Local (hostile abdomen)	8	1 (12%)	7 (87%)
Age > 84	5	3 (60%)	1 (20%)
Hematologic	3	0	3

attributed to the combination of difficult anatomy leading to failure of endovascular exclusion and requiring open intervention, some deaths occurred after uncomplicated procedures. These findings highlight the importance of the need for a controlled trial between best medical treatment and endovascular repair in high-surgical-risk patients who are deemed unfit for open surgery.

DURABILITY OF ENDOVASCULAR REPAIR

Ultimately, the role of endovascular aneurysm surgery will be determined by the long-term efficacy of this method in preventing aneurysm expansion and rupture. Because of limited follow-up data, the long-term results are still unknown. However, intermediate follow-up has provided some understanding of the behavior of these repairs and the factors influencing late success or failure.

Stent migration has been observed in many earlier Chuter bifurcated graft procedures, and some of these migrations were associated with a significant increase in the neck diameter. This increase is a troublesome finding because such neck expansion poses a threat to the long-term integrity of any endoprosthesis. It is anticipated that use of a suprarenal stent, oversizing, and column strength of a fully skeletonized endoprosthesis might prevent stent-graft displacement. However, perhaps the only certain way of preventing such late failures is the development of a more secure means for attachment of the stent-graft to the aneurysm neck. Only time will tell.

In the presence of a side branch endoleak, aneurysms have been observed to continue to expand or at least to fail to undergo a reduction in size (18). Preemptive selective embolization of these potential sources of endoleaks is a demanding procedure. However, we found that selective use of large particulate emboli introduced into the aneurysm sac in the presence of patent lumbar or inferior mesenteric arteries on an intraoperative aneurysmogram can be effective in preventing late side branch endoleaks.

CONCLUSION

Data from our studies show that endovascular surgery is associated with fewer physiologic derangements that, it is possible to treat most aneurysms with an aortouni-iliac endoprosthesis, and endovascular repair of a contained rupture of an aortic aneurysm is feasible. However, endovascular repair may not improve survival in surgically unfit patients. These studies, along with the experiences of other centers, have clarified the questions that need to be addressed in a randomized trial to determine the place of endovascular aortic aneurysm repair.

References

1. Clark NJ, Stanley TH. Anaesthesia for vascular surgery. In: Miller RD, ed. Anesthesia. 4th ed. New York: Churchill Livingstone, 1994.
2. Baxendale BR, Baker DM, Hutchinson A, et al. Haemodynamic and metabolic response to endovascular repair of infra-renal aortic aneurysm. Br J Anaesth 1996;77:581–585.
3. Baker DM, Wenham PW. Renal damage. In: Hopkinson BR, Yusuf SW, Whitaker SC, Veith FJ, eds. Endovascular surgery for aortic aneurysms. London: WB Saunders, 1997.
4. Soong CV, Blair PBH, Halliday MI, et al. Bowel ischaemia and end organ impairment in elective abdominal aortic aneurysm repair. Br J Surg 1994;81:965–968.
5. ElMarasy N, Yusuf SW, Lonsdale RJ, et al. Study on the effect of endovascular repair on colonic perfusion. J Endovasc Surg 1996;3:95–96.
6. Chuter TAM. Transfemoral aneurysm repair. D.M. Thesis, University of Nottingham, 1993.
7. Yusuf SW, Baker DM, Chuter TAM, et al. Transfemoral endoluminal repair of abdominal aortic aneurysm with bifurcated graft. Lancet 1994;344:650–651.
8. Armon MP, Yusuf SW, Lateif K, et al. The anatomical suitability of abdominal aortic aneurysms for endovascular repair. Br J Surg 1997;84:178–180.
9. Yusuf SW, Baker DM, Hind RE, et al. Endoluminal trans-femoral abdominal aortic aneurysm repair with aorto uni-iliac graft and fem-fem cross over. Br J Surg 1995;82:916–917.
10. Yusuf SW, Whitaker SC, Chuter TAM, et al. Early results of endovascular abdominal aortic aneurysm repair with aorto uni-iliac graft and femoro-femoral bypass. J Vasc Surg 1997;25:165–172.
11. Parodi JC. Endovascular repair of abdominal aortic aneurysm and other arterial lesions. J Vasc Surg 1995;21:549–557.
12. May J, White G, Waugh RC, et al. Treatment of complex abdominal aortic aneurysms by a combination of endoluminal and extraluminal aortofemoral grafts. J Vasc Surg 1994;19:924–933.
13. Marin ML, Veith FJ, Cynamon J, et al. Initial experience with transluminally placed endovascular grafts for the treatment of complex vascular lesions. Ann Surg 1995;22:449–469.
14. Callam MJ, Haiart D, Murie JA, et al. Ruptured aortic aneurysm: A proposed classification. Br J Surg 1991;78:1126–1129.
15. Ernst CB. Abdominal aortic aneurysm. N Engl J Med 1993;328: 1167–1172.
16. Armon MP, Yusuf SW, Whitaker SC, et al. The anatomy of abdominal aortic aneurysm: Implications for sizing of endovascular grafts. Eur J Vasc Endovasc Surg 1997;13:398–402.
17. Yusuf SW, Whitaker SC, Chuter TAM, et al. Emergency endovascular abdominal aortic aneurysm repair. Lancet 1994;344:1645.
18. Armon MP, Yusuf SW, Whitaker SC, et al. The fate of the aneurysm sac following endovascular aneurysm repair. Eur J Vasc Endovasc Surg 1998, in press.

Chapter 11

Endovascular Grafting for Abdominal Aortic Aneurysms Using a Modular System[1]

Wolf-J. Stelter

HISTORY OF MODULAR SYSTEMS

In contrast to conventional operations, approximately 90% of abdominal aortic aneurysms (AAAs) suitable for transfemoral treatment require a bifurcated stent-graft (Table 11.1). Whether cylindrical or tapered, tubular stent-grafts require a sufficient "neck" above the bifurcation. To fix a stent-graft into position, the uncovered portion of the proximal fixation device or stent may extend above the renal arteries; however, such extension is not possible at the aortic bifurcation.

The first attempt to deal with aneurysms involving the distal aorta and iliac arteries was the extension of a tapered tube graft from the infrarenal aorta into an enlarged or aneurysmal iliac artery with occlusion of the contralateral iliac artery and restoration of contralateral extremity arterial flow by a femorofemoral bypass (1–4). After this, custom-made bifurcations with self-expanding stents (i.e., the Chuter device and the Endovascular Technologies [EVT] device) were introduced through one femoral or iliac artery, and the contralateral limb was manipulated and pulled across the aortic bifurcation into the opposite iliac artery (5,6).

Soon a modular system was developed in which the components of the bifurcation are introduced separately through both femoral or external iliac arteries and positioned within the vascular tree (Figs. 11.1 and 11.2). A self-expanding framework along the total length of the graft is the most desirable. Oversized graft diameters allow good fixation and sealing at the ends of the device and reduce the need for stockpiling a large number of sizes of grafts. Because a modular system seems to be simpler and more straightforward to insert, its use may reduce the risk of peripheral embolization and shorten or avoid the need for prolonged clamping of one or both of the exposed femoral arteries.

DEVICES

The first widely used modular system was the Mialhe-STENTOR bifurcation graft (Mintec, Inc, Grand Bahama Island), now known as the Vanguard system (Meadox/Boston Scientific Corporation, Natick, MA) (see Fig. 11.1). The Vanguard system appears to be the most versatile, easily adaptable, and practicable bifurcation system, although other similar modular systems have more recently become available (Talent [World Medical Manufacturing Corporation, Sunrise, FL], AneuRx [Medtronic, Eden Prairie, MN], and Corvita [Corvita Corporation, Miami, FL] devices).

During the last 2 years, the latter modular bifurcation devices have been used in more than 1200 patients worldwide (7–9). However, any modular system may have specific weaknesses and develop leaks at the junctions of its components.

In the Vanguard system, the graft consists of self-expanding stents made of nitinol with thermal memory at a transition temperature, 32° to 33° C. These are covered by a thin-walled, woven polyester fabric (Dacron).

[1] Dedicated to the memory of my teacher and father-like friend, George C. Morris Jr, MD, born February 15, 1924, died August 29, 1996, Professor of Surgery at Baylor University, Houston, Texas.

Table 11.1. Patients Receiving Endovascular Treatment of AAAs With Stent-Grafts (N = 123)

Patient characteristics	Number of patients
Male	111
Female	12
Mean age	69 years (range 50–90)
Number receiving bifurcated grafts	109
Number receiving tube grafts	14

This study was carried out at the Stadtische Kliniken Frankfurt-Haöchst between August 1994 and October 1996.

The proximal part of the aortic stent is reinforced and fitted with small hooks on the third row of nitinol. The proximal two rows of nitinol (10 mm in length) are uncovered, allowing this portion of the device to overlap the region of the renal arteries. Platinum markers identify the beginning of the stent and the polyester fabric covering. If needed, the uncovered part of the stent can be constructed with only one row of zigzag nitinol wire. The aortic section of this stent is available in diameters of 22, 24, and 26 mm.

The junction between the main aortic graft and the contralaterally introduced iliac leg is tapered in a trumpet-like fashion. The iliac legs are available in diameters of 12 mm and 10 mm. In the original Mialhe-STENTOR series, only 9-mm legs could be connected to the main graft. The length of these legs could be increased by telescope-like extensions inserted stepwise using a 10-mm or 12-mm (diameter) covered stent. This configuration created additional junctions, which caused subsequent problems (see Secondary Leakage).

Although the Dacron fabric of the Miahle-STENTOR graft had some weak points, the new denser Vanguard graft is sutureless and woven with greater strength. The Vanguard graft is available in diameter lengths of 153 mm and 165 mm. The distal limbs can be extended in a telescope-like fashion by other tubular stent-grafts of var-

Figure 11.1. Vanguard modular bifurcation for the endovascular treatment of AAAs.

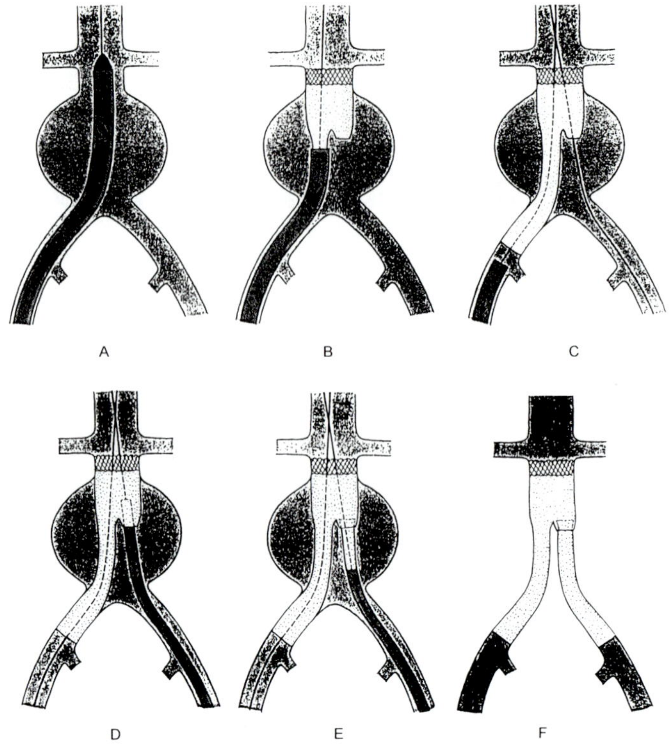

Figure 11.2. Series of steps involved in the implantation of a modular bifurcation system (see text for details).

ious lengths (Passager [Meadox/Boston Scientific Corporation]). Thus, the graft can be extended into more normal-sized external iliac arteries, thereby excluding aneurysmal common iliac arteries.

The main graft is fitted inside an 18F (internal diameter [I.D.]) delivery catheter, the distal end of which is occluded by a short, tapered tip to facilitate passage through the iliac arteries, even when they are tortuous, calcified, or stenosed.

The "pusher" for the main aortic part of the graft functions as the delivery sheath for the iliac limb, which is kept in place during deployment by another pusher. A soft latex balloon that is integrated within the nose cone can be used to occlude the aorta during deployment, although we no longer use this component of the device for this purpose. However, we do use this balloon to expand or "model" the graft against the vessel wall. A 10F (I.D.) catheter system (Passager), which can be introduced percutaneously or via an open arteriotomy, is used for delivery and deployment of the opposite iliac limb.

TECHNIQUE

Standard 8F (I.D.) sheaths are introduced into both femoral arteries—one percutaneously and one after surgical exposure of generally the right femoral artery at the inguinal ligament. Following is the technique for placing endovascular grafts.

1. Proximal angiography with road-mapping is performed to identify and mark the renal arteries and

the aortic neck. A super-stiff wire is introduced through the angiography catheter. The exposed femoral artery is clamped distally, and the 8F sheath is removed. Through an open arteriotomy, the 18F delivery system is advanced under fluoroscopy over the stiff wire (see Fig. 11.2*A*). Using the markers on the graft, the outer sheath is withdrawn when properly positioned below the renal arteries (see Fig. 11.2*B*). This deploys the proximal main part of the bifurcation that is held in place by the appropriate pusher, which also acts as the sheath for the iliac limb. Distal flow is immediately restored through the orifice for the contralateral iliac limb.

2. When the aortic portion is fully expanded and modeled by gentle inflation of the integrated balloon using warm saline with contrast and the proximal hooks have been pushed into the aortic wall, the iliac leg is then fully deployed by retraction of the second pusher sheath together with the outer sheath.
3. As the deflated balloon is withdrawn in stages, it is intermittently inflated to fully expand and model the limb.
4. After retraction of the empty delivery system over the guidewire, the arterial incision is closed, and a standard 8F introducer sheath is tightly placed over the wire. This allows removal of the femoral clamp and restores perfusion to the leg.
5. Through the contralateral percutaneous sheath, the opening in the graft body for the second limb is catheterized using interventional techniques (see Fig. 11.2*C*). Occasionally, a cross-over maneuver is required, which leaves a stiff guidewire in place. We have set a limit of about 7 to 10 minutes for the ipsilateral retrograde catheterization of the orifice before changing to the cross-over maneuver.
6. A transbrachial catheterization of the orifice from the left arm is another option. Using a dilatation balloon that is partially inflated by contrast and saline, the position of the wire within the orifice is thoroughly examined. A dilatation balloon is used because it can be easily pulled through the orifice but not outside between the graft and the vessel wall. One of our earliest conversions was required because of incorrect positioning of the wire outside the orifice.
7. Using a special delivery sheath, the contralateral limb is positioned (see Fig. 11.2*D*) and deployed under fluoroscopy so that the opening with the special markers fits into its counterpart on the main graft (see Fig. 11.2*E*).
8. Using a dilatation balloon, the second limb is then modeled against the main graft. If angiography demonstrates that the limb is too short or too small and further extension is required, other covered stents can be added and introduced in a telescope-like fashion over the same wire.
9. A final digital subtraction angiogram is performed to check for possible leaks before all the sheaths and wires are removed and the exposed artery is fully closed (see Fig. 11.2*F*).

Knowledge of the mode and extent of pressure changes can be very helpful during delivery. One femoral artery is clamped for only a few minutes during the introduction and deployment of the main graft. No occlusion clamps are required during catheterization of the second orifice and deployment of the second limb.

GENERAL CONSIDERATIONS

All procedures performed on high-risk patients are done in the operating room under general anesthesia with the patient prepared for an immediate conversion to a conventional open approach. A mobile, digital C-arm fluoroscope with subtraction and playback capabilities is adequate for this procedure, and high-power fluoroscopy and angiography with fixed installation are not required. We use large sheaths, including angiography catheters, which can be injected by hand and are easier to handle because of their stiffness. However, we have found that pressure injection uses less contrast and produces better images.

Early in our experience on the occasion when the pressure injector and C-arm failed, we used a simple guidewire through the contralateral sheath to effectively mark the renal artery for proper stent deployment. As we gained experience with the implantation of these devices, our operative time and blood loss decreased (Fig. 11.3). In our later experience with the Vanguard bifurcation device, the typical time for uncomplicated insertion is 90 minutes. Less than 80 mL of agent is required, and fluoroscopy time is approximately 10 minutes.

PATIENT SELECTION

Our patients were selected based on anatomic criteria alone. We found that over 50% of our patients were suitable candidates for endovascular grafting procedures. The aortic diameter below the renal ostia must not exceed 24 mm to permit the graft diameter to be 2 mm more than the proximal aortic neck (a limitation that will change as soon as larger-diameter grafts become available). The outer vessel wall is measured on computed tomography (CT) scans as a reference point for the diameter. The minimal length of the aortic neck is 16 mm. Minor thrombus formation on the aortic wall within the neck is acceptable, although long-term studies are needed to document this with certainty. For tube graft insertion, a narrow segment of not less than 16 mm in length above the bifurcation is required.

For bifurcation grafts, the diameter of the iliac vessels should not exceed 12 mm, because the graft limb would then have to be extended into the external iliac artery or other options would have to be used (see Early Persistent

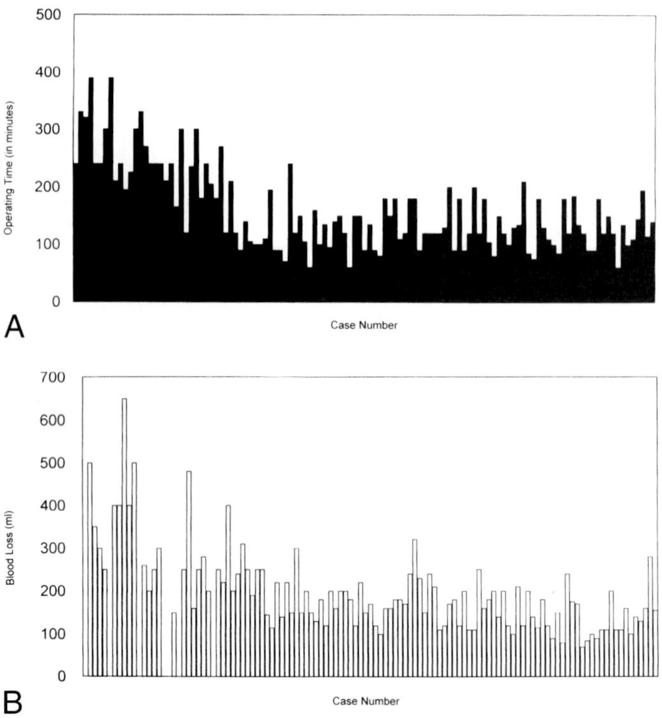

Figure 11.3. Learning curves in the endovascular treatment of AAAs. **A.** Operating time. **B.** Blood loss.

Leakage). Although intentional occlusion of one internal iliac artery is acceptable, bilateral occlusion probably should be avoided. Patent inferior mesenteric and lumbar arteries are not considered contraindications, nor are minor aberrant renal arteries. Moderately tortuous or stenotic iliac arteries are acceptable, because they can usually be straightened and stenosis or occlusion of the iliac vessels can be balloon-dilated or recanalized. When a relatively short proximal neck is present, we prefer to position the standard graft high with the uncovered portion of the device overlapping the renal arteries.

PATIENT REQUIREMENTS

All patients require spiral CT scanning with slices no thicker than 5 to 6 mm and arterial angiography using a calibrated marker catheter. All patients are informed that conversion to a conventional operation is possible and that late results are not known. Each patient must agree to follow-up examination every 3 months, including clinical examination, laboratory studies, a spiral CT scan, and possible angiography.

COMPLICATIONS AND DEATHS

In our series of 123 patients in whom we attempted placement of an endovascular graft to treat an AAA, three patients required conversion to a standard open operation. Two of these patients required conversions because of a technical problem with the integrated balloon, which led to an unanticipated, device-related failure of graft deployment. The third conversion was caused by inexperienced handling, resulting in misplacement of the guidewire outside the orifice of the graft.

In our first 23 patients, 5 perioperative deaths occurred 2 to 3 weeks after technically uneventful operations, with documented symptoms of disseminated intravascular coagulopathy as a contributory cause of death (Table 11.2) (see Coagulation Disorders). We have had no such fatalities since we began to replace one main key protein—antithrombin III—in high-risk patients before surgery, despite similar or increased risk profiles in one subsequent patient.

Embolization

Embolization does not appear to be an important problem with the use of the modular device. We had only four cases of embolization in our series. One major embolization occurred in a patient after vigorous movement and dislocation of the graft by a defective deployment device within an aneurysm filled with thrombus. This patient was converted to a standard aneurysm repair. In two patients without clinical symptoms, small cutaneous embolization (2 cm^2 and 1 cm^2 in size) occurred between the toes. One embolus was detected immediately, and the other was detected 1 week after operation. One other patient returned after 1 week with blue discoloration of one toe, which disappeared in a short time.

Table 11.2. Mortality After Endovascular Treatment of AAAs With Stent-Grafts

Patient number	Age	AAA	ASA Rating	Risk	Cause
5	63	Symptomatic, 46 mm	III	Coronary, pulmonary	Recent myocardial infarction
7	86	Symptomatic, 70 mm	IV	Age, coronary	Pneumonia
8	63	Symptomatic, 45 mm	IV	Liver, pulmonary	Liver, acute pulmonary failure
12	69	Symptomatic, 51 mm	IV	Cardiac	Cardiac
23	73	Asymptomatic, 42 mm	III	Cardiac, metabolic	Pulmonary
101	82	Symptomatic, 59 mm	IV	Congestive heart failure, chronic obstructive lung disease	Hemorrhage

REACTIONS TO ENDOVASCULAR GRAFT DEPLOYMENT

Blood Pressure Changes

Early in our experience we observed a distinct decrease in blood pressure over several minutes immediately after deployment of the main graft as well as after deployment of the second limb or during any extension (Fig. 11.4). In several patients, this was associated with a visible skin flushing reaction, which was noticed by the anesthesiologist. We first attributed this reaction to a "declamping effect" after the use of the balloon for temporary supra-aortic occlusion. However, it happened to a similar extent when the balloon was not inflated and released.

Blood pressure reactions became less pronounced and less frequent during the second year of our series, and open aortic clamping for the same short period of time did not produce similar effects.

Inflammatory Reactions

Fifty to eighty percent of our patients had a characteristic postprocedural febrile course. Sometimes this lasted up to 4 to 6 weeks after endovascular grafting. In addition, many patients complained about diffuse back pain, night sweats, and general malaise. Patients with these symptoms, which we called "postimplantation syndrome," responded well to indomethacin. We now routinely administer indomethacin, and during the past year these inflammatory reactions have decreased in severity and duration.

Two patients whose grafts had already been deployed in the aorta and were then converted to standard open repair had a reaction consisting of edema and fluid accumulation around the aortic wall a few minutes after laparotomy. In addition, we noted generalized subcutaneous edema primarily in the lower body (scrotum, legs, and thighs).

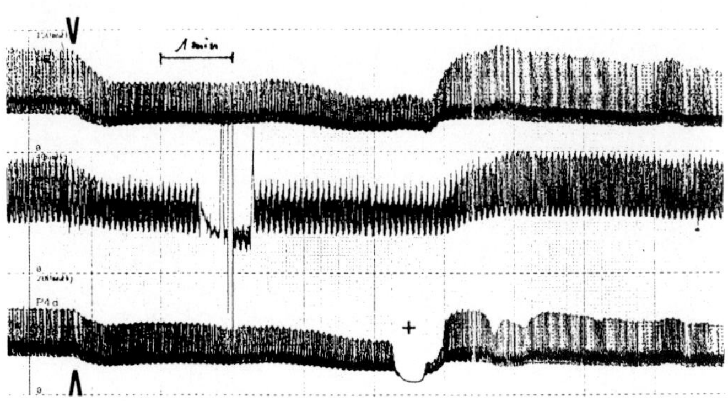

Figure 11.4. Blood pressure changes during and after graft deployment. (*Upper curve*) Blood pressure, systemic (radial artery). (Middle curve) Pulmonary artery pressure. (*Lower curve*) Pressure in the aneurysm sac (see text). At V, deployment of the main graft. At +, inflation of the aortic balloon to model the graft causing a drop of the pressure in the distal aorta (aneurysm sac, *bottom curve*).

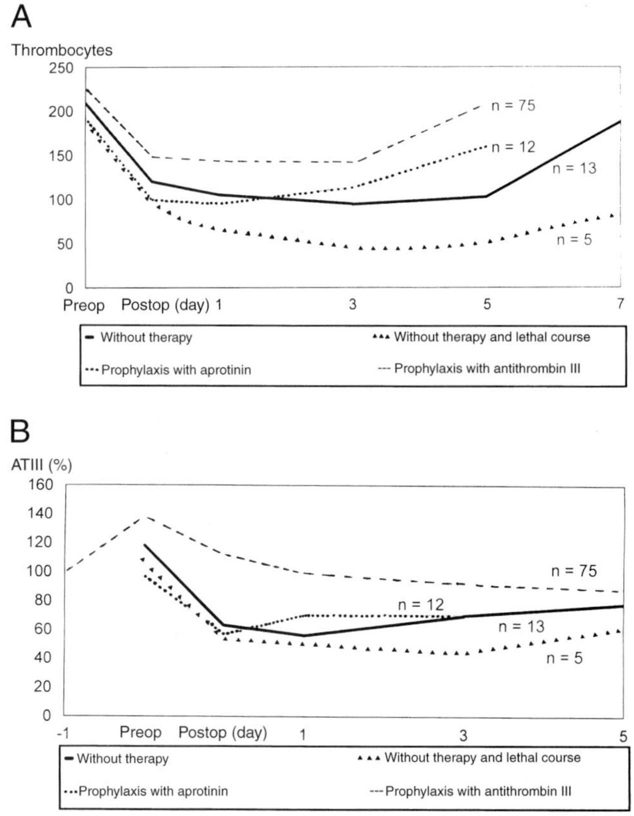

Figure 11.5. **A.** Thrombocytes (in thousands per mm^3) after insertion of endovascular grafts for the treatment of AAAs. **B.** Antithrombin III (ATIII) after insertion of endovascular grafts for the treatment of AAAs.

Coagulation Disorders

When blood samples were obtained 6 hours after the end of the operation, all patients exhibited a distinct drop in antithrombin III and thrombocytes (Fig. 11.5). Leukocyte counts, primarily lymphocytes, also decreased during the first 12 hours after operation and then increased to above normal.

Five high-risk patients with symptomatic aneurysms died from multiorgan failure 2 to 3 weeks after an uncomplicated operation (see Table 11.2). These mortalities were unexpected because other patients at an even higher risk did well. All patients who died exhibited a drop in antithrombin III and platelets beyond that of the other patients in the study, and these parameters did not recover after 24 hours (see Fig. 11.5). We therefore supplied high-risk patients with an excess of antithrombin III before surgery, and we have had no further mortalities with this sequence of events.

Thrombotic Graft Occlusion

Any arterial graft is associated with an increased risk of thrombosis when compared with a natural, endothelialized, although diseased, vessel. Thrombotic graft occlusion, which occurs almost exclusively in only one limb of bifurcation grafts and, fortunately, only rarely in the aor-

Table 11.3. Graft Thrombosis After Endovascular Treatment of AAAs ($N = 123$)

Time period	Number of patients
Perioperative (<24 hr)	4
Early (<30 days)	6
Late (>30 days)	4

tic tube or main graft, may occur perioperatively or postoperatively from a variety of causes (Table 11.3). For effective treatment and prevention of this complication, the underlying cause has to be analyzed and eliminated. However, accurate diagnosis and elimination of the cause are not always possible.

One of our patients had repeated postoperative episodes of thrombosis of one of the limbs of his bifurcation graft. The lumen was percutaneously reconstituted by lysis and balloon dilatations, which had always resulted in adequate angiographic and clinical outcomes. However, after the fifth occlusion, we performed a femorofemoral bypass. This solved the problem.

Based on the latter and other experience, we recommend intravascular ultrasound (IVUS) examination of the graft even after successful insertion, because it will reveal irregularities that cannot be seen arteriographically. A plain abdominal film may also delineate improper stent positioning (Fig. 11.6).

Special attention should be paid to a distal graft that ends in a tortuous vessel. The nitinol stents are flexible and therefore usable in tortuous vessels. If an endovascular graft ends at a bend, its orifice may be in an eccentric position, leading to obstruction. If this situation is seen during the procedure, the limb should be extended by another covered stent that ends coaxially in a straight section of artery.

Routine IVUS exploration has changed our attitude about deliberately excessively oversizing iliac limbs. Large limbs can be compressed and even bent at the iliac orifices or within the small residual lumen of a partially thrombosed aneurysm. The normally circular cross-section is then markedly reduced and deformed with wrinkling and infolding of the graft tissue, which may provoke clot formation.

Shock or other low-flow situations can also provoke thrombosis in a graft that, under normal circumstances,

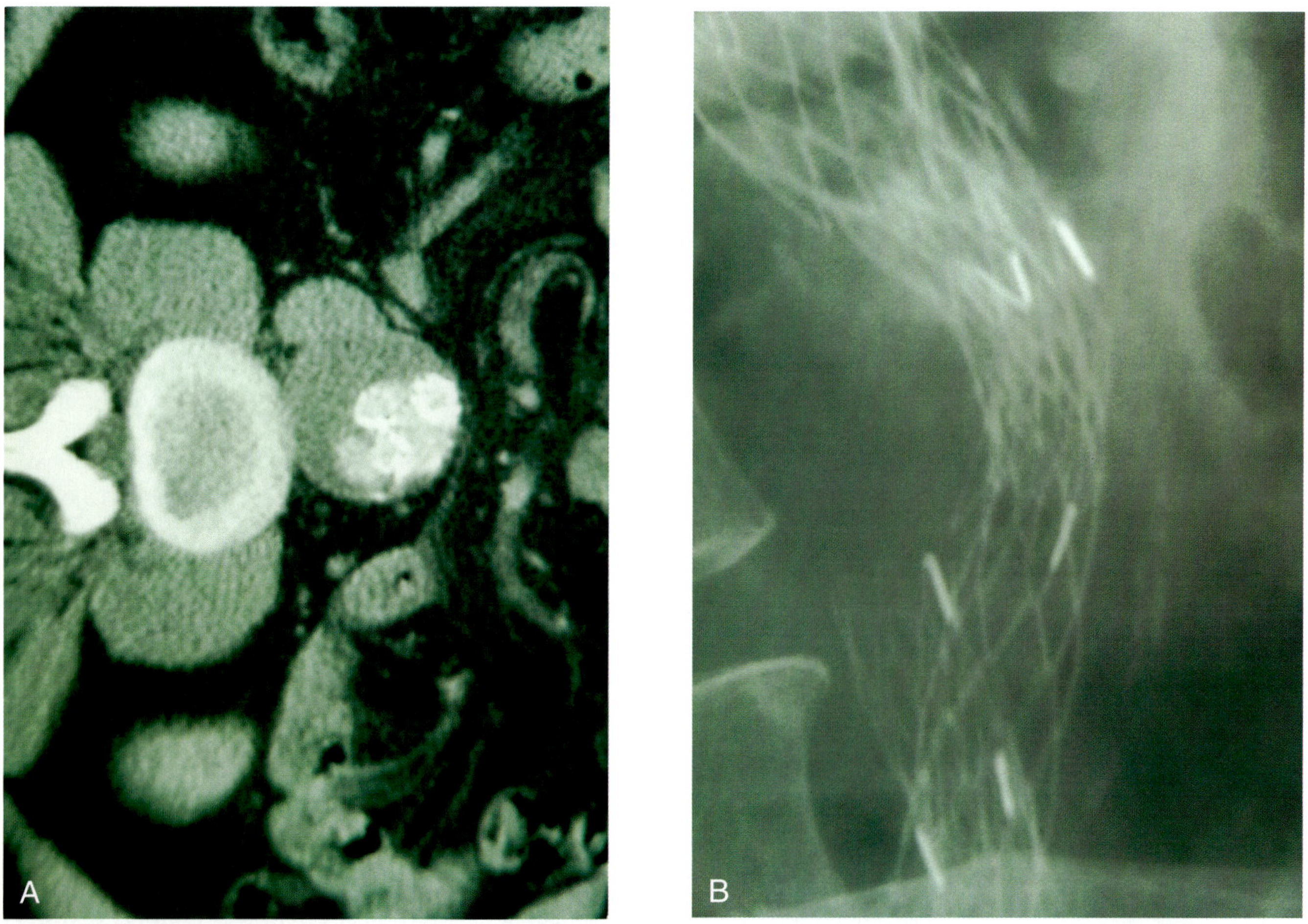

Figure 11.6. Secondary (late) leakage occurring 11 months after the procedure. **A.** CT scan showing contrast endoleak around the disconnected limb. **B.** Disconnected limb is visible on plain lateral abdominal film.

would remain patent. One of our patients had late limb occlusion after a complicated and time-consuming transurethral prostate resection. This resulted in massive bleeding and shock. The patient died with multisystem organ failure. Another patient developed late thrombosis after massive bleeding from a duodenal ulcer, which resulted in shock.

Graft Thrombosis Treatment

We first attempt catheter thrombolysis either by means of a cross-over catheter or, if this is unsuccessful, percutaneous retrograde catheterization. Angiography should show the underlying cause of thrombosis, which can then be corrected. The last and most invasive option after failure is to perform a femorofemoral bypass. Occasionally, surgical correction from a small extraperitoneal approach may be necessary, such as shortening and stretching of a kinked iliac artery or removal of a misplaced obstructing stent combined with banding of a leaking iliac artery.

One patient with a secondary late leak from one limb that was composed of three stent-grafts developed thrombosis of the limb while waiting for repair of the leak. Angiography revealed complete disconnection of the stents with obvious resulting compression and deviation of the orifice of the distal stent; this led to thrombosis of the proximal stump. Because the aneurysm was completely sealed, we performed a simple femorofemoral bypass without attempting to reconstruct the thrombosed limb (Fig. 11.7).

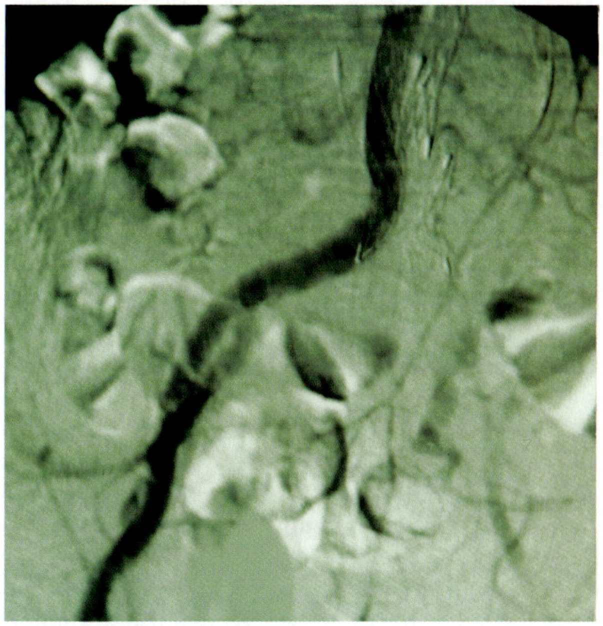

Figure 11.7. Thrombotic occlusion of a modular bifurcation endovascular graft limb after disconnection occurring 9 months after the insertion procedure (see text). The patient was treated with a femorofemoral bypass.

Endoleaks

Early Persistent Leakage

Immediate persistent leakage after endovascular graft deployment is the consequence of graft size mismatch or graft misplacement (Table 11.4). Short proximal necks can also contribute to an increased incidence of leaks.

Proximal leakage at the aortic junction is considered a failure of treatment and appears to be dangerous. Sealing of such a leakage by pure clot formation may suggest an angiographic success, but it does not effectively reduce pressure in the aneurysm sac. A proximal leakage in one of our elderly patients with a short kinked aortic neck was repaired by transfemoral introduction of another short, custom-made tube graft, which was easily positioned because the old stent-graft served as a precise marker.

Distal leakage along the iliac graft limb is, in our opinion, a less serious failure. In such situations, we extend the limb distally to just above the iliac bifurcation and then wait 4 to 6 weeks for clot formation along the elongated iliac graft. Alternatively, the graft can be extended into the smaller external iliac artery, thus depriving the internal iliac artery from prograde flow. Extension into one of the external iliac arteries can also be performed after 6 weeks of observation. Other options include banding of the common iliac artery through a small extraperitoneal approach to preserve internal iliac prograde flow.

Secondary (Late) Leakage

Late leakage developing after 6 months of good function without leakage is the result of graft material failure (n = 3) or graft dislocation (n = 7). During our 2 years of experience with self-expanding, oversized stent-grafts, we have not seen a case in which the vessel wall expanded and led to leakage. Graft material failure can occur because of wear on the polyester texture by a stent wire, especially when the graft is in a tortuous position. Another reason for graft material failure has been a break or tear at the bifurcation suture line. This has not occurred with the Vanguard device.

Graft dislocation occurred in the distal limbs, usually when they were placed in tortuous arteries. Grafts also slipped out of the iliac orifices or became disconnected when the overlap was too short (see Fig. 11.6). Disconnections, which are specific complications of modular systems, occurred early in our series in limbs

Table 11.4. Primary Leakage After Endovascular Treatment of AAAs (N = 123)

Type of leakage	Number of patients
Proximal	2
Distal	10
Repaired	6
Spontaneous sealing	1

that were constructed of multiple telescoping grafts. These failures can be avoided by providing a long overlap at the time of insertion or by construction of larger and longer limbs. We did not observe any late secondary leaks in our last 73 patients.

All our late endoleaks (n = 10) were repaired by percutaneous insertion of appropriate extension grafts.

MEASURING SUCCESS IN THE ENDOVASCULAR TREATMENT OF ABDOMINAL AORTIC ANEURYSMS

When treating an AAA, can our goals of protection from rupture, thromboembolism, or compression of adjacent structures be achieved by endovascular techniques with the same safety and predictability as by conventional graft insertion? It has been assumed but not proved that successful graft insertion must reduce pressure within the aneurysm sac and, therefore, tension on the degenerated wall. But how much pressure reduction is necessary? What is the role of collaterals that are not immediately occluded by the endovascular procedure?

We routinely measure pressure in the aneurysm during and, if necessary, for a short period of time after insertion of the stent-graft. This aids in the determination of adequate sealing at the ends of the graft. However, we did not see a clear correlation between pressure reduction and the status of patent collaterals as determined by the injection of angiographic contrast into the pressure catheter located outside the graft within the aneurysm sac.

Another "safe" parameter of success is assumed to be shrinkage of the aneurysm during follow-up. In our experience, shrinkage could be clearly demonstrated only in small saccular aneurysms, which, in part, seemed to nearly disappear after 2 years. Larger aneurysms treated by bifurcation grafts exhibited much less or no shrinkage (Fig. 11.8), and this could not be detected until after at least 1 year after implantation. To date, in our series no aneurysm has increased even when persistent or secondary distal leakage was present for more than 1 year.

Clearly, more studies are required to determine the effectiveness of this attractive, less invasive modular method of AAA repair. Such research applies to all devices that are being and will be used to treat these lesions.

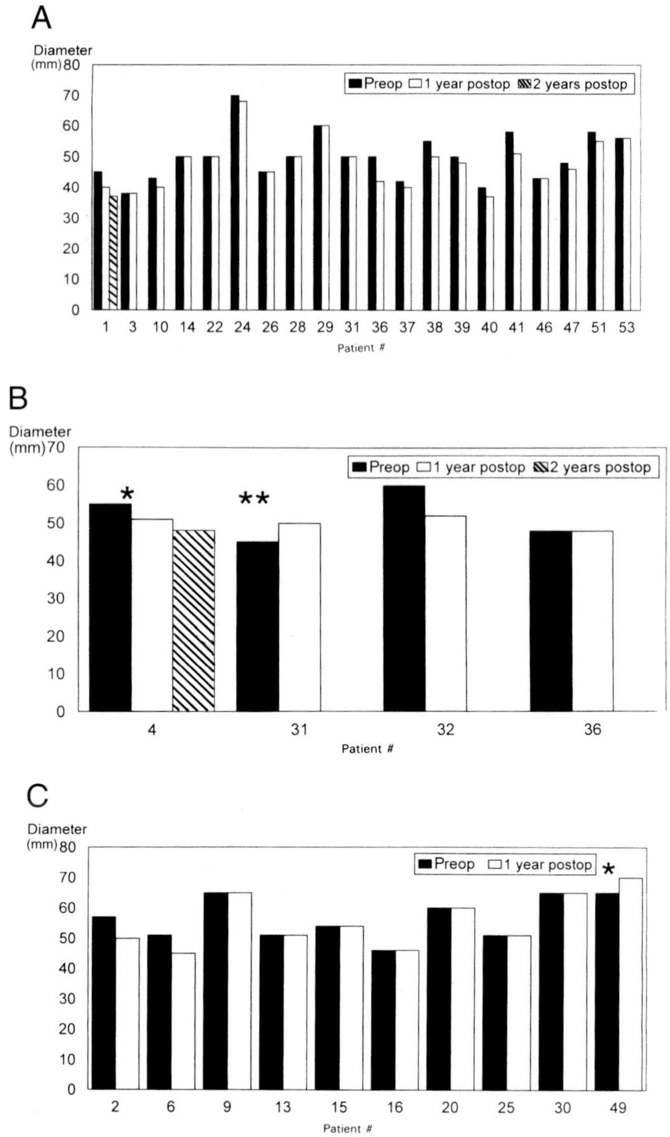

Figure 11.8. Changes in maximum diameter of AAAs after endovascular treatment with bifurcated grafts. **A.** Patients without leakage. **B.** Patients with primary leakage. (* = leakage sealed after 1½ years; ** = diameter has remained unchanged since percutaneous repair.) **C.** Patients with secondary leakage.

References

1. Parodi JC. Endovascular repair of abdominal aortic aneurysms and other arterial lesions. J Vasc Surg 1995;21:549–557.
2. Marin ML, Veith FJ, Cynamon J, et al. Initial experience with transluminally placed endovascular grafts for the treatment of complex vascular lesions. Ann Surg 1995;222:449–469.
3. Yusuf SW, Whitaker SC, Chuter TAM. Early results of endovascular aortic aneurysm surgery with aortouniiliac graft, contralateral iliac occlusion, and femorofemoral bypass. J Vasc Surg 1997;25:165–172.
4. May J, White G, Waugh R, et al. Treatment of complex abdominal aortic aneurysms by a combination of endoluminal and extraluminal aortofemoral grafts. J Vasc Surg 1994;19:924–933.
5. Chuter TAM, Risberg Bo, Hopkinson BR. Clinical experience with a bifurcated endovascular graft for abdominal aortic aneurysm repair. J Vasc Surg 1996;24:655–666.
6. Moore WS. Transfemoral endovascular repair of abdominal aortic aneurysm using the endovascular graft system device. In: Greenhalgh RM, ed. Vascular and endovascular surgical techniques, 3rd ed. London: WB Saunders, 1994:78–91.
7. Ohki T, Veith FJ, Sanchez LA, et al. Varying strategies and devices for endovascular repair of abdominal aortic aneurysms. Semin Vasc Surg 1997;10:242–256.
8. Blum U, Voshage G, Lammer J, et al. Endoluminal stent-grafts for infrarenal abdominal aortic aneurysms. N Engl J Med 1997;336:13–20.
9. Naslund TC, Edwards WH Jr, Neuzil DF, et al. Technical complications of endovascular abdominal aortic aneurysm repair. J Vasc Surg 1997;26:502–510.

Chapter 12

Endovascular Management of Descending Thoracic Aortic Aneurysms

James I. Fann, Michael D. Dake, R. Scott Mitchell, Charles P. Semba, and D. Craig Miller

Aneurysms involving the descending thoracic aorta affect approximately 40% of all patients with thoracic aortic aneurysms (1,2). The estimated incidence of descending thoracic aortic aneurysms is over 2 cases per 100,000 population per year (1,2). Untreated thoracic aortic aneurysms progressively enlarge. In up to 75% of cases, the patients succumb to aneurysm rupture (1–3). Consequently, the 1-year and 5-year actuarial survival estimates for nonsurgically treated patients are low at 60% and 20%, respectively (1,2).

Traditionally, patients with descending thoracic aortic aneurysms are managed by surgical replacement of the aorta with a Dacron graft. Since 1992, we have been investigating an alternative approach, namely, endovascular stent grafting. Compared with open surgical repair, the endovascular technique is less invasive, is potentially less expensive, and may be associated with lower morbidity (4–9).

In this chapter, we review the experience with endovascular stent grafting of descending thoracic aortic aneurysms at Stanford University School of Medicine.

DIAGNOSIS

Because most patients with thoracic aortic aneurysms are asymptomatic, this type of aneurysm is often suspected, based on an abnormal chest radiograph or during routine patient evaluation, (1,3,10,11). The major symptom, if any, is chest pain localized to the aneurysm site. Other manifestations are predominantly the result of compression or erosion of adjacent structures by the aneurysm. Thus, signs and symptoms include the following:

- Hoarseness from recurrent stretching of the left laryngeal nerve
- Tracheal deviation and respiratory symptoms from airway compression
- Hemoptysis from pulmonary or bronchial erosion
- Dysphagia from esophageal compression
- Superior vena cava syndrome from venous compression (3,5,11)

Patients with thoracic aortic aneurysms are typically hypertensive and often have major cardiac and pulmonary comorbidities (1,3,10–14). Associated vascular pathologies include abdominal aortic aneurysms in 20% of patients, previous cerebrovascular accidents in 6%, and peripheral arterial occlusive disease in 4%.

The patient's cardiac, pulmonary, cerebrovascular, and renal status are carefully evaluated before surgical or endovascular intervention is considered. Radiographic imaging of the thoracic aorta includes computed tomography (CT) or magnetic resonance imaging (MRI) and aortography. In planning the surgical strategy, aortography is useful in assessing critical aortic branches, including arch vessels and intercostal, renal, and visceral arteries. CT and MRI provide valuable information regarding the pathoanatomy of the aneurysm and the contiguous structures. Coronary arteriography is obtained in older patients and in those with symptoms suggestive of ischemic heart disease.

THERAPY

Conventional Approach

Therapeutic intervention for descending thoracic aortic aneurysms is considered in all symptomatic patients. In asymptomatic patients, indications for intervention include an aneurysm diameter twice that of a contiguous segment of normal-caliber aorta (or more than 6 cm) or documented aneurysm enlargement (3,12). Conventional surgical repair of the descending thoracic aortic aneurysm is associated with an overall (elective and emergent) operative mortality rate of 11% (13,14–22). The actuarial survival statistics for discharged surgical patients at 5 years and 10 years are 70% to 79% and 40% to 49%, respectively (10–13,21).

Late death as a result of rupture of another aortic aneurysm occurs in about 25% of patients, with most dying of cardiovascular or cerebrovascular events (11,12). In spite of improvements in operative and perioperative management, major surgical complications include paraplegia (estimated risk of 4%) and renal insufficiency (estimated risk of 5%) (11–25). The use of adjunctive, distal circulatory support, including cardiopulmonary bypass, has not eliminated the risk of these complications.

Stent-Graft Placement

Obtaining Measurements

The general preoperative evaluation strategy for endovascular stent grafting is identical with that for the conventional "open" approach because of the possibility that conversion to an open procedure may be necessary if stent grafting is unsuccessful. In regard to imaging, attention is directed toward locating critical side branches using aortography and three-dimensional spiral CT. Localization of patent intercostal arteries is important so that an attempt can be made to preserve critical intercostal vessels at the time of stent-graft placement.

It is possible to incorporate a segment of a noncovered stent body attached to the stent-graft to anchor the prosthesis distally or proximally at the level of a patent intercostal or another branch artery, thus theoretically maintaining flow into the artery. Based on CT or aortography, the relationships of the proximal aneurysm neck to the left subclavian artery and the distal neck to the celiac axis are defined. Measurements of the luminal diameters of the femoral and iliac arteries are obtained to ensure adequate access for the delivery sheath.

Certain selection criteria have been established for endovascular stent grafting, based mainly on anatomic considerations (4,5,7,9). It is important that the aneurysm neck is of relatively small caliber (less than 4 cm) and does not involve the origin of the left subclavian artery or the celiac axis. Ideally, the aneurysm neck should be at least 3 to 4 cm long. Patients with chronic aortic dissection are usually excluded because the true lumen is small and distorted and the false lumen is large and eccentric. Access via the peripheral arterial or abdominal aortic approach needs to be adequate for the 27F or 20F (outer diameter) delivery devices. With increased experience, involvement of the subclavian artery proximally is no longer an absolute contraindication to stent grafting, because a left carotid-subclavian transposition (or bypass) can be performed to permit stent-graft deployment within the proximal neck situated in the distal arch.

Construction and Deployment

After obtaining the aneurysm dimensions, a custom-made stent-graft is constructed. Stent-grafts can be tubular with a constant diameter or tapered, depending on the configuration of the aneurysm. The calculated length of the stent-graft allows for an additional 2 to 3 cm to serve as a friction anchor in the proximal and distal aneurysm necks.

Each stent-graft prosthesis is fabricated of stainless-steel Z-stents (Cook, Inc, Bloomington, IN) covered with a Cooley Verisoft woven Dacron graft (Meadox Medicals, Inc, Oakland, NJ) (Fig. 12.1) (4,5,9). The 20F or 27F (external diameter) delivery sheath is constructed of Teflon and has a plastic valve (Cook Urological, Inc, Bloomington, IN) attached to the end of it to provide hemostasis during angiographic catheter manipulation and stent-graft loading (Fig. 12.2). Following are the steps for deployment of stent-grafts:

1. A tapered dilator is placed within the delivery sheath and advanced over a guidewire.

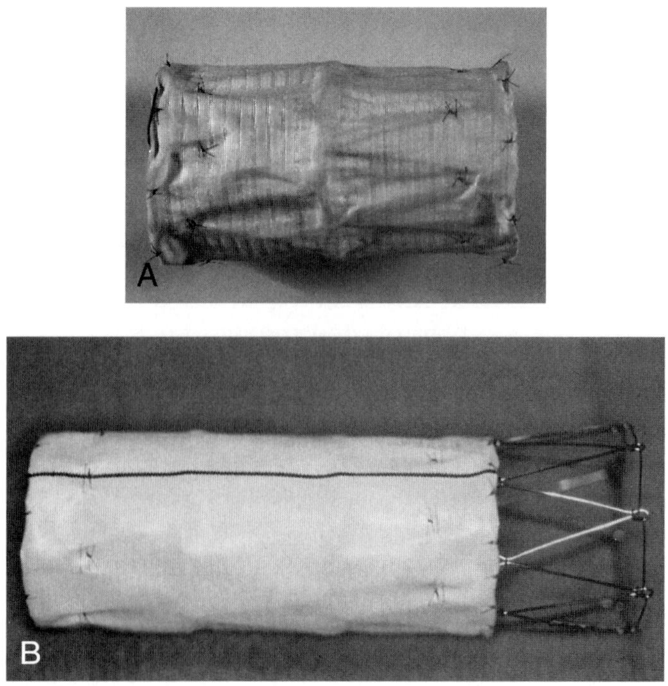

Figure 12.1. A. Endovascular stent-graft constructed of a stainless-steel endoskeleton consisting of Z-shaped stent elements covered with a Cooley Verisoft woven Dacron graft. **B.** Stent-graft with one uncovered stent body to permit anchoring at the neck of the aneurysm while maintaining flow into a patent side branch of the aorta.

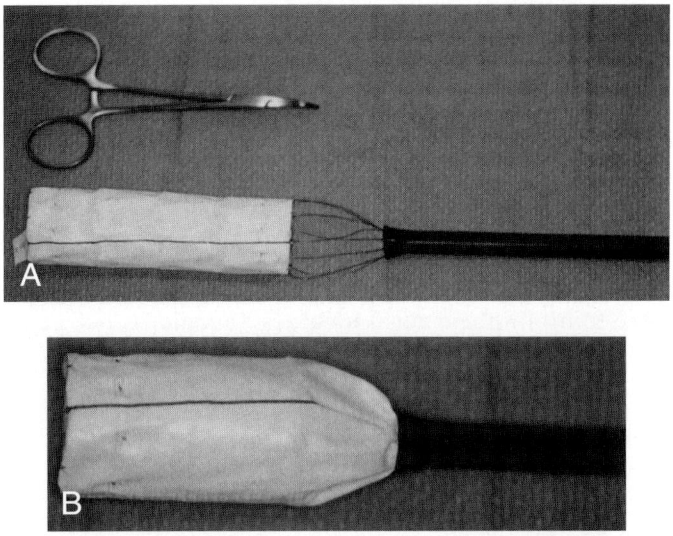

Figure 12.2. **A** and **B.** The stent-graft is compressed and inserted into the delivery sheath with the assistance of temporary nylon traction sutures.

2. After removal of the dilator, a solid Teflon pusher is used to position the compressed stent-graft into the delivery sheath and for stent-graft deployment. Additional stent-grafts of shorter lengths are available if further extensions are necessary because the original endovascular placement is less than optimal.
3. Endovascular stent grafting is performed in the operating room with the patient under general anesthesia. Patient positioning depends on which fluoroscopic projection provides optimal visualization of the descending thoracic aortic aneurysm and the proposed access site.
4. Cardiopulmonary bypass is on standby if an emergency open procedure is required.
5. Access is achieved via the femoral artery, iliac artery, infrarenal aorta, or a side limb graft (usually, a 10-mm Dacron graft) attached to an abdominal aortic graft.
6. Intraoperative aortography with digital road-mapping is obtained with a portable radiographic image intensifier and monitor. Transesophageal echocardiography is used for cardiac monitoring and to guide stent-graft deployment.
7. After intraoperative aortography, the angiographic catheter is exchanged for a super-stiff, 0.035-inch guidewire, which is advanced into the thoracic aorta proximal to the aneurysm. The patient is given heparin (100 to 300 U/kg), and a femoral or iliac arteriotomy, small aortotomy, or transverse incision is made in the side limb graft.
8. With fluoroscopic guidance, the tapered dilator along with the delivery sheath is passed over the guidewire and carefully advanced into the descending thoracic aorta (Fig. 12.3).
9. The dilator and guidewire are removed, leaving the tip of the delivery sheath approximately 1 to 2 cm proximal to the aneurysm neck.

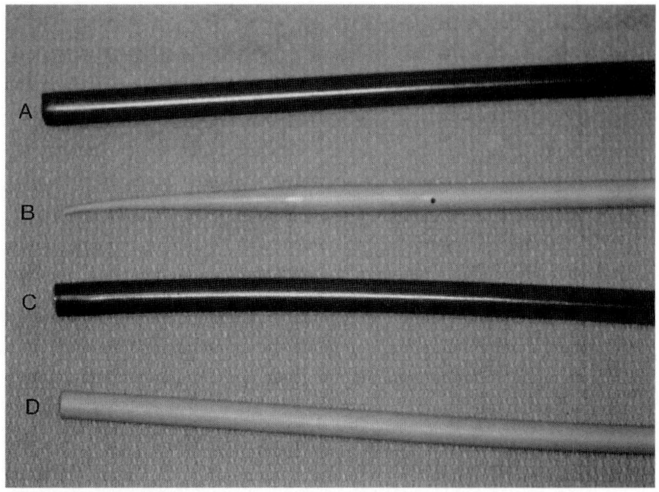

Figure 12.3. The four devices used for stent-graft deployment. After placing the dilator **(B)** within the Teflon delivery sheath **(A),** the dilator-sheath combination is placed over a guidewire and situated in the descending thoracic aorta using fluoroscopic guidance. The dilator is removed followed by placement of the stent-graft loaded within the cartridge **(C)** into the delivery sheath. The proximal end of the stent-graft is positioned at the tip of the delivery sheath using the Teflon pusher **(D)**. After placing the tip of the stent-graft/delivery sheath at the level of the proximal neck of the aneurysm, the stent-graft is deployed by retracting the delivery sheath while holding the pusher in place.

10. The stent-graft is inserted into the delivery sheath via a loading cartridge and advanced to the end of the sheath using the Teflon pusher. Mean arterial pressure is temporarily lowered to 50 to 60 mm Hg (with intravenous esmolol and inhalational isoflurane).
11. The stent-graft is deployed by holding the pusher and retracting the delivery sheath.

After stent-graft deployment, the pusher is removed, and a guidewire is reinserted. A pigtail catheter is passed over the guidewire, and a completion aortogram is obtained (Fig. 12.4). After removal of the catheter and delivery sheath, the arteriotomy or distal aortotomy is repaired. Protamine sulfate is given for heparin reversal. A spiral CT and aortogram are obtained before the patient is discharged (Fig. 12.5).

The protocol for follow-up radiologic studies is shown in Table 12.1.

Additional Modifications

Occasionally, there may be residual blood flow into the aneurysm sac after stent-graft deployment, which is evident on completion angiography. In this case, an additional stent-graft extension segment is deployed (either distal or proximal to the first stent-graft) to exclude flow into the aneurysm sac. In patients with multiple thoracic aortic aneurysms (involving the ascending aorta or arch as well as the descending thoracic aorta), it is possible to surgically replace the ascending aorta and arch using the

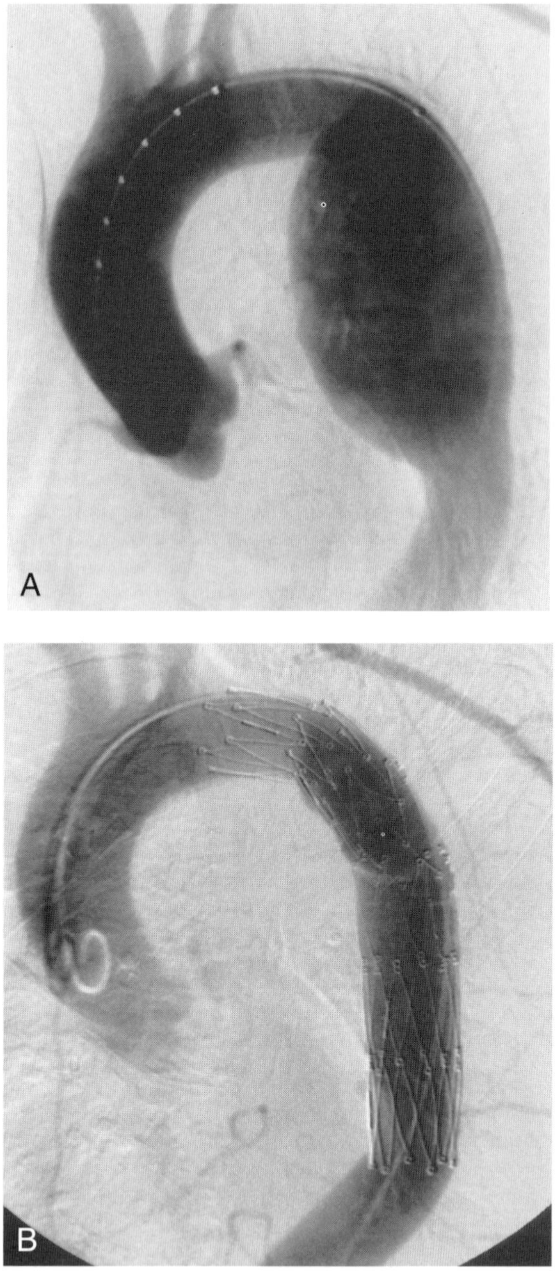

Figure 12.4. **A.** Oblique aortogram demonstrating a large proximal descending thoracic aortic aneurysm. **B.** Oblique aortogram after stent-graft deployment demonstrating exclusion of flow into the aneurysm sac.

"elephant trunk" technique initially described by Borst and colleagues (26); at a later date, stent grafting of the descending thoracic component of the aneurysm is carried out by means of deploying the proximal end of the stent-graft inside the distal opening of the "elephant trunk" graft (Fig. 12.6) (6,26).

Patients with abdominal aortic and thoracic aortic aneurysms (not true thoracoabdominal aortic aneurysms) have been successfully treated with a conventional, open abdominal aortic aneurysm repair combined with descending thoracic aortic stent grafting (8). Thoracic aortic stent-graft placement is facilitated by sewing a side limb (10-mm Dacron graft) onto the abdominal aortic graft; alternatively, if a bifurcation graft is necessary, one limb of this graft may be used. The side limb is used for delivery sheath access, and the stent-graft is deployed in the above manner. At the conclusion of the procedure, the auxillary side limb is oversewn.

CURRENT RESULTS WITH STENT GRAFTING

Our experience with endovascular stent grafting of the descending thoracic aorta includes 81 patients (61 men

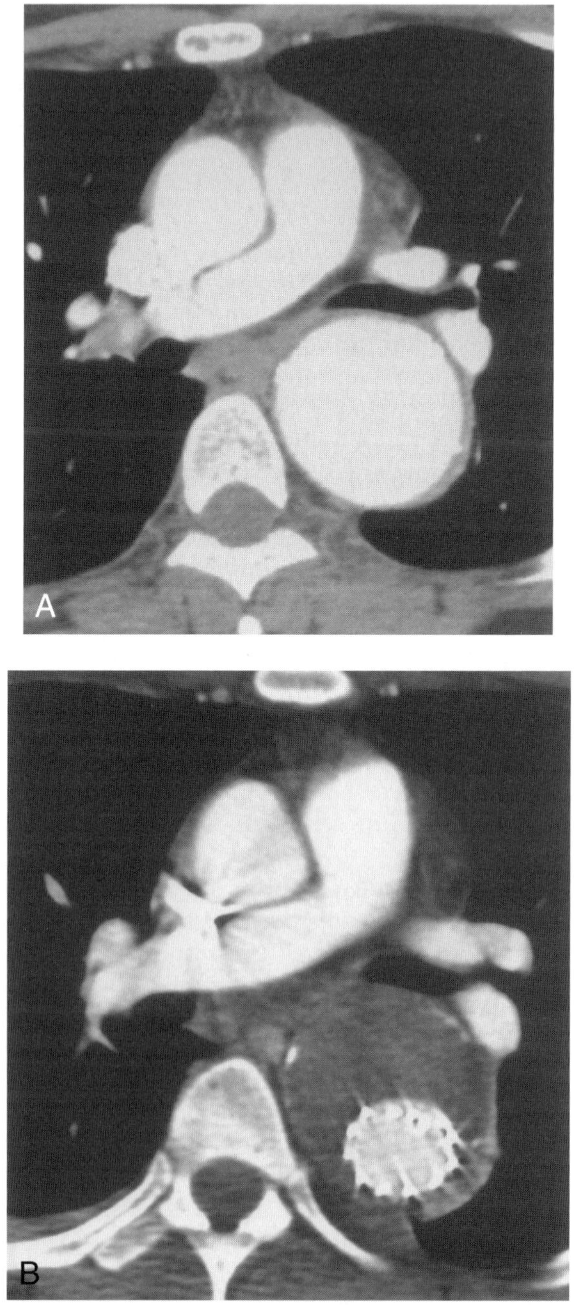

Figure 12.5. **A.** CT demonstrating a large proximal descending thoracic aortic aneurysm. **B.** CT obtained before discharge after endovascular stent grafting of descending thoracic aortic aneurysm. Flow into the aneurysm sac is excluded by the stent-graft.

Table 12.1 Imaging Protocol for Endovascular Stent Grafting

Timing	Imaging procedure		
Preprocedure	CXR	CT	Angiography
Intraoperative			Angiography
Predischarge	CXR	CT	Angiography
Postprocedure			
2 months	CXR		Angiography
6 months	CXR	CT	Angiography
1 year	CXR	CT	Angiography
2 years (and annually thereafter)		CT	Angiography

CXR = chest radiography; CT = spiral computed tomography.

and 20 women) (9). Seventy-six patients underwent stent-graft placement at Stanford University, and five patients at outside institutions by one of the authors (MDD). Approximately 50% of the patients were judged by cardiovascular surgeons not to be candidates for open operation owing to multiple risk factors (7).

The predominant aortic pathology was atherosclerotic or degenerative aneurysmal disease, which occurred in 67 patients. Other pathology was aortic dissection in five patients, post-traumatic aneurysm in seven patients, and healed mycotic aneurysm in two patients.

A total of 52 patients were approached using a femoral arteriotomy, 3 through an iliac arteriotomy, 1 through the aortic arch, and 25 via the abdominal aorta. Among those who required the abdominal aortic approach, seven patients were accessed via a small aortotomy because of inadequate peripheral access, and 18 patients underwent concomitant abdominal aortic aneurysm repair.

Patient follow-up was 100% complete but averaged only 13.2 months, because a large fraction of the patients underwent stent grafting within the last 2 years. There were seven early or procedure-related deaths, yielding an early mortality rate of 9 ± 3% (±70% confidence limit) (9). Paraplegia occurred in three (4 ± 2%) patients; two had complicated intraoperative and postprocedure courses. In one patient, stent grafting was uncomplicated, but the entire descending thoracic aorta had to be excluded. Cerebrovascular events occurred in four (5 ± 3%) patients. Angiographic and stent-graft instrumentation at the aortic arch level were associated with three embolic strokes, and one patient sustained an intracerebral hemorrhage during an otherwise uneventful stent-graft deployment.

To date, no patient has required conversion to an open, surgical approach. In four patients (5 ± 3%), incomplete thrombosis of the aneurysm occurred after initial stent-graft deployment. Three subsequently died, two of whom sustained unexpected late deaths (categorized as possible treatment failures). One developed fatal aneurysm rupture into the esophagus.

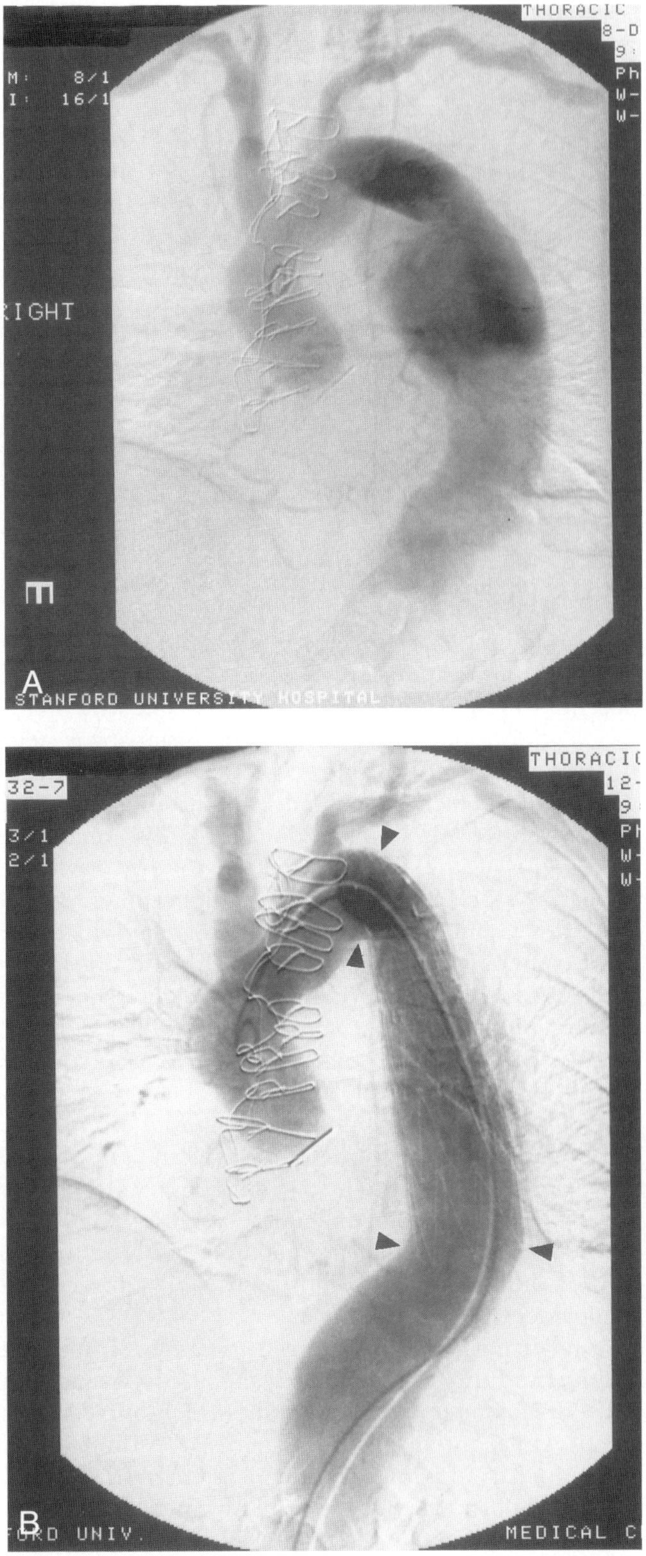

Figure 12.6. A. Oblique aortogram demonstrating the remaining descending thoracic aortic aneurysm after repair of ascending aortic and arch aneurysms using the "elephant trunk" technique. **B.** Oblique aortogram immediately after stent-graft deployment showing virtual exclusion of flow in the descending aortic aneurysm. Arrowheads show the proximal and distal extent of the stent-graft. (Reproduced with permission from Fann JI, Dake MD, Semba CP, et al. Endovascular stent-grafting after arch aneurysm repair using the "elephant trunk." Ann Thorac Surg 1995;60:1102–1105.)

For the 76 patients who underwent endovascular stent grafting at Stanford University, the actuarial survival rate estimates were 87 ± 4% at 1 year, 81 ± 6% at 2 years, and 81 ± 6% at nearly 4 years.

Given the high-risk nature of this specific patient cohort, these results satisfactorily compare with the reported 5-year actuarial survival rate estimate of 70% to 79% in patients who underwent open surgical repair (10–12,16,23). There were five late deaths. Two deaths were sudden, and one was due to aneurysm rupture; thus, these three deaths can be categorized as treatment failure or possible treatment failure. Five patients developed late recanalization or filling of the aneurysm sac over time; three of these were successfully treated with secondary interventional embolization, one underwent open surgical repair, and one is being followed up.

LIMITATIONS AND UNKNOWNS

The endovascular stent grafting technique for treating thoracic aortic aneurysms is associated with real and potential limitations (4,9). Anatomic concerns include the need for adequate proximal and distal aortic necks, the absence of critical aortic branches in the affected aorta, and the fact that curved segments of aorta cannot be as effectively excluded using this technique. Placement of the stent-graft close to the distal arch appears to be associated with a higher incidence of strokes, presumably owing to catheter manipulation in the ascending aorta or arch. Although the thoracic aorta is not cross-clamped during stent-graft deployment, the risks of spinal cord ischemia and paraplegia remain because critical intercostal arteries can be obliterated by the deployed stent-graft.

In addition, reimplantation of intercostal arteries is not possible with the stent-graft technique, as is the case during an open repair. The current delivery system is relatively large, requiring adequate arterial access. The deployment process is somewhat imprecise, and the position of the stent-graft cannot be reliably adjusted after deployment.

A major potential concern is the stability of stent-graft fixation and the ultimate fate of the aneurysm sac (4,9). After deployment, the stent-graft is friction-fixed within the proximal and distal aneurysm necks. Although it has not yet occurred in our experience, stent-graft migration is a potential problem, especially if an aneurysm neck progressively dilates over time. Also conceivable is continued aneurysm expansion over time despite being thrombosed outside the prosthetic stent-graft. In addition, late filling of the aneurysm sac or recanalization around the stent-graft has been observed.

Finally, the long-term outcome of patients who undergo endovascular stent grafting of descending thoracic aortic aneurysms remains unknown. Only 5- to 10-year follow-up along with serial imaging studies of a large number of patients will answer this important question.

SUMMARY

Endovascular stent grafting is an alternative therapeutic approach for high-risk patients with descending thoracic aortic aneurysms. Those likely to benefit most include the very elderly, those with compromised cardiac, pulmonary, or renal status, and persons who have undergone previous complex thoracic aortic procedures. For patients with combined abdominal and descending thoracic aortic aneurysms, the endovascular stent grafting approach provides a useful adjunct. Compared with open surgical repair, current results suggest that this less invasive approach may be associated with less early morbidity. If this method proves to be reliable and reproducible, the economic implications include substantially lower cost in the treatment of these patients. The true efficacy of endovascular stent grafting for patients with descending thoracic aortic aneurysms, however, awaits long-term follow-up.

References

1. Pressler V, McNamara JJ. Thoracic aortic aneurysm: Natural history and treatment. J Thorac Cardiovasc Surg 1980;79:489–498.
2. Bickerstaff LK, Pairolero PC, Hollier LH, et al. Thoracic aortic aneurysms: A population-based study. Surgery 1982;92:1103–1108.
3. McNamara JJ, Pressler VM. Natural history of arteriosclerotic thoracic aortic aneurysms. Ann Thorac Surg 1978;26:468–473.
4. Dake MD, Miller DC, Semba CP, et al. Transluminal placement of endovascular stent-grafts for the treatment of descending thoracic aortic aneurysms. N Engl J Med 1994;331:1729–1734.
5. Fann JI, Miller DC. Descending thoracic aortic aneurysms. In: Baue AE, Geha AS, Hammond GL, et al, eds. Glenn's thoracic and cardiovascular surgery. 6th ed. Stamford, CT: Appleton & Lange, 1995.
6. Fann JI, Dake MD, Semba CP, et al. Endovascular stent-grafting after arch aneurysm repair using the "elephant trunk." Ann Thorac Surg 1995;60:1102–1105.
7. Mitchell RS, Dake MD, Semba CP, et al. Endovascular stent-graft repair of thoracic aortic aneurysms. J Thorac Cardiovasc Surg 1996;111:1054–1062.
8. Moon MR, Mitchell RS, Dake MD, et al. Simultaneous abdominal aortic replacement and thoracic aortic stent-graft placement for multilevel aortic disease. (Abstract). Chicago, Society of Vascular Surgery, 1996.
9. Fann JI, Mitchell RS, Dake MD, Miller DC. The results of endovascular stent-grafting in patients with descending thoracic aortic aneurysm. In: Yao JST, Pearce WH, eds. Vascular surgery: Twenty years of progress. Norwalk, CT: Appleton & Lange, 1996.
10. Pressler V, McNamara JJ. Aneurysm of the thoracic aorta. J Thorac Cardiovasc Surg 1985;89:50–54.
11. DeBakey ME, McCollum CH, Graham JM. Surgical treatment of aneurysms of the descending thoracic aorta. J Cardiovasc Surg 1978;19:571–576.
12. Moreno-Cabral CE, Miller DC, Mitchell RS, et al. Degenerative and atherosclerotic aneurysms of the thoracic aorta. J Thorac Cardiovasc Surg 1984;88:1020–1032.
13. Hamerlijnck RP, Rutsaert RR, DeGeest R, et al. Surgical correction

of descending thoracic aortic aneurysms under simple aortic cross-clamping. J Vasc Surg 1989;9:568–573.
14. Livesay JJ, Cooley DA, Ventemiglia RA, et al. Surgical experience in descending thoracic aneurysmectomy with and without adjuncts to avoid ischemia. Ann Thorac Surg 1985;39:37–46.
15. Borst HG, Jurmann M, Buhner B, Laas J. Risk of replacement of descending aorta with a standardized left heart bypass technique. J Thorac Cardiovasc Surg 1994;107:126–133.
16. Verdant A. Descending thoracic aortic aneurysms: Surgical treatment with the Gott shunt. Can J Surg 1992;35:493–496.
17. Najafi H, Javid H, Hunter J, et al. Descending aortic aneurysmectomy without adjuncts to avoid ischemia. Ann Thorac Surg 1980;30:326–335.
18. von Segesser LK, Killer I, Jenni R, et al. Improved distal circulatory support for repair of descending thoracic aortic aneurysms. Ann Thorac Surg 1993;56:1373–1380.
19. Cooley DA, Baldwin RT. Technique of open distal anastomosis for repair of descending thoracic aortic aneurysms. Ann Thorac Surg 1992;54:932–936.
20. Crawford ES, Walker HSJ, Saleh SA, Normann NA. Graft replacement of aneurysm in descending thoracic aorta: Results without bypass or shunting. Surgery 1981;89:73–85.
21. Hilgenberg AD, Rainer WG, Sadler TR. Aneurysm of the descending thoracic aorta. J Thorac Cardiovasc Surg 1981;81:818–824.
22. Lawrence GH, Hessel EA, Sauvage LR, Krause AH. Results of the use of the TDMAC-heparin shunt in the surgery of aneurysms of the descending thoracic aorta. J Thorac Cardiovasc Surg 1977;73: 393–398.
23. Najafi H. 1993 Update: Descending aortic aneurysmectomy without adjuncts to avoid ischemia. Ann Thorac Surg 1993;55: 1042–1045.
24. Cartier R, Orszulak TA, Pairolero PC, Schaff HV. Circulatory support during crossclamping of the descending thoracic aorta. J Thorac Cardiovasc Surg 1990;99:1038–1047.
25. Schepens MA, Defauw JJ, Hamerlijnck RP, Vermeulen FE. Risk assessment of acute renal failure after thoracoabdominal aortic aneurysm surgery. Ann Surg 1994;219:400–407.
26. Borst HG, Frank G, Schaps D. Treatment of extensive aortic aneurysms by a new multiple-stage approach. J Thorac Cardiovasc Surg 1988;95:11–13.

Chapter 13

Endovascular Grafts for Vascular Injuries

Takao Ohki, Frank J. Veith, and Luis A. Sanchez

The surgical repair of vascular injuries may be complicated by the inaccessibility of the vascular lesion when trauma occurs to a vessel within a central body cavity, by distorted anatomy due to a large hematoma and a false aneurysm, and by venous hypertension when an arteriovenous fistula is present (1). These conditions make the use of endovascular techniques appealing, because repair can be performed from a remote access site, obviating the need for direct surgical exposure of the hostile injury site and reducing the morbidity and mortality rates of the repair.

Endovascular techniques for the treatment of vascular trauma include the use of coil embolization, intravascular stents, and, more recently, the use of stented grafts.

This chapter describes our experience at the Montefiore Medical Center in New York and briefly reviews the literature on the use of endovascular grafts for the treatment of traumatic vascular lesions.

LIMITATIONS OF EMBOLIZATION COILS AND INTRAVASCULAR STENTS

Embolization coils are wires with integrated fabric materials that are used to induce thrombosis. They can be deployed through a catheter that is placed at the target site (Fig. 13.1). Because coil embolization results in the sacrifice of the treated vessel, it has been used predominantly to treat relatively small traumatic arteriovenous fistulas and pseudoaneurysms involving nonessential vessels, such as the lumbar artery, the internal mammary artery, or branches from the hypogastric or deep femoral arteries (2–4). However, in some instances coil embolization has been used to treat lesions involving major vessels including the femoral (5,6), internal carotid (7,8), and vertebral arteries (8).

The most common complication involving the use of coil embolization is the inadvertent embolization of a distal or pulmonary artery. This may occur during the treatment of an injured artery or arteriovenous fistula that lacks a suitable segment of vessel within which the coils can be stabilized (5). Long-term follow-up of successfully coil-treated lesions has proved to be favorable (3).

The placement of intravascular stents is efficacious for the repair of intimal flaps (9–12) (Fig. 13.2). However, because of their porous nature, stents are not indicated for treating arteriovenous fistulas or pseudoaneurysms because they require complete exclusion from the circulation (13); the exception is a single case report in which a Wallstent and coils were used in combination to embolize a traumatic iliac arteriovenous fistula (14).

Although both embolization coils and intravascular stents have proved effective in selected cases, most patients with vascular trauma are not amenable to such therapy.

ENDOVASCULAR GRAFTS

In 1969, Dotter initially conceptualized the use of endovascular grafts for the treatment of traumatic arterial lesions (15). The first clinical application of this technology was performed by Volodos and colleagues in Russia in 1986, who used a Dacron graft and a self-expanding stent to treat a thoracic aortic pseudoaneurysm (16).

Endovascular grafts are new techniques, which com-

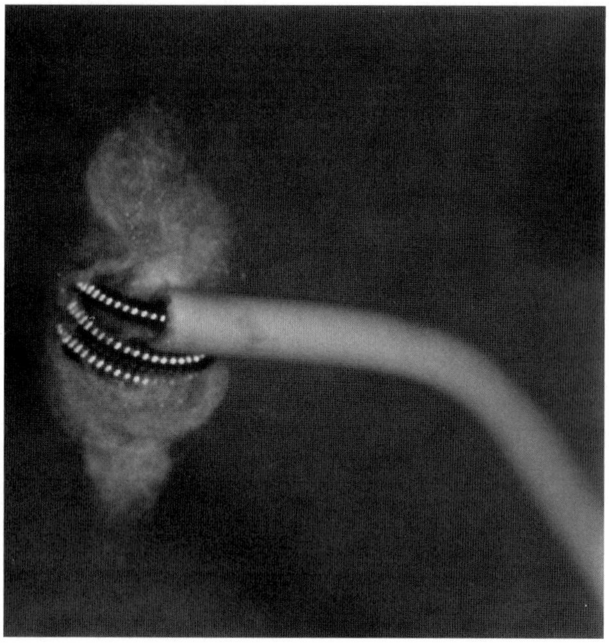

Figure 13.1. Embolization coil (Cook Inc, Bloomington, IN). The coil that is delivered through a catheter has synthetic fibers attached to induce thrombosis and hemostasis of the bleeding site. (Reprinted with permission from Ohki T, Veith FJ, Kraas C, et al. Endovascular therapy for upper extremity injury. Semin Vasc Surg, 1998;11:106–115.)

bine the structural and therapeutic features of both metallic stents and prosthetic vascular grafts. The grafts are inserted within the vessel wall from a remote access site, for example, the femoral or brachial arteries. The device is anchored to noninjured portions of the artery, thereby completely excluding the injured portion of the artery from the circulation. Fixation of the graft is accomplished by means of a metallic stent. The procedure is performed under fluoroscopic guidance.

Endovascular grafts have significantly extended the potential for the endovascular therapy of vascular trauma. These grafts have been used to treat almost every type of injury in a variety of locations. Most of these grafts were used in circumstances in which coil embolization or stent placement was deemed inappropriate (17–42). In some instances, endovascular grafts have also been used to treat life-threatening, acute arterial hemorrhage (34,40).

Most clinical experience reported in the literature used a device consisting of a Palmaz stent and an expanded polytetrafluoroethylene (ePTFE) graft. Because the traumatized field is often contaminated, some investigators used a vein graft in combination with a Palmaz stent (20,30). The various types of stented grafts used are summarized in Table 13.1 and shown in Figure 13.3.

The characteristics of arterial injuries, sites of arterial access, technical success rates, and the outcome of endovascular stented grafts in the treatment of vascular trauma are summarized in Table 13.2. These results are encouraging, demonstrating a high technical success rate with a midterm patency of 83% to 100% and a complication rate of 0% to 17%. The results are especially impressive when one considers the difficulties that almost certainly are encountered in the direct surgical repair of these injuries. The fact that most trauma patients have nonatherosclerotic vessels facilitates the proximal and distal graft fixations and contributes to the high success rate.

This is not the case with abdominal aortic aneurysms

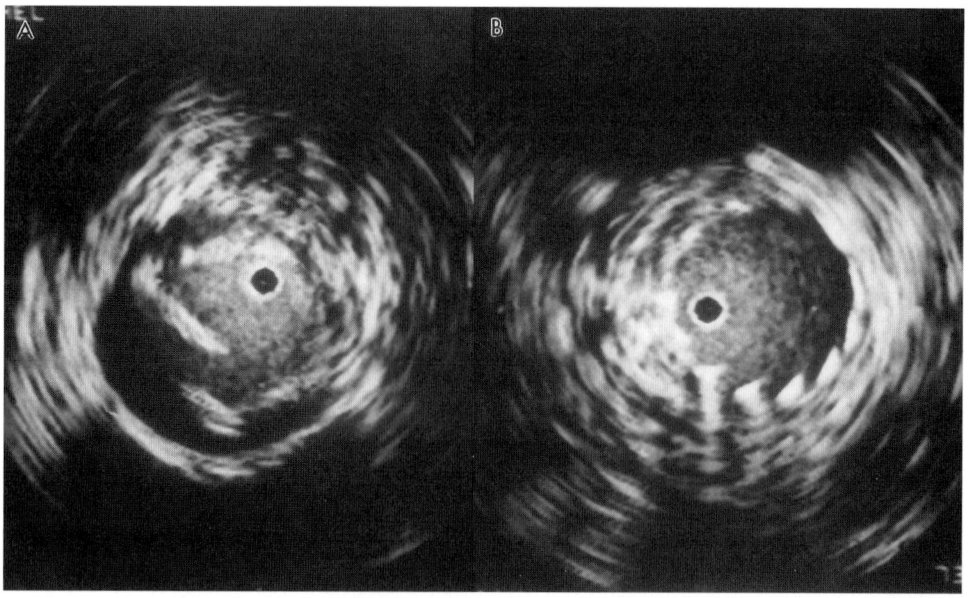

Figure 13.2. **A.** Intravascular ultrasound image showing iatrogenic dissection of an external iliac artery secondary to passage of a large introducer sheath containing a stent-graft. **B.** The dissection was repaired by the placement of a Wallstent. (Reprinted with permission from Ohki T, Veith FJ, Sanchez LA. Five year experience with endovascular grafts for the treatment of aneurysmal, occlusive and traumatic arterial lesions. J Cardiovasc Surg, 1998, in press.)

Table 13.1. Various Devices Used for Repair of Arterial Trauma

Type	Combination of Palmaz Stent and various grafts				Cragg Endopro	Corvita Graft
Stent	Palmaz Stent				Nitinol	Self-Expanding Braided Stent
Graft material	PTFE	Dacron	Vein	Silicone	Ultrathin, woven polyester fabric	Polycarbonate urethane
Arterial access	1[a] or 2[b]	2[b]	2[b]	1[a] or 2[b]	2[b]	2[b]
Reference	17–28	18,29	20,30–33	20,34	35–38	18,39

PTFE = polytetrafluoroethylene.
[a]Open arteriotomy; [b]percutaneous.

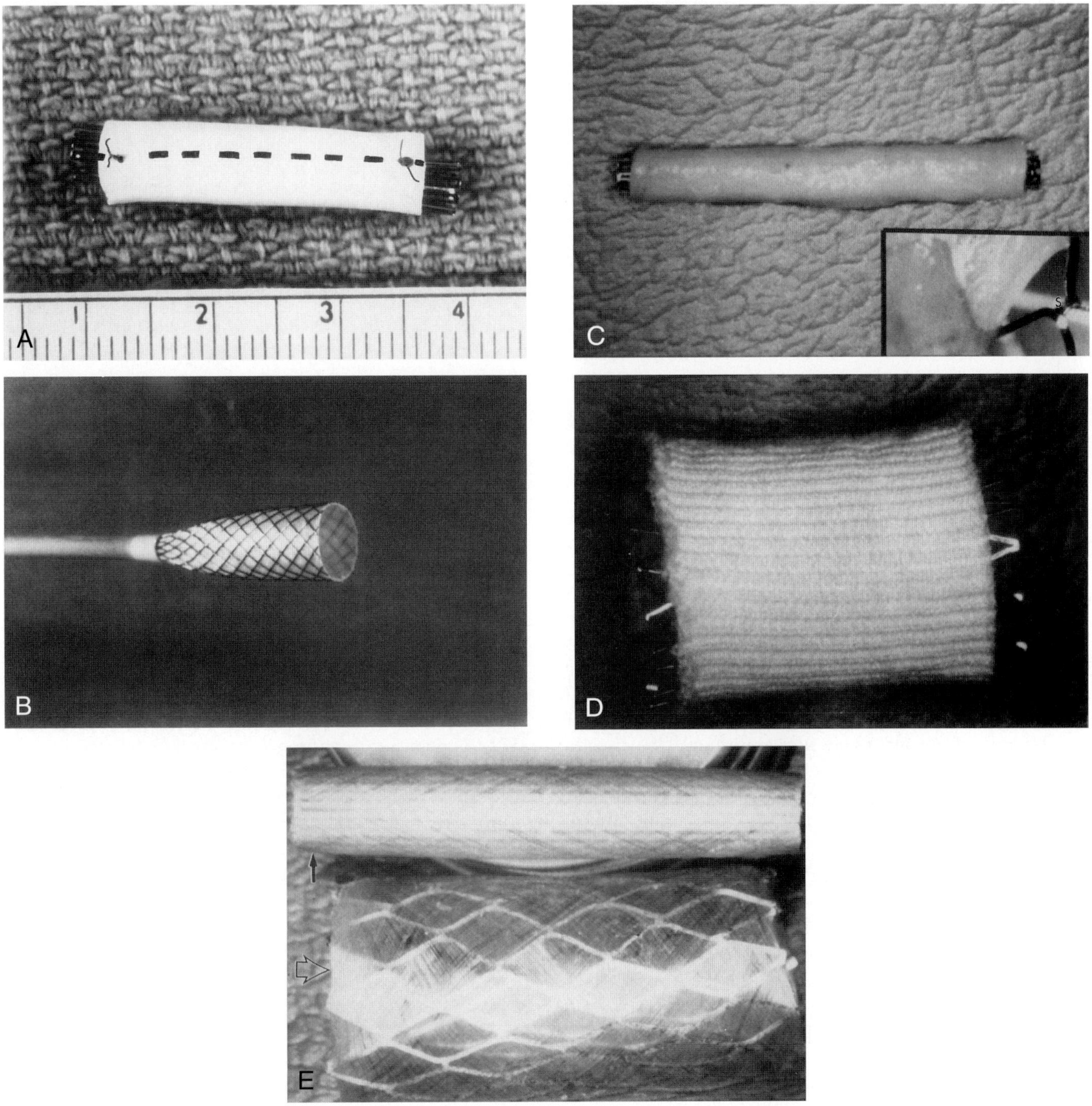

Figure 13.3. Endovascular stented grafts. **A.** Picture of a Palmaz stent sewn to an ePTFE graft. **B.** Corvita endovascular graft. This polycarbonate elastima-covered stent structurally resembles a Wallstent. A Corvita graft is manually crimped into a 9F or 10F delivery sheath. **C.** Autologous vein can be used to cover the stent, creating a biologic stented graft. The collapsed stent-graft assumes a small profile that effectively covers the struts of the stent following deployment (*inset*). **D.** Dacron graft material may be used to cover a balloon-expandable stent. **E.** Polyurethane can be fabricated directly onto a stent. This material has elastic properties, permitting a closed stent (*arrow*) to remain well covered following deployment (*open arrow*). (Reprinted with permission from Ohki T, Veith FJ, Marin ML, et al. Endovascular approaches to vascular injuries. Semin Vasc Surg 1997;10:272–285.)

Table 13.2. Lesion Characteristics and Outcome by Location of Injury

Location of trauma	Axillary-subclavian artery	Aorta or iliac artery	Femoral artery	Other locations[a]
Number of cases	18	20	7	6
Cause of injury	Bullet, 55%; catheterization 22%; other, 22%	Surgical, 44%; catheterization, 13%; bullet, 6%; other, 37%	Bullet, 57%: catheterization, 43%	Percutaneous transluminal angioplasty, 33%; other, 67%
Presence of pseudoaneurysm	61%	60%	57%	83%
Presence of arteriovenous fistula	44%	70%	57%	33%
Arterial access	Brachial arteriotomy, 39%; brachial percutaneous, 39%; femoral percutaneous, 22%	Femoral arteriotomy, 56%; femoral percutaneous, 44%	Femoral arteriotomy, 67%; femoral percutaneous, 33%	Femoral percutaneous, 83%; other, 17%
Technical success rate	94% (17/18)[b]	100%	100%	83%[e]
Complications: Minor	0%	5%	0%	0%
Complications: Major	6%[c]	5%	0%	0%
Mean length of stay	3.3 days	4 days	4.4 days	7.5 days
Mean follow-up	13 months	11.3 months	15.5 months	15.6 months
Primary patency	83%[d]	100%	100%	100%
References	17,18,20,22,23,25,34 39,40	17,18,20,21,25,29,32, 35,38,41	17,18,20,26,30,40	27,29,31,33,42

Reprinted with permission from Ohki T, Veith FJ, Bertucci WR, et al. New technologies for vascular injury. Curr Opin Crit Care 1997;3:465–469.
[a]Includes coronary artery, internal mammary artery, aortopulmonary, and aortobronchial lesions.
[b]One failure due to misdiagnosis.
[c]Brachial artery injury during device insertion.
[d]Two failures due to stent deformity.
[e]Failure to introduce the device.

(AAAs) and, therefore, the endovascular repair of AAAs has suffered from a higher endoleak rate and migration of the fixation stents. In addition, the minimal access nature of the procedure allows for a shorter hospital length of stay (3.3 to 7.5 days) and probably results in increased cost-effectiveness (see Table 13.2).

The use of intravascular ultrasound (IVUS) for the endovascular repair of traumatic lesions has been advocated by some investigators (43,44). IVUS provides certain details that are otherwise difficult to obtain, including the exact location and size of the injury or fistula (Fig. 13.4). These details are especially important when a coexisting arteriovenous fistula makes the angiographic interpretation difficult, and they may be essential when performing an endovascular repair (Fig. 13.5).

THE MONTEFIORE EXPERIENCE WITH ENDOVASCULAR GRAFTS FOR ARTERIAL TRAUMA

Technique and Devices

We have been using either a surgeon-made device or the Corvita endoluminal graft (Corvita Corporation, Miami, FL). The surgeon-made device was constructed by suturing a Palmaz stent (Cordis [Johnson & Johnson Interventional Systems, Warren, NJ]) to a thin-walled ePTFE graft (see Fig. 13.3*A*) (43). The stents varied between 2 and 3 cm in length (Palmaz P-204, 294, 308) and were fixed inside 5 to 6 mm Gore-tex grafts (W.L. Gore and Associates, Flagstaff, AZ) by 4 "U" stitches (see Fig. 13.3*A*). The stented graft was then mounted on a balloon angioplasty catheter with a tapered dilator tip firmly attached to

Figure 13.5. A. Preinterventional angiogram of an iatrogenic arteriovenous fistula (*AVF*) involving the right common iliac artery due to lumbar disk surgery. The left common iliac vein (*C*) and the inferior vena cava (*I*) are visualized by the contrast flowing through the AVF (arrow). **B.** Completion angiogram. A Corvita graft (10 mm × 6 cm) was used to repair the AVF. Because the location of the fistula was only 0.5 cm proximal to the origin of the internal iliac artery, this artery was embolized with multiple embolization coils (*E*) at the time of angiogram. Note the preservation of iliac flow and the obliteration of the AVF. (Reprinted with permission from Ohki T, Veith FJ, Marin, ML, et al. Diagnostic and therapeutic strategies for vascular injuries. In: Baert AL, ed. Medical radiology—diagnostic imaging and radiation oncology. Heidelberg: Springer-Verlag, 1997.)

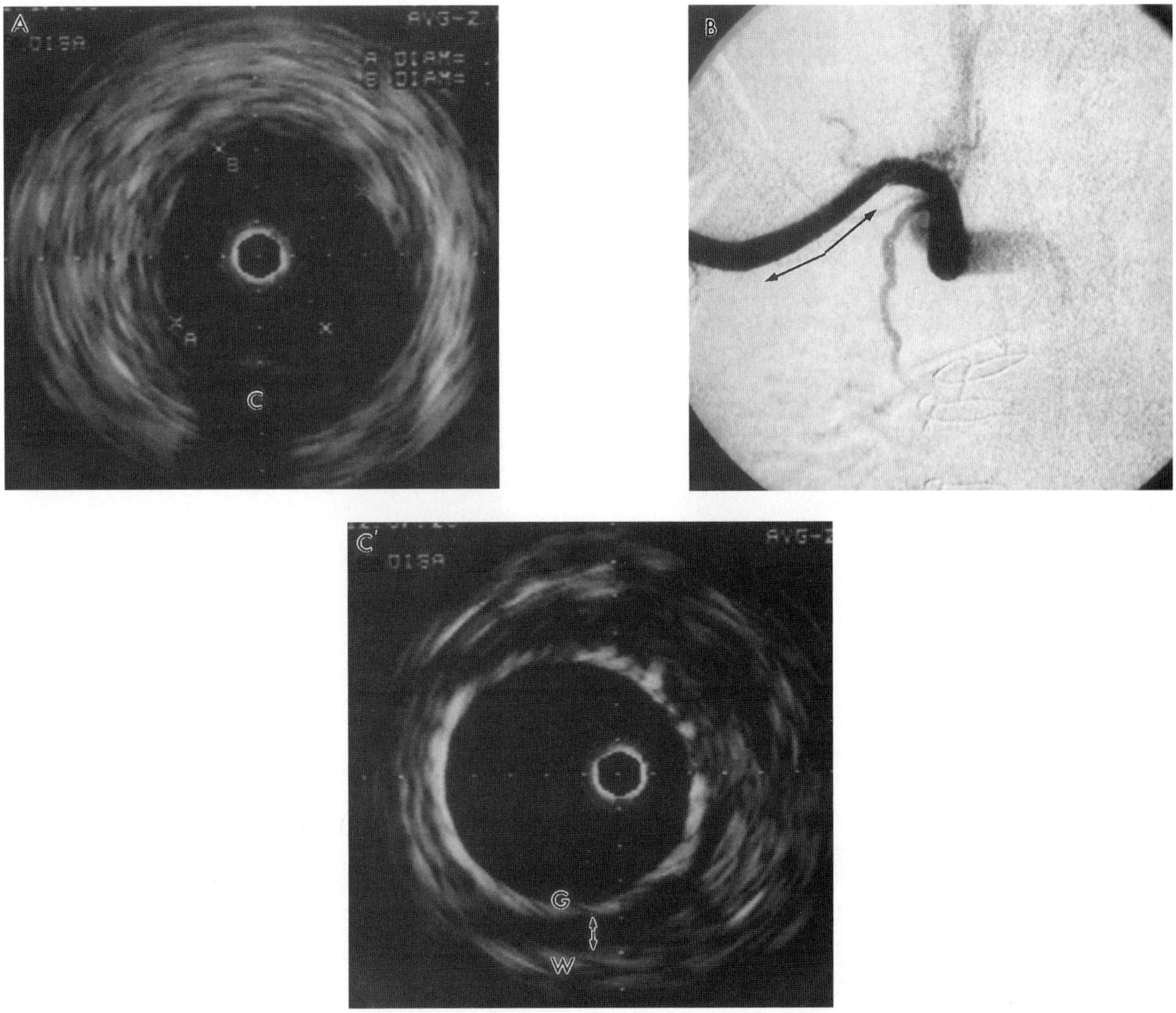

Figure 13.4. **A.** Preinterventional Intravascular ultrasound (IVUS) image of a right subclavian artery with an iatrogenic arterial injury due to central venous catheter placement. The exact insertion site of the catheter is seen (*C*) (by identifying the location of the probe of the IVUS under fluoroscopy), which was difficult to obtain using fluoroscopy because of the overlapping of the catheter and the artery. In addition, IVUS was able to precisely measure the diameter of the vessel, which dictates the size of the stent-graft to be used. **B.** Fluoroscopic image after Corvita graft deployment. There are no signs of hemorrhage through the injury site, and the graft *(arrows)* appears to be well deployed. **C.** IVUS imaging showing post-Corvita graft deployment. Although the graft seemed to be well deployed on the fluoroscopic image, clearly it is underdeployed on IVUS. This problem was managed by further dilatation with an angioplasty balloon. Arrows denote the space between the graft (*G*) and the arterial wall (*W*). (Reprinted with permission from Ohki T, Veith FJ, Marin ML, et al. Endovascular approaches to vascular injuries. Semin Vasc Surg 1997;10:272–285.)

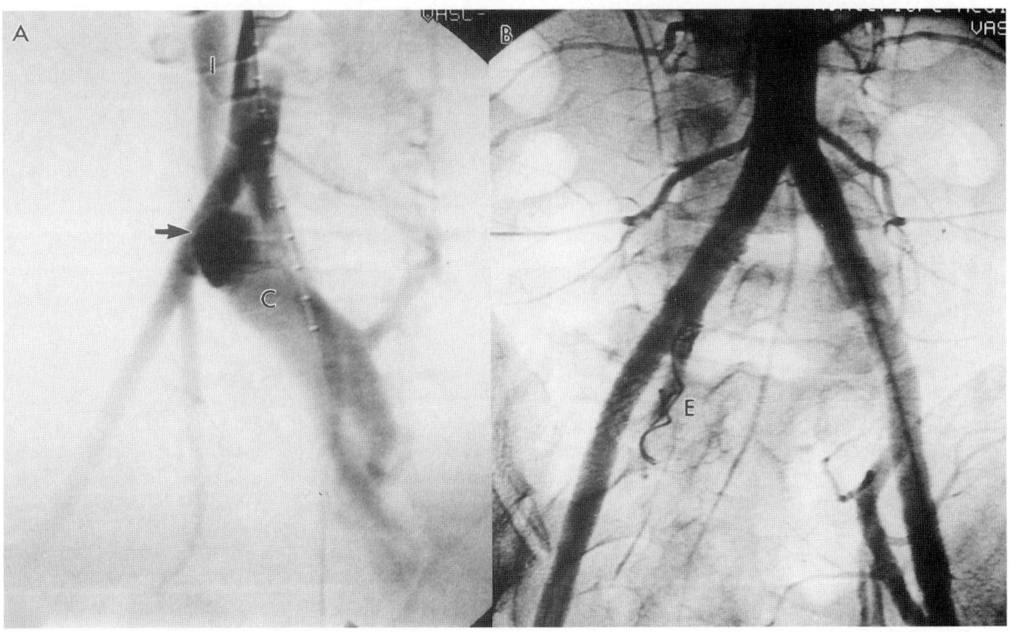

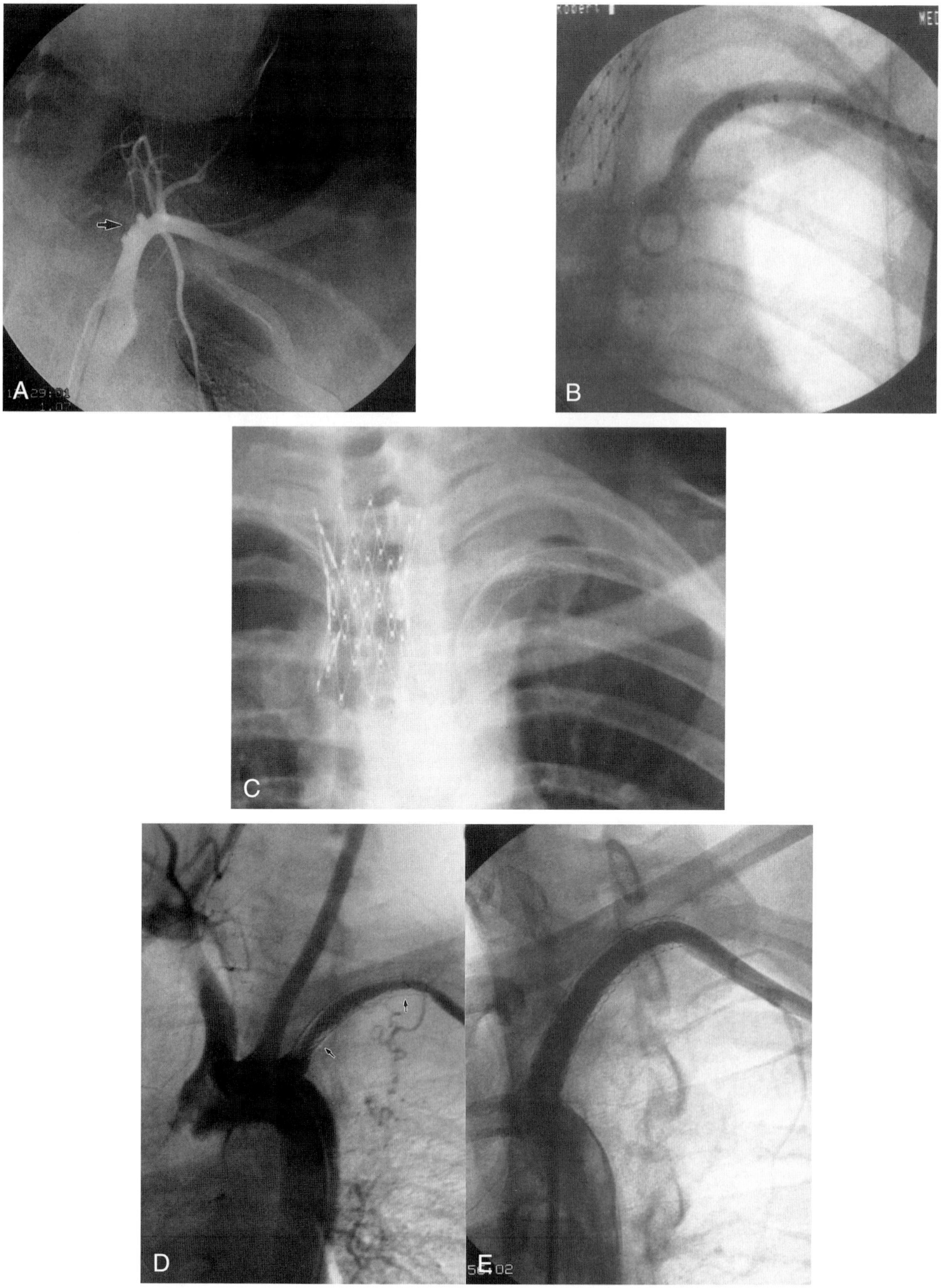

Figure 13.6. A 19-year-old man presented status after a gunshot wound to the chest, which traversed the mediastinum from right to left injuring his esophagus, trachea, and left subclavian artery. **A.** The initial arteriogram showed occlusion of the left vertebral artery and a small pseudoaneurysm at that site (*arrow*). He was transferred to our institution for endovascular treatment after placement of a covered esophageal stent to repair his tracheoesophageal fistula. **B.** A Corvita endovascular graft was placed across the lesion with excellent flow through it. **C.** A plain x-ray film demonstrated the esophageal stent and the Corvita graft in the left subclavian artery. **D.** At 4 months after graft insertion, his left radial pulse was diminished, although he remained asymptomatic. An angiogram revealed intimal hyperplasia throughout the graft, more prominent at both ends of the graft (*arrows*). **E.** The patient had angioplasty for the lesions, and a Palmaz stent was placed. After the procedure, the patient had a strong radial pulse. (Reprinted with permission from Ohki T, Veith FJ, Kraas C, et al. Endovascular therapy for upper extremity injury. Semin Vasc Surg 1998;11:106–115.)

its end. The entire device was contained within a 9F to 12F delivery system for over-the-wire insertion either percutaneously or through an open arteriotomy.

The Corvita endovascular graft is fabricated from a self-expanding stent of braided wire, which is covered with polycarbonate elastima fibers (see Fig. 13.3*B*). The stent-graft can be cut to the desired length in the operating room using a wire cutting scissors and then loaded into a specially designed delivery sheath. This sheath has a central pusher catheter, which is used for maintaining the graft in position while the outer sheath is being retrieved.

RESULTS

All procedures but one were performed in the operating room under fluoroscopic (Phillips, BV 212, The Netherlands) and intravascular ultrasonographic control (Hewlett Packard Company, Paramus, NJ). One Corvita graft was inserted in the angiography suite. To date, a total of 17 stented grafts or covered stents were used to treat 17 patients with traumatic arterial lesions (see Table 13.2). Seven injuries occurred as a result of gunshot wounds (Fig. 13.6), one as a result of a knife wound. Four were iatrogenic catheterization injuries (Fig. 13.7), two were iatrogenic arterial trauma (gynecological surgery or lumbar disk surgery) (see Fig. 13.5), and three were a result of arterial graft disruptions possibly associated with infection.

All but one injury were associated with an adjacent pseudoaneurysm. In five instances, the arterial injury formed a fistula to an injured adjacent vein. Associated injuries were present in eight patients with arterial trauma (see Table 13.2). Most of the cases were performed under either local or epidural anesthesia. One patient who had an axillary pseudoaneurysm repaired with a stented graft required a vein patch to close a small brachial artery insertion site. Procedural complications were limited to one distal embolization, which was treated with suction embolization, and one wound hematoma, which resolved without further intervention. Graft patency was 100% with no early or late graft occlusions (mean follow-up of 30 months [range 6 to 46 months]).

One patient with a left axillary-subclavian stent-graft developed compression of the stent at 12 months and was treated with balloon angioplasty. This problem recurred 3 months later, and no intervention was required. The device has not thrombosed with follow-up over 3 years. Another patient who received a Corvita graft required balloon angioplasty and stent placement for intimal hyperplasia (see Fig. 13.6*D* and *E*).

CONCLUSIONS

Coil embolization and stent placement are an effective treatment in a limited number of cases. Endovascular grafts have greatly extended the potential of endovascular therapy for vascular trauma. These grafts have been used to treat all types of vascular injuries at various locations in the body.

Regardless of the type of device used, stented grafts appear to be associated with a low morbidity rate, a high success rate, and reduced anesthesia requirements; they can limit the need for an extensive dissection in an already traumatized field. These advantages are especially important in patients with central arteriovenous fistulas or false aneurysms, especially those who are critically ill from concomitant injuries or other medical comorbidities. In such instances, the use of stented grafts

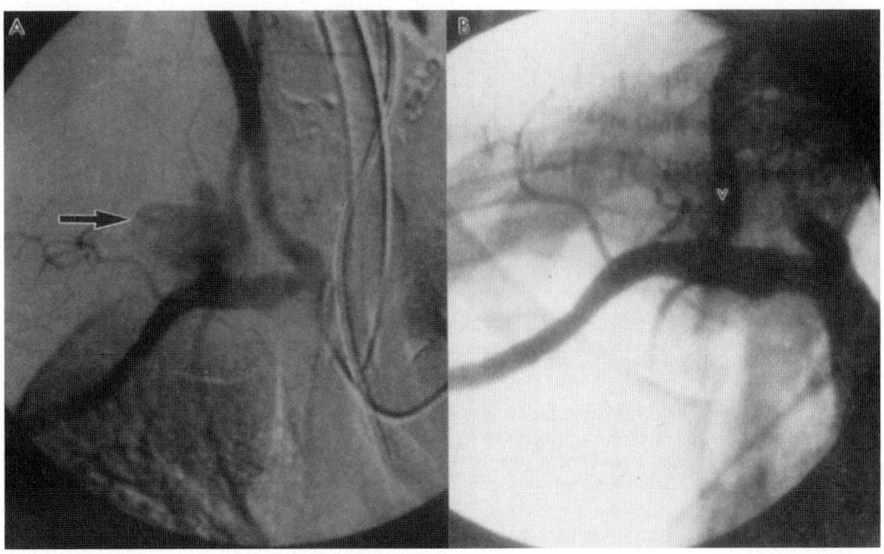

Figure 13.7. **A.** An arteriogram showing a large pseudoaneurysm of a subclavian artery (*arrow*) just distal to the right vertebral artery, which occurred after an attempted subclavian vein catheter insertion. **B.** After stented graft (ePTFE and Palmaz stent) placement through the right brachial artery, the pseudoaneurysm was excluded from the circulation. Vertebral artery flow was maintained (v). (Reprinted with permission from Marin ML, Veith FJ, Panetta TF, et al. Transluminally placed endovascular stented graft repair for arterial trauma. J Vasc Surg 1994;20:466–473.)

appears to be justified. In addition to vascular trauma, endovascular grafts have been used to treat a variety of other vascular pathologies, including aortic aneurysms, peripheral aneurysms, and occlusive disease (45,46). Of these, traumatic arterial lesions appear to be one of the indications for endovascular grafting.

The ideal composition of the external stent covering and the best mechanism for stent deployment have not as yet been determined. Nevertheless, endovascular grafts are important tools for the treatment of vascular trauma and should be included in the armamentarium of the vascular surgeon.

References

1. Perry MO, Thal ER, Shires GT. Management of arterial injuries. Ann Surg 1971;173:403–408.
2. Rosch J, Dotter CT, Brown MJ. Selective arterial embolization: A new method for control of acute gastrointestinal bleeding. Radiology 1972;102:303–306.
3. Panetta T, Sclafani SJA, Goldstein AS, Phillips TF. Percutaneous transcatheter embolization for arterial trauma. J Vasc Surg 1985;2:54–64.
4. Cekirge S, Oguzkurt L, Saatci I, et al. Embolization of a high-output postnephrectomy aortocaval fistula with Gianturco coils and cyanoacrylate. Cardiovasc Intervent Radiol 1996;19:56–58.
5. Lemaire JM, Dondelinger RF. Percutaneous coil embolization of iatrogenic femoral arteriovenous fistula or pseudo-aneurysm. Eur J Radiol 1994;18:96–100.
6. Jain SP, Roubin GS, Iyer SS, et al. Closure of an iatrogenic femoral artery pseudoaneurysm by transcutaneous coil embolization. Cathet Cardiovasc Diagn 1996;39:317–319.
7. Horowitz MB, Miller G, Meyer Y, et al. Use of intravascular stents in the treatment of internal carotid and extracranial vertebral artery pseudoaneurysms. Am J Neuroradiol 1996;17:693–696.
8. Sclafani SJA, Scalea TM, Hoffer EK, et al. Interventional radiology in the treatment of internal carotid artery gunshot wounds. J Vasc Intervent Radiol 1995;6:857–861.
9. Michaels AJ, Gerndt SJ, Taheri PA, et al. Blunt force injury of the abdominal aorta. J Trauma 1996;41:105–109.
10. Whigham CJ, Bodenhamer JR, Miller JK. Use of the Palmaz stent in primary treatment of renal artery intimal injury secondary to blunt trauma. J Vasc Intervent Radiol 1995;6:175–178.
11. Marty-Ané CH, Alric P, Prudhomme M, et al. Intravascular stenting of traumatic abdominal aortic dissection. J Vasc Surg 1996;23:156–161.
12. Ohki T, Veith FJ, Sanchez LA. Five year experience with endovascular grafts for the treatment of aneurysmal, occlusive and traumatic arterial lesions. J Cardiovasc Surg 1998, in press.
13. Geremia G, Bakon M, Brennecke L, Haklin M. Experimental arteriovenous fistulas: Treatment with porous metallic stents. Am J Neuroradiol 1995;16:1965–1973.
14. Rogerts JA, Hays DA, McFarland DR. Coil embolization of an internal iliac artery arteriovenous fistula with a Wallstent. J Vasc Intervent Radiol 1996;7:611–613.
15. Dotter CT. Transluminally-placed coilspring endarterial tube grafts: Long-term patency in canine popliteal artery. Invest Radiol 1969;4:329–332.
16. Volodos NL, Karpovich IP, Troyan VI, et al. Clinical experience of the use of self-fixing synthetic prostheses for remote endoprosthetics of the thoracic and the abdominal aorta and iliac arteries through the femoral artery and as intraoperative endoprosthesis for aorta reconstruction. Vasa 1991;33(Suppl):93–95.
17. Marin ML, Veith FJ, Cynamon J, et al. Initial experience with transluminally placed endovascular grafts for the treatment of complex vascular lesions. Ann Surg 1995;222:449–469.
18. Marin ML, Veith FJ, Ohki T. Endovascular stent-grafts for treatment of traumatic pseudoaneurysms and arteriovenous fistulas. In: Yao JST, Pearce WH, eds. Vascular surgery: 20 Years of progress. Norwalk, CT: Appleton & Lange, 1996.
19. Marin ML, Veith FJ, Panetta TF, et al. Transluminally placed endovascular stented graft repair for arterial trauma. J Vasc Surg 1994;20:466–473.
20. Parodi JC. Endovascular repair of abdominal aortic aneurysms and other arterial lesions. J Vasc Surg 1995;21:549–557.
21. Becker GJ, Katzen BT, Benenati JF, et al. Endografts for the treatment of aneurysm and traumatic vascular lesions: MVI experience. (Abstract). J Endovasc Surg 1995;2:380–382.
22. Schmitter SP, Marx M, Bernstein R, et al. Angioplasty-induced subclavian artery dissection in a patient with internal mammary artery graft: Treatment with endovascular stent and stent-graft. AJR 1995;165:449–451.
23. Marston WA, Criado E, Mauro M, Keagy BA. Transbrachial endovascular exclusion of an axillary artery pseudoaneurysm with PTFE-covered stents. J Endovasc Surg 1995;2:172–176.
24. Zajiko AB, Little AF, Steed DL, Curtiss EI. Endovascular stent-graft repair of common iliac artery-to-inferior vena cava fistula. J Vasc Intervent Radiol 1995;6:803–806.
25. Criado E, Marston WA, Ligush J, et al. Endovascular repair of peripheral aneurysms, pseudoaneurysms, and arteriovenous fistulas. Ann Vasc Surg 1997;11:256–263.
26. Uflacker R, Elliot BM. Percutaneous endoluminal stent-graft repair of an old traumatic femoral arteriovenous fistula. Cardiovasc Intervent Radiol 1996;19:120–122.
27. Dorros G, Avula S, Fox P, et al. Endovascular covered stent repair of an intercostal artery patch dehiscence from a descending thoracic aortic aneurysm graft. J Endovasc Surg 1996;3:299–305.
28. Brunkwall J, Lindblad B, Ivancev K, et al. Iatrogenic AV-fistula treated by a graft-covered self-expandable stent. Eur J Vasc Endovasc Surg 1996;12:243–245.
29. Chuter TA, Ivancev K, Lindblad B, et al. Endovascular stent-graft exclusion of an aortobronchial fistula. J Vasc Intervent Radiol 1996;7:357–359.
30. Dorros G, Joseph G. Closure of a popliteal arteriovenous fistula using an autologous vein-covered Palmaz stent. J Endovasc Surg 1995;2:177–181.
31. Wong CS, Kent KM, Mintz GS, et al. Percutaneous transcatheter repair of a coronary aneurysm using a composite autologous cephalic vein-coated Palmaz-Schatz biliary stent. Am J Cardiol 1995;76:990–991.
32. Kerns DB, Darcy MD, Baumann DS, Allen BT. Autologous vein-covered stent for the endovascular management of an iliac-ureteral fistula: Case report and review of the literature. J Vasc Surg 1996;24:680–686.
33. Dorros G, Jain A, Kumar K: Management of coronary artery rupture: Covered stent or microcoil embolization. Catheter Cardiovasc Diagn 1995;36:148–154.
34. Becker GJ, Benenati JF, Zemel G, et al. Percutaneous placement of a balloon-expandable intraluminal graft for life-threatening subclavian arterial hemorrhage. J Vasc Intervent Radiol 1991; 2:225–229.
35. Allgayer B, Theiss W, Naundorf M. Percutaneous closure of an arteriovenous fistula with a Cragg endoluminal graft. AJR 1996; 166:673–674.
36. Christensen L, Justesen P, Larsen KE. Percutaneous transluminal treatment of an iliac pseudoaneurysm with endoprosthesis implantation. Acta Radiol 1996;37:542–544.
37. McCarter DHA, Johnstone RD, McInnes GC, et al. Iliac arteriovenous fistula following lumbar disc surgery treated by percutaneous endoluminal stent grafting. Br J Surg 1996;83:796–797.
38. Johnstone RD, McCarter DH, McInnes GC, Reid AW. Stent grafting in iliac arteriovenous fistulas. AJR 1996;167:1597–1602.
39. Gomez-Jorge JT, Guerra JJ, Scagnelli T, et al. Endovascular management of a traumatic subclavian arteriovenous fistula. J Vasc Intervent Radiol 1996;7:599–602.
40. Patel AV, Marin ML, Veith FJ, et al. Endovascular graft repair of penetrating subclavian artery injuries. J Endovasc Surg 1996;3:382–388.
41. Terry PJ, Houser EE, Rivera FJ, et al. Percutaneous aortic stent placement for life threatening aortic rupture due to metastatic germ cell tumor. J Urol 1995;153:1631–1634.

42. Campagna AC, Wehner JH, Kirsch CM, et al. Endovascular stenting of an aortopulmonary fistula presenting with hemoptysis. J Cardiovasc Surg 1996;37:643–646.
43. Ohki T, Marin ML, Veith FJ, et al. Endovascular approaches for traumatic arterial lesions. Semin Vasc Surg, 1998, in press.
44. White RA, Donayre CE, Walot I, et al. Preliminary clinical outcome for endovascular prosthesis development in high-risk patients who have aortoiliac and traumatic arterial lesions. J Vasc Surg 1996;24:556–571.
45. Parodi JC, Palmaz JC, Barone HD. Transfemoral intraluminal graft implantation for abdominal aortic aneurysms. Ann Vasc Surg 1991;5:491–499.
46. Ohki T, Marin ML, Veith FJ. Use of endovascular grafts to treat non-aneurysmal arterial disease. Ann Vasc Surg 1997;11:200–205.

Chapter 14

Endovascular Grafting for Aortoiliac Occlusive Disease

Reese A. Wain, Michael L. Marin, and Frank J. Veith

Aortoiliac atherosclerotic disease is a significant cause of lower extremity ischemic symptomatology both as a distinct pathologic entity and when combined with infrainguinal disease. The appropriate treatment of lesions in the aortoiliac segment depends on the local extent of the lesion, the presence of bilateral or infrainguinal disease, the degree of symptomatology and the patient's medical comorbidities. The goals of treatment are similar to those for all types of ischemic arterial disease and include long-term relief of symptoms at an acceptable surgical risk to the patient. Therapeutic options available to the vascular surgeon for treating aortoiliac occlusive disease were developed over the last half century and include endarterectomy, anatomic and extra-anatomic bypass, and the newer percutaneous techniques, such as angioplasty and arterial stenting.

The recent implementation of endovascular grafting techniques for aortoiliac occlusive disease affords another treatment alternative of profound potential for which the indications are only now being realized. With the available range of current treatment options, the comparative advantages and disadvantages of each procedure must be prudently considered before undertaking a therapeutic intervention.

As a new technique such as endovascular grafting becomes technically feasible, it must be carefully scrutinized both in terms of its inherent limitations and its purported advantages over competing procedures. Typically, new techniques have been devised to improve upon existing technologies. However, only when the new approach can outperform its predecessors in terms of ease of use, wider applicability, reduced cost, or decreased patient morbidity will it usurp the role of the more established procedure.

This chapter first undertakes a brief recapitulation of significant developments in the treatment of aortoiliac occlusive disease as they presage the inception of endovascular grafting. Next, a description of the current experience with endovascular grafting techniques is presented highlighting graft design, deployment, and clinical experience. Finally, the merits and disadvantages of each of the established procedures are compared with one another and with endovascular grafting to better define the increasing role of endovascular grafting in clinical practice.

TREATMENT OF AORTOILIAC OCCLUSIVE DISEASE

Overview

Technical innovations over the last four decades have provided the impetus for change in the overall management of aortoiliac occlusive disease and the trend toward less invasive interventions. The first procedure to enjoy widespread use for the treatment of such lesions was thromboendarterectomy, which was initially described in 1947 by dos Santos (1) and in 1952 by Wylie (2) in the United States. Around the same time, aortic excision and homograft replacement as well as prosthetic aortic grafting techniques were being developed for aneurysmal and occlusive lesions (3–6). Although endarterectomy was probably the more commonly performed procedure in the 1960s, aortofemoral grafting became the most frequently performed intervention over the following two decades (7,8). The ease with which the procedure could be performed, a broader applicability, and perceived improvements in long-term patency were responsible for the ascendancy of aortobifemoral grafting.

Limited endarterectomies are often undertaken as an adjunct to bypass grafting to enhance inflow and outflow or to treat localized stenoses; however, endarterectomy is largely of historical significance as a singular procedure to manage aortoiliac occlusive lesions (9).

The development of prosthetic arterial grafts in the 1950s made possible the description of "extra-anatomic" bypasses in the early 1960s (10,11). Axillofemoral and femorofemoral bypasses became important in the treatment of infected aortic grafts and have been increasingly used in high-risk operative candidates with severe coexisting medical problems.

Although transluminal dilatation of atherosclerotic lesions was first performed by Dotter and Judkins in 1964 (12), it wasn't until 10 years later when Gruntzig and Hopff (13) described the double-lumen balloon catheter that angioplasty became a popular technique.

Although without the widespread applicability of aortobifemoral grafting, balloon angioplasty has proved useful in the treatment of aortoiliac occlusive disease for certain well-defined patterns of disease (14–20). Improvements in angioplasty results using intraluminal stents served to further increase the acceptance of this less invasive alternative. In the 1980s, angioplasty and stenting became the procedures of choice for short-segment common iliac artery stenoses. They also became important adjuncts to infrainguinal bypasses in the presence of multilevel disease.

Long-term follow-up results of aortic and iliac artery stenting have proved the procedure to be efficacious in treating disease patterns, which in the past were amenable only to conventional operative management (21–27). In addition, endovascular techniques such as angioplasty and stenting have broadened the indications for another extra-anatomic approach to aortoiliac occlusive disease—femorofemoral bypass.

In the past, femorofemoral bypass had been possible only in patients with unilateral iliac artery disease and a normal contralateral donor iliac artery. Now it can be used in a larger group of patients with bilateral disease.

Endovascular grafts that were initially conceived of and deployed to treat abdominal aortic aneurysms (28–31) have successfully been adapted to the treatment of aortoiliac occlusive disease in clinical practice. Although still in its formative stages, the technique of endovascular grafting combines the advantages of arterial stenting and extra-anatomic procedures to treat aortoiliac occlusive disease when a direct approach to the abdominal aorta and aortobifemoral grafting are not desirable.

Additional advantages of endovascular grafting over both extra-anatomic bypasses and intravascular stents may make it the procedure of choice in other circumstances as well. Just as time largely obviated the need for aortoiliac endarterectomy, it is conceivable that tomorrow endovascular grafts may supplant aortobifemoral grafting, extra-anatomic bypasses, and balloon angioplasty and stenting as the preferred treatment for symptomatic aortoiliac occlusive disease.

ENDOVASCULAR GRAFTING

General Considerations

The basic tenets of endovascular grafting include the ability to treat a lesion distant from the site of arterial access with a device that can be deployed endoluminally. This concept has been satisfied by using novel devices constructed from conventional prosthetic grafts and intravascular stents. The stent portion of an endovascular graft affixes the prosthetic portion of the graft to the vessel wall in a blood tight fashion while the graft itself lines the inner wall of the vessel. Proponents of endovascular grafting hypothesize that by lining the lumen of a vessel with a prosthetic graft, both intimal hyperplasia and tissue ingrowth may be inhibited and, consequently, the restenosis rate may be attenuated.

Two different techniques have recently enjoyed clinical usefulness while addressing the basic concepts of endovascular grafting. One of the techniques involves the percutaneous insertion of a Dacron-covered, self-expanding nitinol stent called the Cragg EndoPro System 1 (Mintec, Inc, Grand Bahama Island) (32–34). After radiographic visualization of the target lesion, balloon angioplasty of the lesion is undertaken and an appropriate-sized covered stent is used to traverse the length of the lesion. When the deployment system is extracted, blood flows through the lumen of the graft, whereas arterial wall recoil is opposed by the stent component of the device.

At the Montefiore Medical Center in New York, our experience with endovascular grafting has relied on an operative rather than a percutaneous approach. In the operating room, access to the vasculature is obtained by means of a standard arteriotomy. Using fluoroscopically guided, catheter-based techniques, a device consisting of a conventional prosthetic graft mated proximally to a Palmaz balloon-expandable stent (Johnson & Johnson Interventional Systems, Warren, NJ) is inserted. The stent is deployed proximal to the lesion, and the graft is retrieved from the arteriotomy site. Individualized management of the distal anastomosis is then undertaken.

The key difference between our approach and the aforementioned one is that in our approach the graft is not supported along its entire course by a stent endoskeleton. More important, in contrast to a covered stent, our graft can bypass a long segment of diseased artery and does support tailoring of the distal anastomosis to the patient's pattern of disease.

Clinical Experience

Covered Stents

Two groups have recently reported their initial clinical experiences with the Cragg EndoPro System. Henry and coworkers (34) in Nancy, France, treated 15 symptomatic patients with iliac artery stenoses measuring between 3 and 20 cm in length and 4 patients with iliac artery occlusions measuring between 5 and 20 cm. The patients in this study had the typical risk factors for atherosclerotic disease, were primarily claudicants, and underwent covered stent procedures for either residual stenoses or dissections after angioplasty, recurrent stenoses, or ulcerated lesions.

With the patient given local anesthesia, percutaneous insertion of the endoluminal prosthesis was undertaken by means of an antegrade femoral artery puncture, and balloon angioplasty of the lesion was accomplished. Next, angiography was used to select the appropriate-sized covered stent, and the preloaded device was inserted through a 9F, 50-cm long Cragg introducer sheath. As the introducer sheath was removed, the covered stent was held in place with a positioning catheter and allowed to expand. To achieve improved fixation of the device to the vessel wall, balloon dilatation of the stent was always undertaken.

Technical success was achieved in all patients as documented by angiographic decreases in percent stenoses of the lesions, decreased pressure gradients across the lesions, and postoperative improvements in symptomatology and noninvasive vascular studies.

One patient with a common iliac artery occlusion who was treated with two separate covered stents experienced a partially occluded contralateral common iliac artery when the proximal-most device protruded into the aorta. Placement of a stent-graft on the contralateral side fixed the problem. Patency was maintained in all patients with up to 8 months of follow-up. The investigators concluded that the technique is appropriate for the treatment of long-segment iliac artery lesions, but that they will await long-term results before hazarding comparisons with operative bypass grafting.

Pernes and his group (33) in Antony, France, have also enjoyed clinical success with the Cragg EndoPro System. In their experience, 10 patients with iliac artery stenoses greater than 6 cm in length in arteries of at least 7 mm in diameter were treated with 15 covered stents. In this study, the indications for treatment were claudication in nine patients and rest pain in one. An approach similar to Henry's was used to deploy all the stent-grafts. Immediate technical success was achieved in all but one patient in whom the external iliac artery ruptured during balloon angioplasty. This was fixed with a second covered stent, but the iliac artery thrombosed the next day and an axillary-to-femoral artery bypass was required. Two patients became symptomatic during the follow-up period. The first underwent repeat dilatation for a stenosis seen at the proximal portion of the graft and did well; the second patient's graft became occluded and his fate is unknown. In addition, in one patient an asymptomatic restenosis of the graft developed, which was also treated with balloon angioplasty.

The Montefiore Experience

Between November 1992 and April 1995, 42 patients underwent 47 endovascular grafting procedures to treat 47 combined aortoiliofemoral artery occlusive lesions (35–44). Twenty-three of the patients were men and 19 were women; their average age was 65 years. Eighty-six percent of the patients had known coexisting coronary artery disease, 52% had diabetes mellitus, and 21% and 17% suffered from chronic obstructive pulmonary disease and renal insufficiency, respectively. Our indications for intervention were gangrene in 39 patients (83%) and pain at rest in the remaining 8 (17%).

The endovascular grafts we use are constructed from 30-mm Palmaz balloon-expandable stents and 6-mm polytetrafluoroethylene (PTFE) thin-walled grafts (W.L. Gore and Associates, Flagstaff AZ). The stents are sutured to the proximal aspect of the graft with four CV-6 PTFE sutures (W.L. Gore and Associates) so that the graft overlaps one-half of the stent's predeployment length. The graft is then coaxially mounted on an 8 mm × 3 cm or 4 cm angioplasty balloon (Blue Max or PMT, Medi-tech, Inc, Watertown, MA).

The endovascular graft and the balloon catheter are loaded into a delivery sheath, which also houses a balloon catheter shaft culminating in a "tip balloon." When inflated, the tip balloon partially protrudes from the distal portion of the delivery sheath to provide a smooth taper, facilitating insertion of the entire device. In addition, inflation of the tip balloon occludes the distal aspect of the delivery sheath, allowing the sheath to be pressurized and facilitating its delivery. A modification of this device with two balloons mounted on the same shaft was also used (Figs. 14.1 through 14.3).

Although the covered stent approach to endovascular grafting is conceptually similar to intra-arterial stenting, our approach to endovascular grafting is more akin to operative bypass grafting. Under general, epidural, or local anesthesia, exposure of the femoral vessels was undertaken in the operating room. After arteriotomy, stenotic lesions were traversed with a guidewire under fluoroscopic control, and occluded arterial segments were recanalized. Recanalization was accomplished by means of a percutaneously inserted guidewire through the contralateral femoral artery (if it was patent).

By advancing the guidewire up and over the aortic bifurcation, a recanalization plane could often be successfully achieved within the lumen of the native vessel. If the contralateral iliac artery was similarly occluded, a

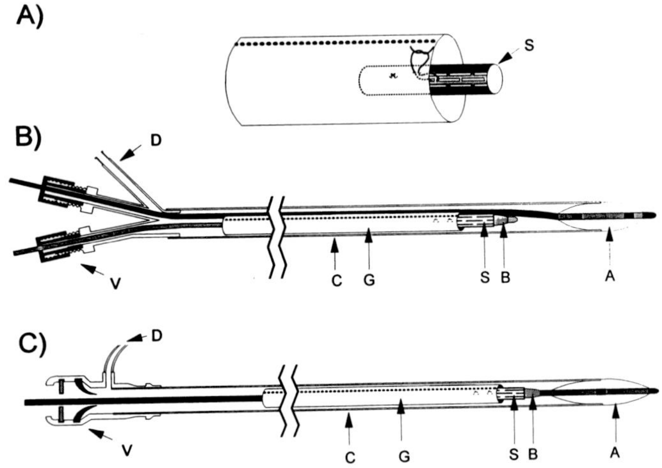

Figure 14.1. Schematic drawings of an endovascular graft and its delivery systems. **A.** An endovascular thin-walled PTFE graft. A Palmaz balloon-expandable stent (*S*) is sutured to the graft using four "U" sutures, which permit 50% of the stent to protrude from the graft. **B.** Double-balloon catheter introducer/delivery system used for the delivery and deployment of endovascular stented grafts. In this system, the introducer catheter is equipped with two separate balloon catheters. Balloon "A" functions as the "tip balloon" to form a tapered end to the catheter system and allow pressurization of the flexible sheath (*C*) after saline is injected from port (*D*). The second balloon (*B*) functions to deploy the overlying Palmaz stent (*S*). (V = Hemostatic valve mechanism; G = graft.) **C.** An alternative delivery system that consists of a single-balloon catheter, which has two balloons on a single shaft. The distal balloon serves as a tip balloon (*A*), whereas the stented graft is mounted on a proximal deployment balloon (*B*). (Reprinted with permission from Ohki T, Marin ML, Veith FJ, et al. Endovascular aortounifemoral grafts and femorofemoral bypass for bilateral limb-threatening ischemia. J Vasc Surg 1996;24:984–997.)

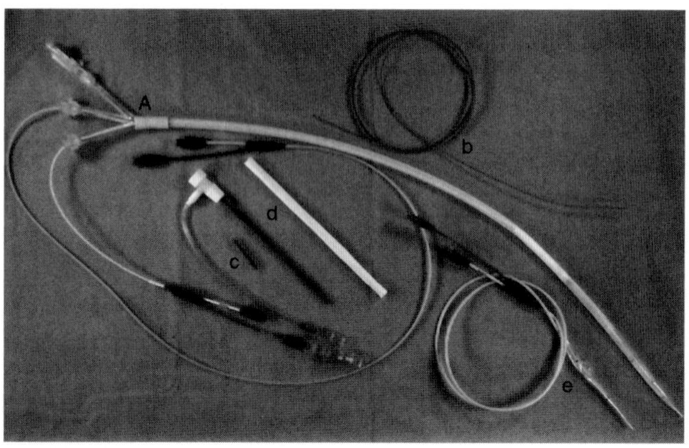

Figure 14.2. Photograph of the components of an endovascular graft and its delivery system. **A.** The fully assembled endovascular graft and deployment system housed within a delivery sheath. The hemostatic trifurcated valve system at the proximal end of the sheath allows for independent inflation of the "tip" and deployment balloons as well as for pressurization of the sheath itself. After the arteriotomy, an introducer (*a*) is inserted into the vessel and a guidewire (*b*) is placed. Diagnostic catheters or the endovascular device can be inserted over the guidewire and advanced to the appropriate level. The components of the endovascular grafting system include the Palmaz stent (*c*), the PTFE graft (*d*), and the tip balloon (*e*). (The deployment balloon is not shown.)

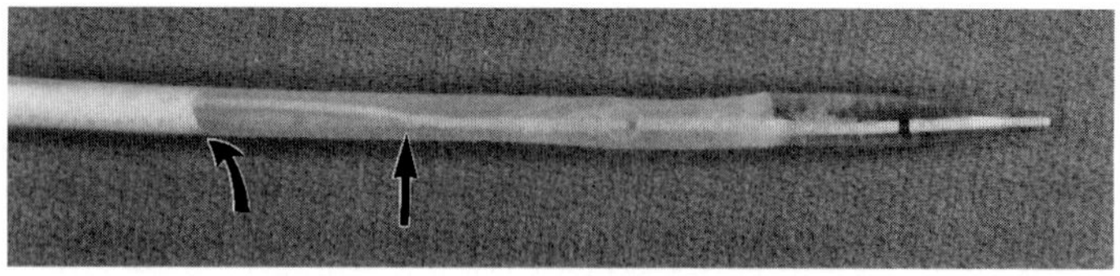

Figure 14.3. Close-up photograph of the distal end of the endovascular graft delivery system. Note the smooth, tapered end to the delivery catheter tip. In this configuration of the system, a fenestration in the tip balloon's catheter (*straight arrow*) permits coaxial (single-wire) integration between this balloon and the deployment balloon. The PTFE graft and Palmaz stent can be visualized within the sheath assembly (*curved arrow*). (Reprinted with permission from Marin ML, Veith FJ, Sanchez LA, et al. Endovascular repair of aortoiliac occlusive disease. World J Surg 1996;20:679–686.)

retrograde approach was used (Fig. 14.4). Recanalizations were attempted through an occluded artery in 34 patients and a diffusely stenotic vessel in 8 patients using a hydrophilic guidewire and an angled directional catheter. Thirty-nine patients (93%) were successfully recanalized, and the remaining three required conversion to a conventional operative procedure.

After the diseased segment had been traversed by a guidewire, the entire length of the diseased iliac artery and the femoral artery proximal to the arteriotomy were dilated. Dilatation with an 8-mm angioplasty balloon was performed so that the graft deployment system could be inserted without excessive friction (Fig. 14.5).

After the endovascular graft had been appropriately positioned, the tip balloon was deflated and the sheath partially withdrawn, permitting expansion of the proximal balloon and stent. As the remainder of the sheath was withdrawn, the terminal aspect of the graft was manually retrieved from within the arteriotomy (Fig. 14.6).

Management of the distal anastomosis was dependent on each individual patient's pattern of disease. Thirty-two of the grafts were endoluminally anastomosed at the level of the common femoral artery, and the rest were extraluminally anastomosed to a distal vessel or a separate bypass graft. The arteriotomy overlying the endoluminal anastomosis was either closed primarily with a patch or covered by the hood of a second bypass graft (Fig. 14.7).

Routine completion angiography was performed, which revealed midgraft stenoses in 8 (21%) of the 39 patients successfully treated. These lesions were felt to be secondary to either inadequate expansion of the prosthetic graft or compression of the graft by the excluded vessel wall. They were corrected by repeat balloon dilatation of the stenotic segment and in four instances by placement of an additional Palmaz stent within the graft.

Successful endovascular graft placement occurred in 37 (88%) of the 42 patients (Figs. 14.8 through 14.10). In addition to the three patients whose occluded iliac vessels could not be recanalized, two patients had thrombosed grafts within 48 hours, which were considered procedural failures. An extensive follow-up protocol was developed consisting of pulse volume recordings and

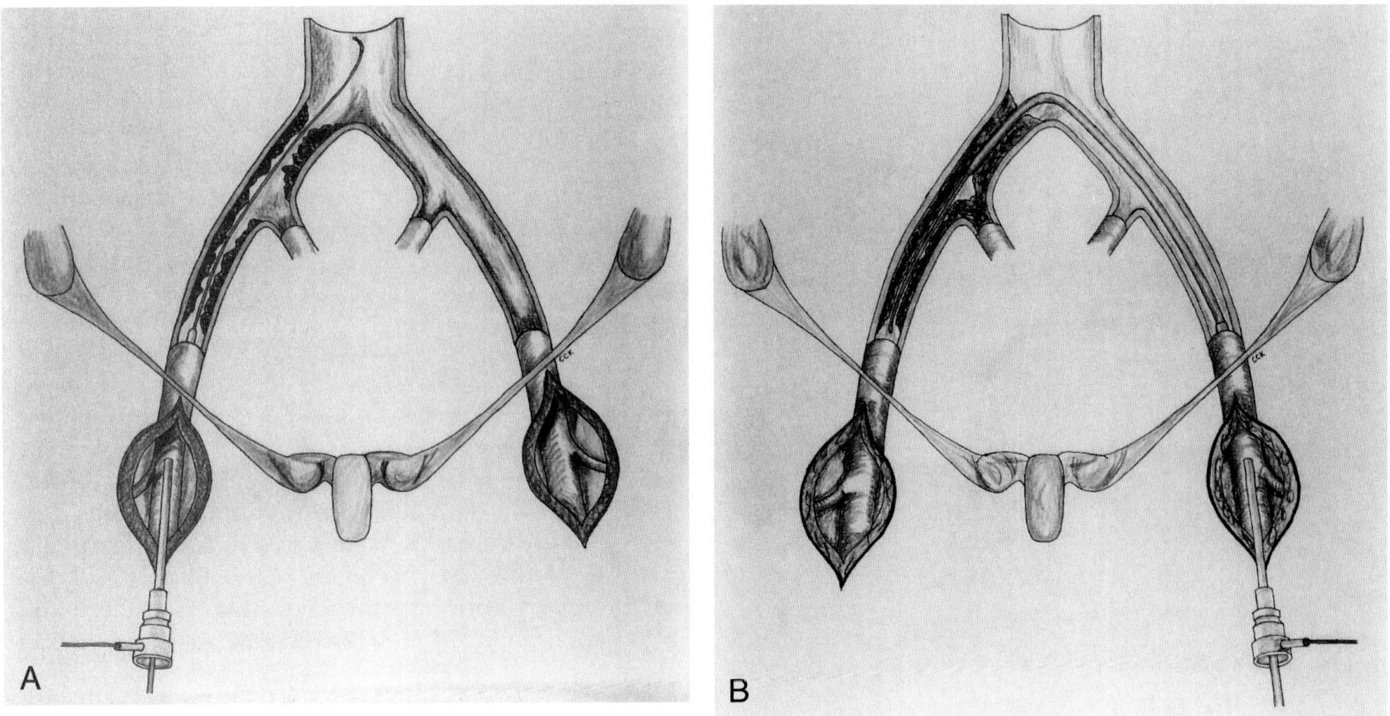

Figure 14.4. **A.** Depiction of a guidewire and catheter traversing a severely stenotic iliac artery. **B.** Illustration of the "up and over" technique of recanalization of an occluded iliac segment.

duplex studies as well as angiography when abnormalities were suspected on the basis of the noninvasive studies or physical examination.

Eighteen-month primary and secondary cumulative life-table patency rates for this cohort of patients were 89% and 100%, respectively. Limb salvage was achieved in 94% of the patients at up to 2 years' follow-up. One patient died of congestive heart failure after placement of an endovascular graft, and four patients accrued minor complications consisting of lymphoceles and

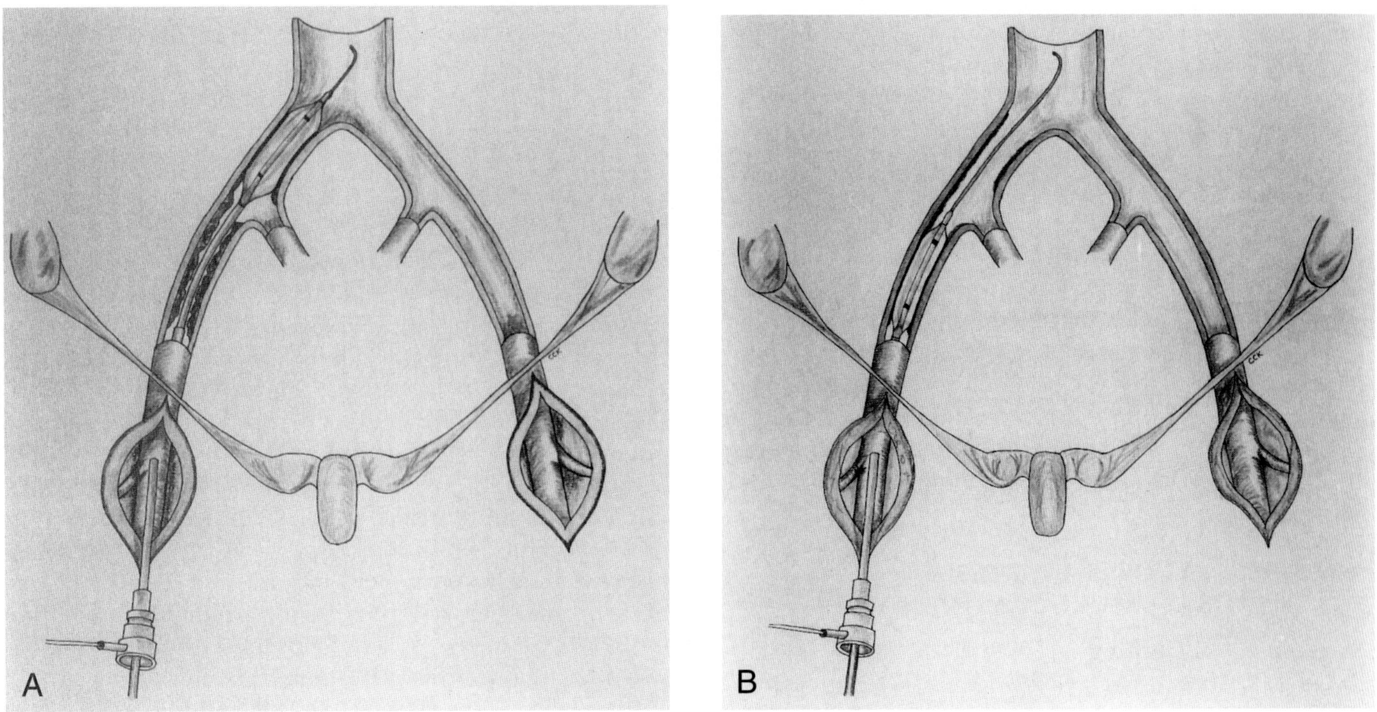

Figure 14.5. **A.** Depiction of the initial balloon dilatation of the proximal aspect of the common iliac artery. **B.** The iliac and common femoral vessels have been sequentially dilated in a proximal-to-distal fashion until the arteriotomy site is reached.

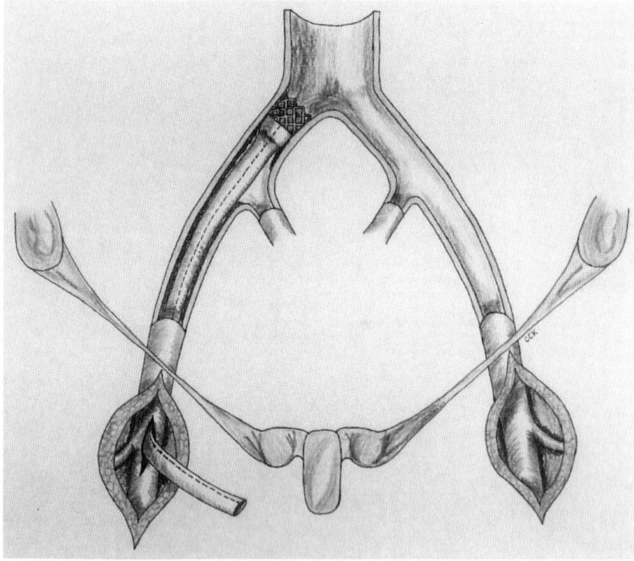

Figure 14.6. The endovascular graft has been deployed within the iliac artery. The stent anchors the proximal aspect of the graft within the common iliac artery, and the distal aspect of the graft is manually retrieved from the arteriotomy.

groin hematomas. One patient suffered a mild, asymptomatic subendocardial myocardial infarction.

We are encouraged by the high technical success rate and the low rate of morbidity that were achieved in this difficult group of patients with multilevel atherosclerotic disease, threatened limbs, and severe medical comorbidities. We believe that our method of endovascular grafting must be considered a viable alternative to more conventional vascular interventions for the treatment of patients with aortoiliac occlusive disease. Covered stents are useful in select circumstances, but their use cannot be generalized to the majority of patients with aortoiliac occlusive disease. As we will describe, covered stents are subject to many of the limitations of conventional stents, whereas the potential applicability of endovascular grafts is not similarly constrained.

The following sections compare the benefits and limitations of aortobifemoral grafting, extra-anatomic grafting, and balloon angioplasty and stenting with our current method of endovascular grafting for the treatment of aortoiliac occlusive disease. We have chosen to omit comparisons with aortoiliac endarterectomy and axillounifemoral, aortounifemoral, and iliofemoral grafting for the sake of brevity and because these procedures are of less widespread applicability and are less frequently performed.

ENDOVASCULAR GRAFTING VERSUS CONVENTIONAL VASCULAR INTERVENTIONS

Aortobifemoral Grafting

Advantages

The procedure that has had the most widespread success in the treatment of aortoiliac occlusive disease is the aortobifemoral bypass. This success is partially explained by the applicability of the approach not only to patients with disease limited to the aorta and common iliac arteries but also to those with more widespread disease extending to the femoral vessels and beyond. Although some still debate the relative merits of a proximal end-side versus end-end anastomosis, in most cases the distal limbs of the graft are extended to the femoral arteries. Not only is exposure superior at this level, but correction of profunda femoris origin disease is possible.

In addition, in bypassing the iliac artery, many of the effects of disease progression in this segment can be rendered clinically irrelevant (45–48). An end-side proximal anastomosis may be indicated if the surgeon wishes to preserve prograde blood flow to an inferior mesenteric artery, an accessory renal artery, or the internal iliac arteries to preserve adequate pelvic blood flow. The versatility of aortobifemoral grafting is underscored by its applicability in these infrequent but important settings.

The ability to perform a standardized procedure to treat a wide variety of disease types and anticipate and correct for disease progression in concert with an excellent source of inflow contribute to the impressive long-term patency of aortobifemoral grafting.

A sampling of studies reveals 5-year cumulative patency rates of 75% to 90% after aortobifemoral grafting (7,8,49,50). Aortobifemoral grafting has become the gold standard against which newer procedures must be compared based on such superior patency data collected over extended intervals.

In a historical context, however, note that the indications for which these procedures were performed differed significantly from those currently in vogue. Large numbers of patients were operated on for varying degrees of claudication and constitute a significant part of the early experience. Today, many of these patients would be treated by alternative, less invasive means and would likely never experience aortobifemoral grafting. Patients who are operated on for limb salvage would most assuredly not enjoy the same degree of patency as their claudicant cohorts. Therefore, care must be taken when inserting such grafts in the expectation of achieving patency rates indigenous to a different subgroup of patients.

Superior inflow of the magnitude attained with aortobifemoral grafting is achieved with endovascular grafting as the proximal limb of the graft is secured at the aortic bifurcation. This placement minimizes the effect of progression of atherosclerotic disease proximal to a bypass graft. Disease progression could decrease patency if the proximal anastomosis were more distal in the iliac artery. Just as the aortobifemoral graft typically terminates below the inguinal ligament at the level of the femoral artery, so, too, can the endovascular graft. In both cases,

Figure 14.7. Tailoring of the distal endovascular graft anastomosis according to the patient's pattern of disease. **A.** The distal anastomosis has been constructed in an endoluminal fashion (*inset*), and the arteriotomy is primarily closed. **B.** The distal anastomosis has also been constructed in an endoluminal fashion, but the arteriotomy is closed with a patch (*a*). View of the endoluminal anastomosis before patch closure of the arteriotomy (*b*). Cutaway view of the endoluminal anastomosis and the overlying patch closure (*c*). **C.** The endoluminal anastomosis is covered by the hood of a distal PTFE bypass graft.

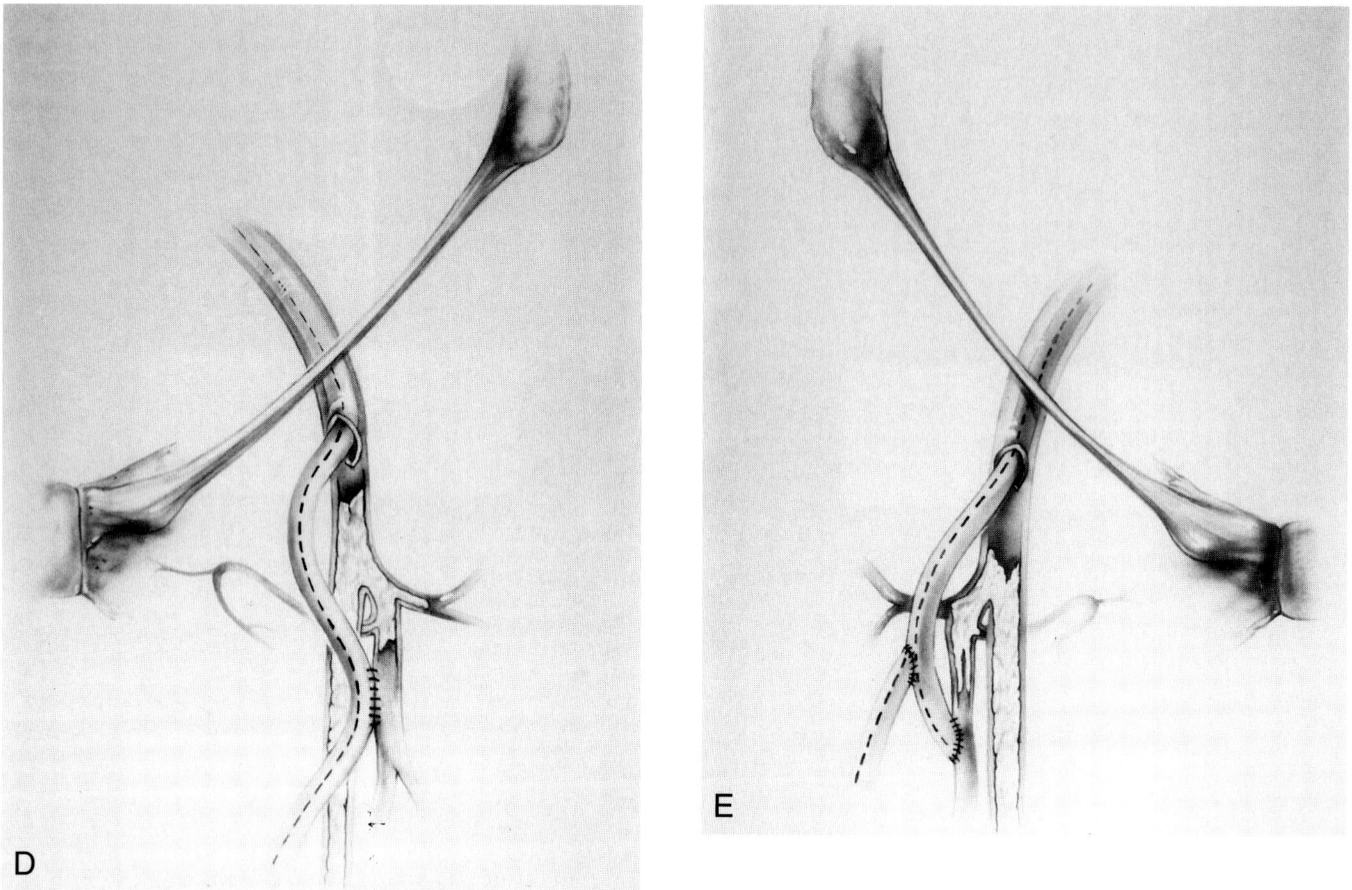

Figure 14.7 (continued). **D** and **E.** Examples of how the endoluminal graft can be brought out via the arteriotomy and anastomosed in a conventional fashion to patent distal vessels or to separate bypass grafts.

coexistent disease within the femoral artery or its branches can be directly addressed at the time of the primary procedure.

In addition, endovascular grafting can also be used in a standardized fashion to treat different patterns of disease. With the exception of the management of the distal anastomosis, recanalization of the diseased segment and graft deployment are identical for all lesions. Bilateral disease can also be readily addressed with the addition of a femorofemoral bypass after unilateral aortofemoral grafting (39). This technique most closely approximates the functional result of an aortobifemoral graft.

Our initial patency data for patients with limb-threatening ischemia are similar to those seen on early follow-up after aortobifemoral grafting. However, results of long-term surveillance must be obtained before definitive conclusions can be reached comparing the durability of the two procedures.

Direct aortobifemoral reconstruction may be the operation of choice in the presence of peripheral atheroemboli originating from an ulcerated lesion in the aorta or when the correction of coexisting renal artery disease is indicated (51). In the past, both of these conditions have mandated operative exposure of the abdominal aorta—to bypass the culprit lesion in the former condition and to fix the diseased renal vessel or vesssels in the latter condition.

We have not performed endovascular grafting for the primary indication of symptomatic atheroemboli of aortic origin. However, the procedure can be adapted to treat such lesions. If the proximal end of a prosthetic graft is predilated to the diameter of the infrarenal aorta and affixed by a stent of a similar deployment diameter, the graft can be inserted beneath the renal arteries in the abdominal aorta. Using the same technique we use to endovascularly treat abdominal aortic aneurysms, the distal limb of the graft can be endoluminally anastomosed to the femoral artery, and femorofemoral bypass grafting can be performed to perfuse the contralateral extremity (Fig. 14.11).

To prevent retrograde filling of the aortic sac, an occluder "windsock" device can be deployed within the contralateral iliac artery. The ulcerated aortic plaque is thereby excluded from the circulation, and prograde flow is restored. Coexisting renal artery disease cannot presently be addressed with endovascular grafting techniques. However, if a suitable lesion exists, balloon angioplasty with or without stenting may be attempted via a percutaneous approach after the patient's aortic disease has been appropriately treated.

The manner in which endovascular grafting is performed in no way compromises blood flow through accessory renal arteries or the inferior mesenteric artery. In contrast, a conventional bypass procedure may be further complicated by the need to preserve these branches when the patient is dependent on them for end-organ perfusion.

Our technique of common iliac to femoral artery bypass grafting involves sacrificing the ipsilateral hypogastric artery when its orifice is covered by the side of the prosthetic graft. In our experience, sacrifice of this vessel has not been a problem, and we believe that bilateral hypogastric artery occlusion may not carry the morbidity that has been attributed to this condition in the past and can be tolerated in patients with contralateral hypogastric artery disease or those requiring bilateral lower extremity endovascular procedures.

Finally, aortobifemoral grafting with concomitant aortic endarterectomy plays an important role in the treatment of patients with infrarenal aortic occlusion. If the level of thrombosis has progressed to the level of the renal arteries, endovascular grafting would be contraindicated because of possible thrombus embolization to the kidneys. Unless a means can be devised to protect the renal arteries from this potentially harmful sequelae, aortobifemoral grafting with endarterectomy or alternative procedures should be used to treat this condition. On the other hand, aortic occlusions distal to the level of the renal arteries can be treated with endovascular grafting if the occlusion can be successfully recanalized (Table 14.1).

Limitations

Direct transperitoneal aortobifemoral grafting is an invasive procedure, which must be performed with the patient under general anesthesia and with the aorta cross-clamped most of the time. The subset of patients with aortoiliac occlusive disease are also commonly afflicted with occlusive disease of the coronary vessels. Consequently, hemodynamic stressors such as general anesthesia and aortic clamping may exact a toll on a patient with limited cardiac reserve.

Several series have shown that up to 30% of patients operated on for aortoiliac disease may be dead within 5 years of the procedure and 60% within 10 years (52). Not unexpectedly, most of these patients die from complica-

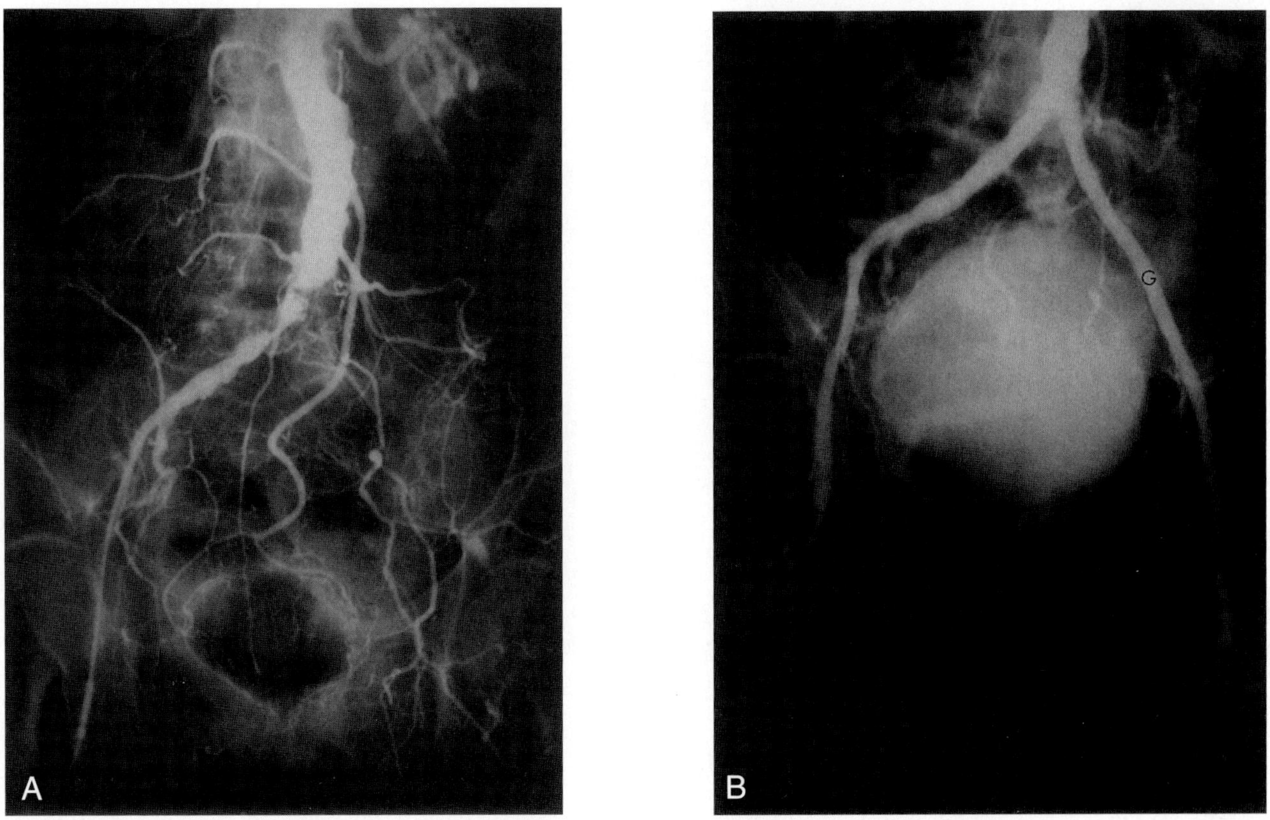

Figure 14.8. Transfemoral iliofemoral bypass for limb-threatening ischemia. **A.** Before insertion of an endovascular graft, severe aortoiliac disease is demonstrated by this preoperative transfemoral arteriogram. Disease in the right common iliac artery system would preclude effective standard femorofemoral bypass to reestablish circulation to the left lower extremity in this 86-year-old woman. The left common and external iliac arteries are completely occluded. **B.** After long-segment balloon dilatation of the left common and external iliac arteries and insertion of an endovascular graft (*G*), vascular continuity is established to the left lower extremity. A percutaneous balloon-expandable stent has also been inserted into the right common iliac artery to treat the symptomatic high-grade common iliac artery stenosis and to protect the origin of the right common iliac artery during endovascular graft insertion. (Reprinted with permission from Marin ML, Veith FJ, Cynamon J, et al. Initial experience with transluminally placed endovascular grafts for the treatment of complex vascular lesions. Ann Surg 1995;222:449–469.)

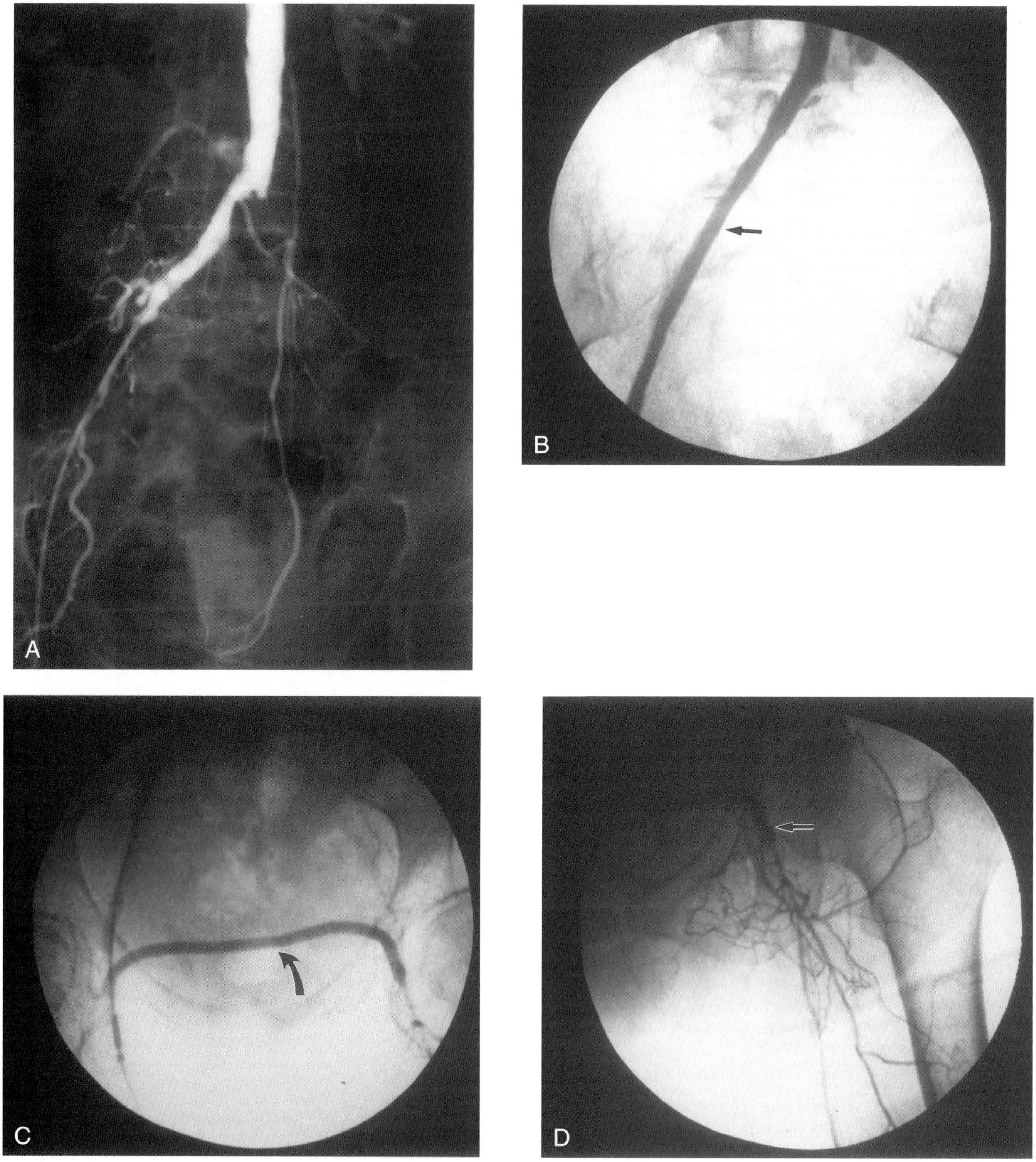

Figure 14.9. Angiograms of a 77-year-old hypertensive female smoker who sought treatment for a nonhealing, gangrenous left ankle ulcer of traumatic origin. **A.** Preoperative angiogram demonstrating chronic occlusions of both common iliac arteries. The superficial femoral arteries were reconstituted on later films (not shown). **B.** As seen on an intraoperative film, an endovascular graft (*arrow*) was inserted between the right common iliac artery and the common femoral artery. **C.** A crossover femorofemoral bypass graft (*arrow*) completed the revascularization of the affected extremity. **D.** Close-up view of the anastomosis (*arrow*) of the left limb of the femorofemoral bypass demonstrating improved outflow of a magnitude sufficient to heal the patient's ulcer within 2 months.

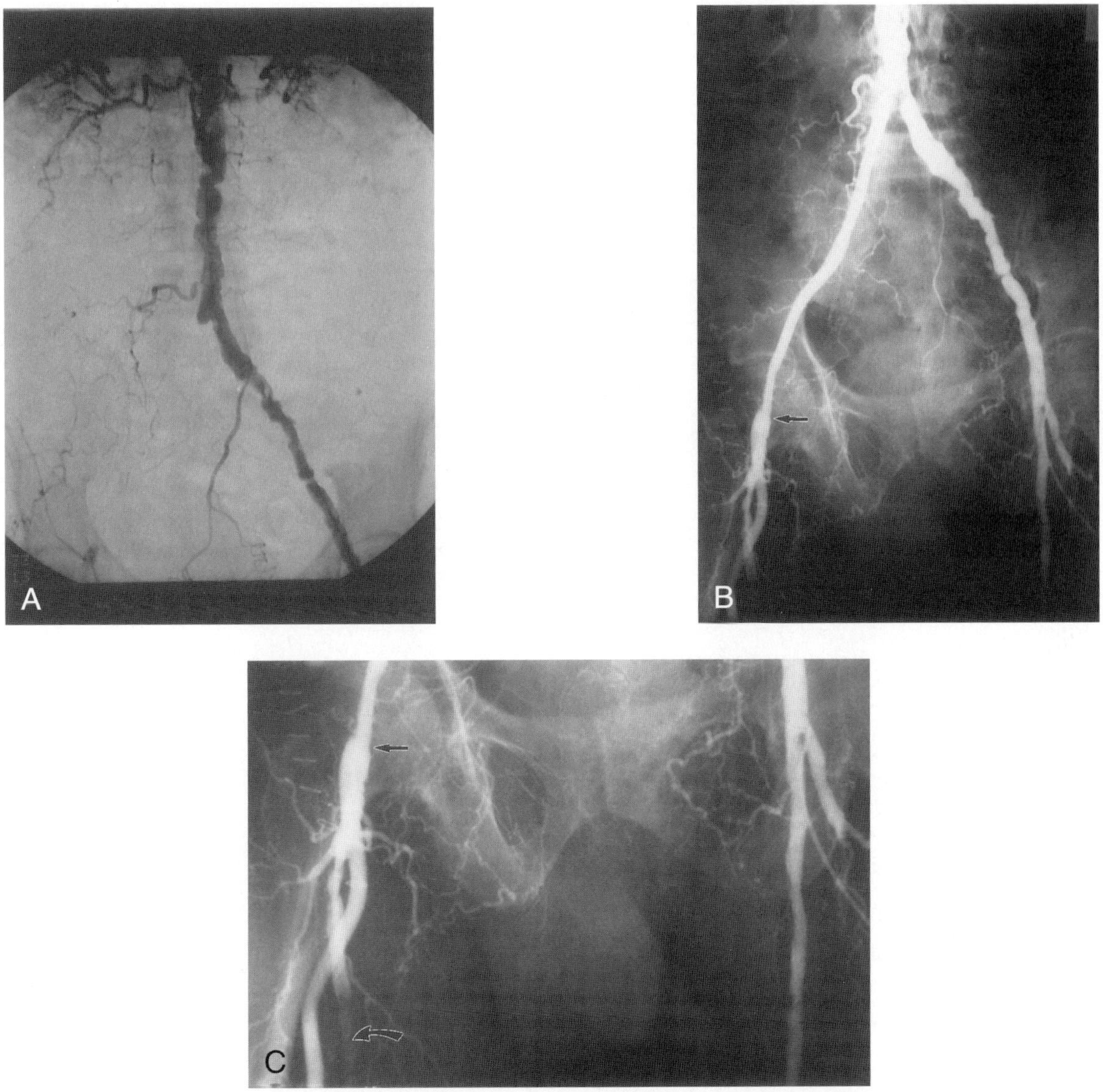

Figure 14.10. A 77-year-old hypertensive man on hemodialysis for end-stage renal disease presented with gangrene of his right foot. **A.** Diagnostic arteriography revealed an occlusion of the entire right iliac, common, and superficial femoral arteries (not shown). The patient subsequently underwent placement of an endovascular graft between the right common iliac and the right common femoral arteries after the diseased segment had been recanalized and diffusely dilated. To treat the superficial femoral artery occlusion, a conventional PTFE femoropopliteal bypass was performed and proximally anastomosed to the femoral arteriotomy site. **B.** Postoperative angiogram demonstrating the iliofemoral endovascular graft and the femoropopliteal graft originating at the site of the common femoral arteriotomy (arrow). **C.** Close-up view of the femoral artery anastomotic segment. The technical result achieved by overlying an endoluminal anastomosis with the hood of a distal bypass graft (*straight arrow*) is shown. Note the occlusion in the superficial femoral artery (*curved arrow*).

tions of coronary artery disease. In addition, cardiac complications represent the most common sequelae in both the peri- and the postbypass period. Although still a matter of vigorous debate, recent series support reductions in morbidity and mortality when surgical patients can be managed with regional rather than general anesthesia (53–55).

Of necessity, the patient's comorbid conditions, especially the cardiac status, weigh significantly in choosing not to undertake aortobifemoral grafting and, rather, to proceed with an alternative procedure.

Similarly, patients with severe pulmonary disease may be denied aortobifemoral grafting because of the need for endotracheal intubation and the anticipated inability to successfully wean the patient from mechanical ventilation. Other preoperative patient characteristics such as a "hostile abdomen" may contraindicate aortobifemoral grafting or even preclude patients from undergoing this procedure in the presence of a multiply reoperated abdomen, infection, stomas, or prior radiation treatment. Finally, a limited life expectancy based on age, comorbid medical conditions such as those previously discussed, or cancer may cement the decision not to pursue a conventional aortic approach to revascularization.

One of the most significant advantages of endovascular grafting is the ability to insert the graft from a groin

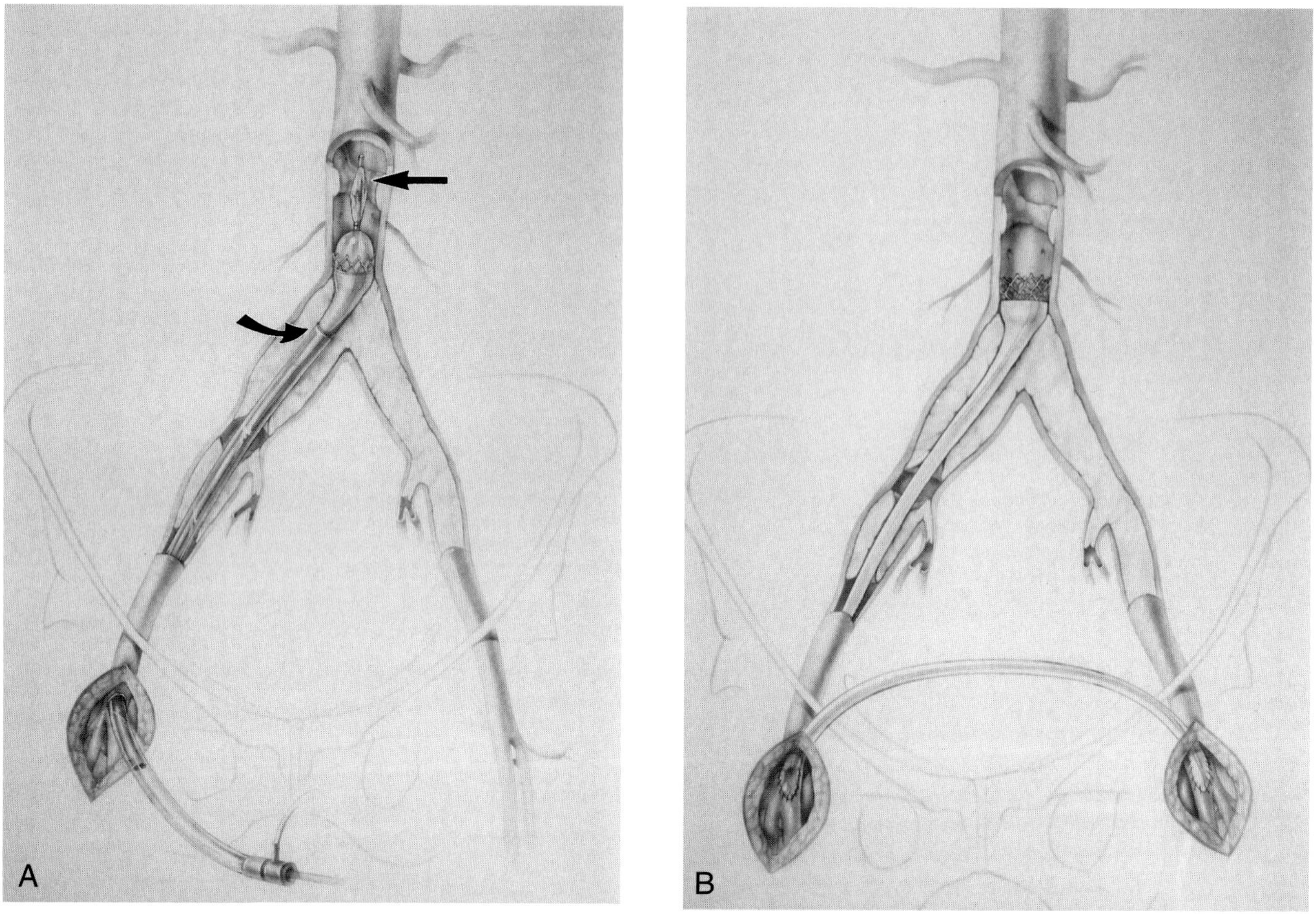

Figure 14.11. **A.** Infrarenal aortic endovascular grafting for the treatment of abdominal aortic aneurysms or possibly atheroembolic disease. The proximal aspect of the graft has been appropriately positioned in the infrarenal aorta. The tip balloon (*straight arrow*) has been deflated and the delivery sheath (*curved arrow*) withdrawn, permitting balloon expansion of the stent. **B.** After balloon expansion, the endovascular graft has been affixed to the aorta, where it can exclude an aneurysm or ulcerated plaque (not shown). The distal aspect of the graft has been endoluminally anastomosed to the common femoral artery and the femorofemoral graft has been fashioned. (Reprinted with permission from Ohki T, Marin ML, Veith FJ, et al. Endovascular aortounifemoral grafts and femorofemoral bypass for bilateral limb-threatening ischemia. J Vasc Surg 1996;24:984–997.)

Table 14.1. Comparison of the Advantages of Aortobifemoral Bypass Grafting With Those of Endovascular Grafting

Advantage	Aortobifemoral grafts	Endovascular grafts
Widespread applicability	Yes	Yes
Acceptable long-term patency	Yes	?
Can treat deep femoral artery lesions	Yes	Yes
Can treat renal artery lesions	Yes	No
Treatment of aortic atheroembolic disease	Yes	Yes
Treatment of all infrarenal aortic occlusions	Yes	No
Treatment of bilateral disease	Yes	Yes
Preservation of blood flow to major aortic branches	Yes	Yes
Aortic inflow	Yes	Yes

cut-down under regional or local anesthesia. In our series, we were able to avoid general anesthetics in nearly 80% of the patients treated. This allowed us to minimize both the cardiac and the pulmonary stressors attendant with general anesthesia. We believe the avoidance of general anesthesia significantly contributed to our low rate of morbidity. Patients who might be denied an aortic procedure because of contraindications to general anesthesia can safely undergo an endovascular procedure made even less stressful because aortic cross-clamping is unnecessary.

Finally, because endovascular grafting appears to be safe in high-risk patients such as those we have treated, one could make the decision to proceed with such a procedure in a patient with a limited life expectancy in whom aortobifemoral grafting would not be entertained.

A hostile abdomen is not a contraindication to endovascular grafting as it might be for an abdominal procedure. Access to the femoral vessels through the groin is required, but a trans- or retroperitoneal approach to the aorta is unnecessary. Patients with multiple prior abdominal procedures, multiple stomas, or other inflammatory conditions can be safely treated endoluminally.

In our series of patients treated for aortoiliac occlusive disease, 31% of the patients had at least one previous abdominal dissection, scarring, or infection in the region of the vascular lesion, which could have complicated or precluded an open repair.

Long-term follow-up of patients undergoing aortobifemoral grafting has revealed a significant rate of secondary and even tertiary procedures that were performed to address complications of the primary procedure. Most often, these procedures were performed for either graft thrombosis, an anastomotic pseudoaneurysm, or, less frequently, graft infection. Up to 20% of patients may experience a graft limb occlusion over a 10-year period (56), 5% an anastomotic aneurysm (usually at the femoral limb of the graft), and less than 2% an aortic graft infection.

In reviewing a series of 1647 patients who had operative treatment for aortoiliac occlusive disease, Szilagyi and associates (49) found that 26% of this considerable group of patients who underwent aortobifemoral grafting required a secondary repair to treat one of the aforementioned late complications. Although good results in terms of continued graft patency were achieved, the need for an additional hospitalization and procedure must be recognized as a limitation of the primary intervention.

Patients undergoing endovascular grafting can also be expected to have their course complicated by graft thrombosis. The presence of disease progression distal to the bypass graft and compromised outflow are factors common to both aortobifemoral grafting and endovascular grafting that cannot be prevented. Long-term endovascular graft follow-up is necessary before significant numbers of graft occlusions would be expected to occur to facilitate a meaningful comparison. The long-term effect of diffuse dilatation of an arterial segment is also unknown. It is conceivable that this maneuver may lead to a high rate of arterial recoil or progressive disease within the native artery, both of which may cause constriction of the endovascular graft and a hemodynamically significant stenosis or occlusion.

One would not expect to see the complication of anastomotic pseudoaneurysms with endovascular grafts, because in most cases the anastomoses are endoluminal. The graft-anastomotic factors leading to false aneurysms would not be expected to occur at either the endoluminal stented proximal anastomosis or the distal limb of the graft, which is endoluminally sutured to the intimal side of the femoral artery.

Blood-borne factors that may predispose a patient to graft infection cannot be avoided in patients being treated endovascularly. However, some protection would likely be afforded to patients whose graft infections are secondary to an intra-abdominal process, because the endovascular grafts are deployed entirely within a native vessel and do not penetrate the peritoneum.

A potential limitation of aortobifemoral grafting is that there is a subgroup of patients whose symptoms will not be relieved by this procedure and will require another bypass to treat multilevel disease. In a series of 812 patients undergoing either bilateral aortoiliac or bilateral aortofemoral bypasses, Crawford and colleagues (8) reported that 108 (13%) required femoropopliteal bypassing within 30 days of the first procedure. The risk involved in undertaking the second major procedure within a short time frame compounds the potential jeopardy to the patient.

In our group of patients who were treated for limb-threatening situations rather than for claudication, an additional distal bypass was necessary in 20% of all cases. This was uniformly performed in conjunction with endovascular grafting and required only a modest increase in procedure time rather than a second visit to the operating room. In addition, patients requiring a bypass graft to the popliteal level may be candidates for a second endovascular procedure. The distal end of a second graft can be stent-anastomosed in a minimally invasive fashion to the popliteal artery, and the two endovascular grafts can be extraluminally anastomosed in a conventional fashion.

We have also performed extraluminal distal anastomoses at the popliteal level for grafts exiting the common femoral arteriotomy site in the face of severe superficial and deep femoral artery disease (Fig. 14.12).

Although somewhat difficult to adequately quantify, the extensive exposure and mobilization of the peritoneal surface that are necessary to perform an aortic bypass procedure may lead to postoperative disturbances in sexual function after damage to pelvic autonomic nerves. Complicating the efforts to truly define the frequency of this development is the incidence of patients with preoperative dysfunction and the considerable contribution of hypogastric and pelvic blood flow to sexual function. The change in pelvic blood flow postoperatively may either compromise or, in some cases, improve overall function (57). Brewster (50) has suggested a 15% to 25% rate of postoperative sexual dysfunction in patients undergoing aortobifemoral grafting. We have not extensively addressed this issue in our patients who have undergone endovascular grafting for aortoiliac occlusive disease. However, sacrifice of hypogastric blood flow attendant to the endovascular graft procedure may offset the benefit of not performing a dissection around the pelvic autonomic nerves. More detailed study is necessary to indicate which procedure is better suited to the

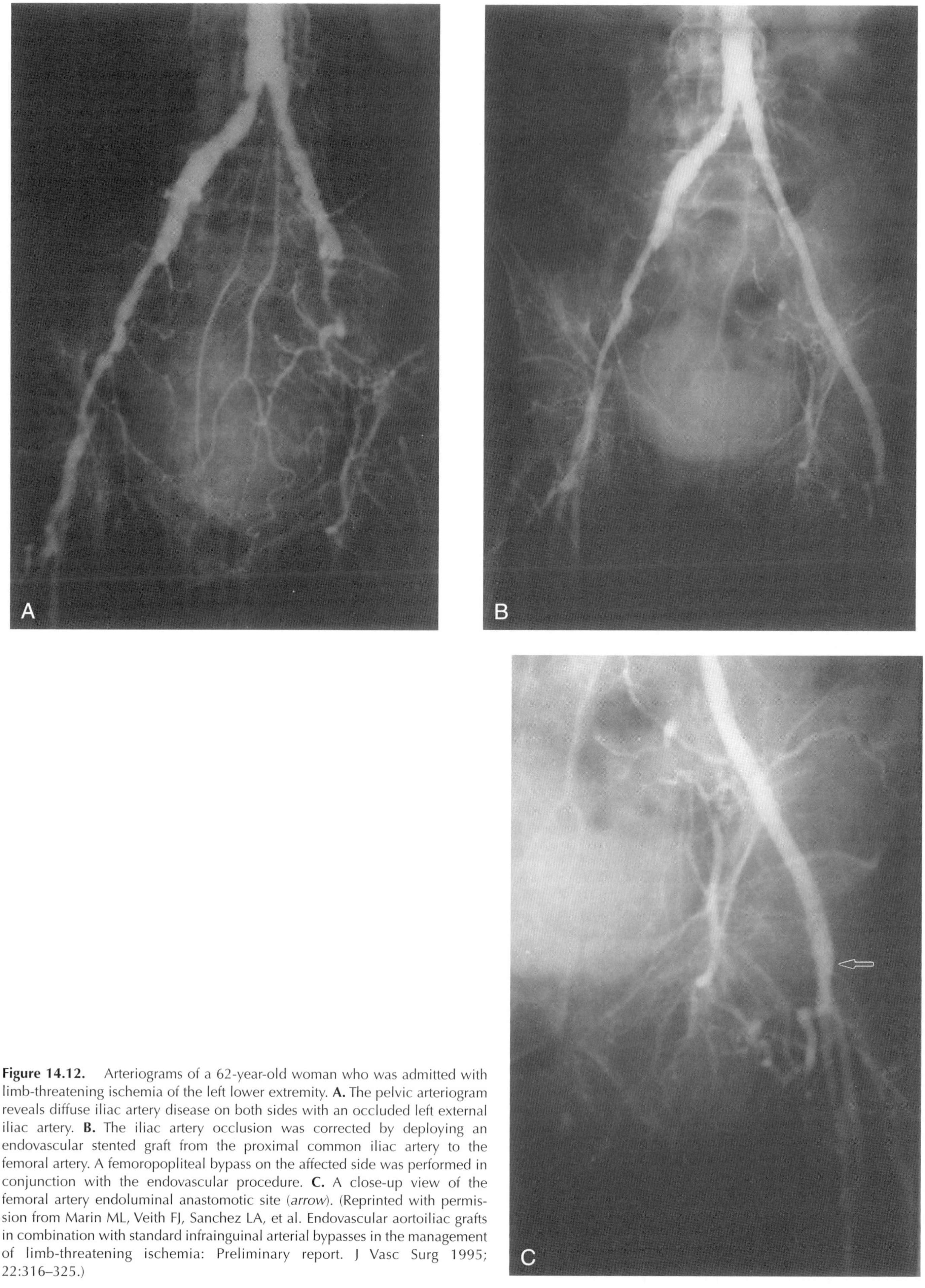

Figure 14.12. Arteriograms of a 62-year-old woman who was admitted with limb-threatening ischemia of the left lower extremity. **A.** The pelvic arteriogram reveals diffuse iliac artery disease on both sides with an occluded left external iliac artery. **B.** The iliac artery occlusion was corrected by deploying an endovascular stented graft from the proximal common iliac artery to the femoral artery. A femoropopliteal bypass on the affected side was performed in conjunction with the endovascular procedure. **C.** A close-up view of the femoral artery endoluminal anastomotic site (*arrow*). (Reprinted with permission from Marin ML, Veith FJ, Sanchez LA, et al. Endovascular aortoiliac grafts in combination with standard infrainguinal arterial bypasses in the management of limb-threatening ischemia: Preliminary report. J Vasc Surg 1995; 22:316–325.)

Table 14.2. Comparison of the Limitations of Aortobifemoral Bypass Grafting With Those of Endovascular Grafting

Limitation	Aortobifemoral Grafts	Endovascular Grafts
May be contraindicated in the presence of:		
Comorbid medical conditions	Yes	No
Hostile abdomen	Yes	No
Limited life expectancy	Yes	No
Need for general anesthesia	Yes	No
Complications:		
Graft limb thrombosis	Yes	Yes
Anastomotic aneurysms	Yes	No
Aortic graft infections	Yes	?
Second procedure needed to treat multilevel occlusive disease	Yes	No
Postoperative sexual dysfunction	Yes	?

preservation of postoperative sexual function. A comparison of the limitations of aortobifemoral grafting and endovascular grafting is given in Table 14.2.

Axillobifemoral and Femorofemoral Grafting

Advantages

The most commonly performed extra-anatomic procedures are the femorofemoral and the axillobifemoral bypasses. These procedures were initially created to treat patients with infected aortic grafts and those who were felt to be a prohibitive operative risk based on a hostile surgical field or the presence of severe, coexisting medical problems. The most important advantage of femorofemoral and axillobifemoral bypass grafting is that these procedures do not require an abdominal procedure with its concomitant risks.

Perhaps the most commonly invoked reason for using an extra-anatomic bypass is the high-risk patient. A femorofemoral bypass can be expeditiously performed under local anesthesia in a patient with limb-threatening ischemia who may have coexisting medical problems that directly contraindicate general anesthesia. In addition, although usually performed under general anesthesia, axillobifemoral bypass grafting does not present the patient with the magnitude of hemodynamic stressors attendant with an aortic procedure. Patients with severe cardiac, pulmonary, or renal disease constitute the majority of patients undergoing placement of such grafts. In addition, these procedures may be desirable in morbidly obese patients or in patients with a limited life expectancy.

Because femorofemoral and axillobifemoral bypass grafting have been relegated to the treatment of a high-risk group of patients, comparison of operative mortality with aortobifemoral grafting is probably unjust. However, a large number of series confirm an acceptable operative mortality rate between 0% and 8% in these difficult to manage patients (58–64).

Similarly, we have demonstrated that endovascular grafting can be performed with 0% perioperative mortality in patients with severe and often multiple coexisting disease processes. Until endovascular grafting for the treatment of aortoiliac occlusive disease has been widely embraced, it will continue to be used in a subset of patients similar to those who have undergone axillobifemoral bypass grafting in past years. Such high-risk patients may be ideal for this procedure because of its minimally invasive nature and because general anesthesia is not mandatory.

The usefulness of axillobifemoral and femorofemoral bypass grafting is perhaps best realized in the patient who has undergone many prior abdominal procedures, has abdominal wall stomas, or has undergone radiation therapy to an intraperitoneal structure. Most surgeons would pursue these approaches in the face of an infected aortic graft over an in situ salvage procedure.

As already discussed, endovascular grafting is also ideal for patients in whom a direct approach to the aorta is thought to be of prohibitive risk. Endovascular grafting does not currently play a role in the treatment of patients with infected aortic grafts, and extra-anatomic grafting will likely remain the gold standard for managing this difficult problem.

Finally, extra-anatomic bypasses may be used to treat patients with aortoiliac occlusive disease and a failed aortic bypass whose operation would be riskier in the redo setting and who might now have progressive systemic disease that would preclude major aortic reconstruction.

Endovascular grafting may become the procedure of choice in the treatment of failed aortic bypasses or in patients who have experienced complications of aortobifemoral grafting. In these patients, endovascular grafts may be deployed after recanalization and balloon dilatation of a chronically thrombosed bifurcation graft limb has been performed. Thrombectomy is often ineffective in this setting, and the endovascular graft can be placed within the previous graft limb so that salvage of graft function can be achieved and the need for a redo abdominal procedure or an extra-anatomic bypass can be avoided. Similarly, if a previous graft has thrombosed to the level of the aorta, direct aortic inflow can be reestablished and endovascular grafting can be performed for lower extremity salvage without resulting to alternative, more invasive, measures.

Limitations

The most frequently cited limitation of extra-anatomic approaches to lower extremity revascularization has been sacrificed patency when compared with aortobifemoral grafting. A direct comparison between the two groups is difficult to undertake because of selection biases in patients undergoing the two procedures. This comparison is further complicated by the large cohort of patients undergoing aortobifemoral grafting for claudication who are obviously healthier than most of the patients who have undergone axillobifemoral or femorofemoral bypass grafting for limb-threatening ischemia. Patency rates are obviously skewed in favor of the former group. Large studies undertaken in the 1970s and 1980s cite 5-year primary patency rates between 40% and 85% for axillobifemoral bypass grafting, with the overwhelming trend toward poorer primary patency rates than those for aortobifemoral grafting (61,64–67).

Schneider and associates (68) recently compared low-risk patients receiving either femorofemoral or aortobifemoral bypasses and found 3-year primary patency rates of 61% and 87%, respectively, in favor of the latter procedure. Again, our initial patency results compare favorably with those obtained in large series of patients undergoing axillobifemoral and femorofemoral bypass grafting, but longer follow-up in a larger cohort of patients is necessary before lasting conclusions can be made.

Some authors have produced results showing that inferior hemodynamic performance of extra-anatomic grafts may contribute to decreased patency and ineffective resolution of symptomatology (68–70). The inferior hemodynamics may be due to graft length for axillobifemoral bypasses and to inflow artery characteristics, such as the size and presence of flow-limiting lesions in the axillary or femoral vessels. In fact, patients may be denied one or the other type of extra-anatomic bypass in the presence of significant donor vessel disease (Fig. 14.13). Endovascular grafting revascularizes the lower extremities with direct aortic inflow, with none of the hemodynamic limitations that may be seen when more peripheral smaller-caliber vessels such as the axillary and femoral arteries are used. Progression of disease in the femoral or axillary segment may have a heavy impact on distal flow, whereas changes in aortic inflow would not be expected to occur to a similar extent.

The overall length of the endovascular graft is considerably shorter than that required for axillobifemoral grafting, and we currently use up to 8-mm diameter grafts to limit the effects of graft resistance.

Another potential disadvantage of axillobifemoral and femorofemoral bypass grafting is the use of an unaffected extremity as inflow for the diseased limb. These approaches put an additional extremity at risk for procedure-related complications such as distal embolization, graft infection, perigraft seroma, steal syndrome, and structural failure of the graft. Although the incidence of complications is relatively low, devastating consequences, including amputation of a previously unaffected limb, may result. Endovascular grafts originate in vessels that themselves harbor disease and do not put unaffected limbs at risk.

The treatment of extra-anatomic graft thrombosis as well as any of the aforementioned complications contributes to a high rate of secondary and tertiary proce-

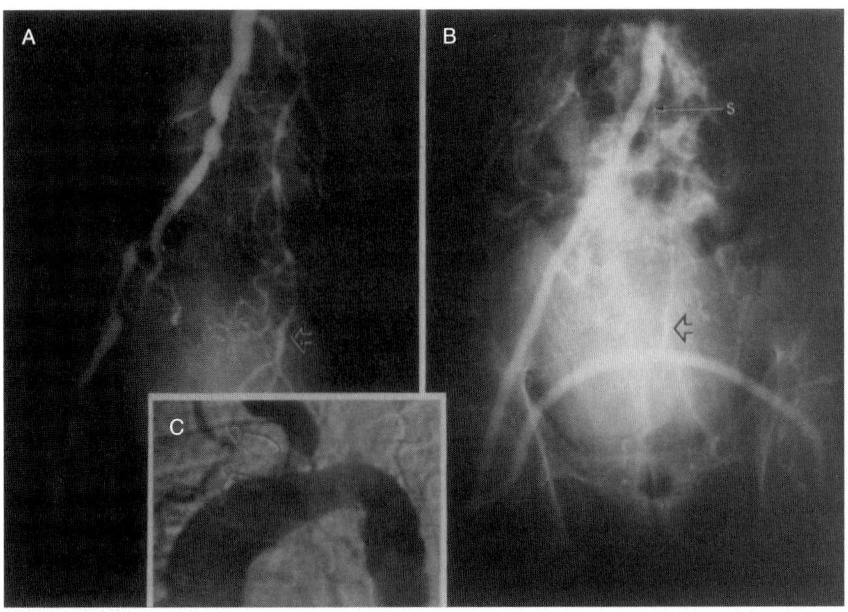

Figure 14.13. **A.** Preoperative arteriogram demonstrating a diffusely diseased right iliac artery and an occluded left common iliac artery. The pelvic circulation is maintained via the inferior mesenteric artery (*open arrow*). **B.** Completion arteriography demonstrates the right iliofemoral and femorofemoral grafts. Note that the stent (*S*) is above the lesion (within the distal aorta). The pelvic circulation is maintained via the inferior mesenteric artery (*open arrow*). **C.** The patient's preoperative arch arteriogram demonstrated a 90% stenosis at the origin of the innominate artery and occluded left common carotid and subclavian arteries, which prohibited axillofemoral grafting. (Reprinted with permission from Ohki T, Marin ML, Veith FJ, et al. Endovascular aortounifemoral grafts and femorofemoral bypass for bilateral limb-threatening ischemia. J Vasc Surg 1996;24:984–997.)

Table 14.3. Comparison of the Advantages and Limitations of Extra-anatomic Bypasses With Those of Endovascular Grafting

	Extra-anatomic bypasses	Endovascular grafts
Advantages		
Avoids a "hostile" abdominal field	Yes	Yes
Treatment of infected aortic grafts	Yes	No
Salvage of failed aortic reconstructions	Yes	Yes
Less stressful in high-risk patients	Yes	Yes
Limitations		
Compromised hemodynamic performance vs aortobifemoral bypass grafting	Yes	No
Potential for donor vessel unsuitability for inflow source	Yes	No
Places normal donor vessel at risk	Yes	No
High rate of subsequent procedures to treat complications	Yes	?

dures in patients undergoing axillobifemoral and femorofemoral bypass grafting. Reoperation rates have been reported in up to 15% to 44% of all patients in several study groups (71–73). Reoperation after endovascular grafting has been necessary; however, the limited number of cases does not lend itself to comparison with existing techniques (Table 14.3).

BALLOON ANGIOPLASTY AND PERCUTANEOUS STENTING

Advantages

Although balloon angioplasty and stenting are not equivalent to aortobifemoral, axillobifemoral, or femorofemoral bypass grafting in applicability to all types of aortoiliac occlusive disease, they are nonetheless important both as stand-alone procedures and as adjuncts to other procedures. Angioplasty followed by stenting is the procedure of choice for hemodynamically significant lesions in the common iliac arteries that are of relatively short length. In addition, as with endovascular grafting, balloon angioplasty and stenting can make feasible an infrainguinal or femorofemoral bypass, which in the past would have mandated aortic or axillary inflow followed by a second operative procedure.

Perhaps the biggest attraction to balloon angioplasty and stent placement is their minimally invasive nature. These procedures can be undertaken in an angiography suite or a suitably equipped operating room with the patient under local anesthesia and in a percutaneous fashion. This approach is seldom associated with major systemic complications of the magnitude of myocardial infarction, pulmonary embarrassment, or coagulation derangements. In addition, the intervention may be applied at a site distant from the vessel that is used for access. Lesions such as those at the aortic bifurcation and iliac vessels that would require considerable effort to access operatively may be approached via a percutaneous groin puncture. Such techniques are desirable in patients with suitable disease anatomy who have contraindications to a transperitoneal or retroperitoneal approach to the aortoiliac vessels.

Although endovascular grafting is considerably less invasive than conventional operative approaches, operative exposure of the femoral vessels is still required, and a hand-sewn distal anastomosis is usually performed. In the future, technologic refinements may allow endovascular grafting to be performed via an entirely percutaneous approach in patients with anatomically favorable disease patterns.

Primary patency rates for balloon angioplasty and stenting in the treatment of stenotic lesions in the iliac arteries are between 80% and 95% at 2 years (24,26,74,75). Recently, primary stenting of the iliac vessels has been performed after the recanalization of occluded arteries. If results continue to be favorable, occlusive lesions formerly amenable only to operative bypass may become an indication for nonoperative intervention.

Primary patency rates of 78% and secondary patency rates approaching 86% were recently reported by Vorwerk and colleagues (76) when treating occluded iliac artery segments. As we have seen, endovascular grafting is also ideally suited for the treatment of occlusive iliac disease and enjoys initial patency rates comparable to balloon angioplasty and stenting.

Limitations

A patient's general medical condition may dictate the type of operative approach that is undertaken, but the pattern of atherosclerotic disease most commonly affords the patient an attempt at balloon angioplasty and stenting.

Only a small percentage of the patients seen by a vascular surgeon are likely to have ideal indications for stent placement. Most often, long-segment or multilevel disease is encountered, and the results of balloon angioplasty and stenting prove less durable. Two-year primary patency rates are poor in the femoropopliteal segment and average approximately 50% (77,78). Similarly, patency suffers when long-segment balloon angioplasty is undertaken or when multiple stents are required (16,79). In many cases, intimal hyperplasia and disease progression proximal and distal to the stented lesion are responsible for the sacrificed patency.

In contrast to aortobifemoral and axillobifemoral bypass grafting in which disease progression in the iliac arteries is largely inconsequential, in balloon angioplasty and stenting progressive iliac disease can have limb-threatening consequences. These approaches treat intrasegmental arterial disease rather than the entire susceptible length of artery, such as occurs during an operative bypass procedure. In addition, intrastent hyperplasia is a recognized cause of both early and late treatment failure.

On the other hand, endovascular grafting has been successfully used to treat a wide range of patients with multilevel disease. The prosthetic graft traverses the entire iliac artery and covers both the lesion itself and the artery at potential risk for disease progression. Therefore, the progression of disease proximal to the endovascular graft cannot limit graft flow or facilitate graft thrombosis.

Intimal hyperplasia may also be of less concern in patients undergoing endovascular grafting, because the prosthetic graft might limit intraluminal cellular adherence and the coincident hyperplastic response when compared with uncovered stents.

Other relative contraindications to the placement of intra-arterial stents include severely calcified vessels, lesions coexisting with iliac artery aneurysmal disease, and contrast extravasation and arterial rupture after balloon dilatation. Some interventionalists shy away from difficult stent and balloon angioplasty procedures because of the possibility of incurring a higher than expected rate of complications and instead refer the patient for surgery.

An overall complication rate of 10% after stenting is commonly reported and includes groin and retroperitoneal hematomas, allergic reactions to the contrast dye, and renal failure. Of particular interest to the surgeon is the incidence of pseudoaneurysm formation, puncture site laceration, and subintimal dissection, all of which may require operative repair. Distal embolization, which is more likely to occur when intervention is being undertaken for an occluded vessel, can be particularly troublesome.

Extensive calcification has not precluded endovascular graft placement in any of the patients we have treated despite the presence of circumferential calcific deposits in several patients. Arterial rupture and contrast extravasation complicating balloon dilatation are readily treated after device deployment and prosthetic graft coverage of the site of injury. Complications after endovascular grafting are similar to those experienced with balloon angioplasty and stenting, but the rate of distal embolization can be expected to be less for the former procedure because surgical control of femoral artery branches limits the potential for distal emboli. This built-in safety mechanism does not exist during percutaneous procedures (Table 14.4).

Table 14.4. Comparison of the Advantages and Limitations of Balloon Angioplasty and Stenting With Those of Endovascular Grafting

	Balloon angioplasty and stenting	Endovascular grafting
Advantages		
Allows for infrainguinal or femorofemoral procedure	Yes	Yes
Minimal systemic stressor	Yes	Yes
Access site different from site of pathology	Yes	Yes
Can be performed under local or regional anesthesia	Yes	Yes
Can treat iliac artery occlusions	Yes	Yes
Percutaneous approach	Yes	No
Limitations		
Limited indications	Yes	No
Poor results with long-segment disease	Yes	No
Segmental disease progression of concern	Yes	No
Many relative contraindications	Yes	No
Distal embolization of major concern	Yes	No
Intrastent hyperplasia an issue	Yes	No

CONCLUSIONS

We have successfully used endovascular grafts in the treatment of a high-risk subset of patients with limb-threatening aortoiliac occlusive disease. The technique is certainly feasible and may offer significant advantages over conventional procedures for a wide range of patients suffering from aortoiliac and multilevel occlusive disease. Improvements in endovascular graft design and deployment techniques that will occur in the future should promote the viability of this evolving field and encroach upon the indications for which aortobifemoral and axillobifemoral bypass grafting have traditionally been performed.

References

1. dos Santos JC. Sur la desobstruction des thromboses artrievelles anciennes. Memoires de l'Academie de Chirurgie 1947;73: 409–411.
2. Wylie EJ. Thrombo-endarterectomy for arteriosclerotic thrombosis of major arteries. Surgery 1952;32:275–284.
3. Oudot J. La greffe vasculaire dans les thromboses du carrefour aortique. Presse Med 1951;59:234–236.
4. Dubost C, Allary M, Oeconomos N. Resection of an aneurysm of the abdominal aorta: Reestablishment of the continuity by a preserved human arterial graft, with result after five months. Arch Surg 1952;64:405–408.
5. Debakey ME, Creech O Jr, Cooley DA. Occlusive disease of the aorta and its treatment by resection and homograft replacement. Ann Surg 1954;140:290–310.
6. Julian OC, Dye WS, Olwin JH, Jordan PH. Direct surgery of arteriosclerosis. Ann Surg 1952;136:459–474.
7. Brewster DC, Darling RC. Optimal methods of aortoiliac reconstruction. Surgery 1978;84(4):739–748.
8. Crawford ES, Bomberger RA, Glaeser DH, et al. Aortoiliac occlusive disease: Factors influencing survival and function following reconstructive operation over a twenty-five-year period. Surgery 1981;90(6):1055–1067.
9. Nicolaides AN, Grigg MJ. Are there any indications for aortic and iliac endarterectomy? In: Greenhalgh RM, ed. Indications in vascular surgery. Philadelphia: WB Saunders, 1988.

10. Louw JH. Splenic-to-femoral and axillary-to-femoral bypass grafts in diffuse atherosclerotic occlusive disease. Lancet 1963;1: 401–402.
11. Blaisdell WF, Hall AD. Axillary-femoral artery bypass for lower extremity ischemia. Surgery 1963;54:563–568.
12. Dotter CT, Judkins MP. Transluminal treatment of arteriosclerotic obstruction: Description of a new technic and a preliminary report of its application. Circulation 1964;30:654–670.
13. Gruntzig A, Hopff, H. Perkutane Rekanalisation chronischer arterieller verschlusse mit einem neuen Dilatationskatheter: Modifikation der Dotter Technik. Deutsch Med Wochenschr 1974;99:2502–2511.
14. Tegtmeyer CJ, Hartwell GD, Selby JB, et al. Results and complications of angioplasty in aortoiliac disease. Circulation 1991; 83(Suppl I):I-53–I-60.
15. Kotb MM, Kadir S, Bennett JD, Beam CA. Aortoiliac angioplasty: Is there a need for other types of percutaneous intervention? J Vasc Intervent Radiol 1992;3:67–71.
16. Johnston KW. Iliac arteries: Reanalysis of results of balloon angioplasty. Radiology 1993;186:207–212.
17. Waltman AC, Greenfield AJ, Novelline RA, et al. Transluminal angioplasty of the iliac and femoropopliteal arteries. Arch Surg 1982; 117:1218–1221.
18. van Andel GJ, van Erp WFM, Krepel VM, Breslau PJ. Percutaneous transluminal dilatation of the iliac artery: Long-term results. Radiology 1985;156:321–323.
19. Johnston KW. Factors that influence the outcome of aortoiliac and femoropopliteal percutaneous transluminal angioplasty. Surg Clin North Am 1992;72(4):843–850.
20. Blankensteijn JD, van Vroonhoven TJ, Lampmann L. Role of percutaneous transluminal angioplasty in aorto-iliac reconstruction. J Cardiovasc Surg 1986;27:466–468.
21. Sigwart U, Puel J, Mirkovitch V, et al. Intravascular stents to prevent occlusion and restenosis after transluminal angioplasty. N Engl J Med 1987;316:701–706.
22. Gunther RW, Vorwerk D, Antonucci F. Iliac artery stenosis or obstruction after unsuccessful balloon angioplasty: Treatment with a self-expandable stent. AJR 1991;156:389–393.
23. Palmaz JC, Garcia OJ, Schatz RA, et al. Placement of balloon-expandable intraluminal stents in iliac arteries: First 171 procedures. Radiology 1990;174:969–975.
24. Long AL, Page PE, Raynaud AC, et al. Percutaneous iliac artery stent: Angiographic long-term follow-up. Radiology 1991;180:771–778.
25. Gunther RW, Vorwerk D, Bohndorf K, et al. Iliac and femoral artery stenoses and occlusions: Treatment with intravascular stents. Radiology 1989;172:725–730.
26. Palmaz JC, Laborde JC, Rivera FJ, et al. Stenting of the iliac arteries with the Palmaz stent: Experience from a multicenter trial. Cardiovasc Intervent Radiol 1992;15:291–297.
27. Wolf YG, Schatz RA, Knowles HJ, et al. Initial experience with the Palmaz stent for aortoiliac stenoses. Ann Vasc Surg 1993;7:254–261.
28. Parodi JC, Palmaz JC, Barone HD. Transfemoral intraluminal graft implantation for abdominal aortic aneurysms. Ann Vasc Surg 1991;5:491–499.
29. Marin ML, Veith FJ. Transfemoral repair of abdominal aortic aneurysm. N Engl J Med 1994;331:1751.
30. Laborde JC, Parodi JC, Clem MF, et al. Intraluminal bypass of abdominal aortic aneurysm: Feasibility study. Radiology 1992;184:185–190.
31. Parodi JC. Endovascular repair of abdominal aortic aneurysms and other arterial lesions. J Vasc Surg 1995;21:549–557.
32. Cragg AH, Dake MD. Percutaneous femoro-popliteal graft placement. Radiology 1993;187:643–648.
33. Pernes JM, August MA, Hovasse D, et al. Long iliac stenosis: Initial clinical experience with the Cragg endoluminal graft. Radiology 1995;196:67–71.
34. Henry M, Amor M, Ethevenot G, et al. Initial experience with the Cragg Endopro System 1 for intraluminal treatment of peripheral vascular disease. J Endovasc Surg 1994;1:31–43.
35. Marin ML, Veith FJ, Sanchez LA, et al. Endovascular repair of aortoiliac occlusive disease. World J Surg 1996;20:679–686.
36. Veith FJ, Marin ML. The present status of endoluminal stented grafts for the treatment of aneurysms, traumatic injuries and arterial occlusions. Cardiovasc Surg 1996;4(1):3–7.
37. Wain RA, Marin ML, Veith FJ, Levine BA. The Montefiore Medical Center experience with endovascular stented grafts. In: Szabo Z, Lewis JE, Fantini G, eds. Surgical technology international IV. San Francisco: Universal Medical Press, 1995.
38. Marin ML, Veith FJ, Cynamon J, et al. Transfemoral endovascular stented graft treatment of aorto-iliac and femoropopliteal occlusive disease for limb salvage. Am J Surg 1994;168:156–162.
39. Ohki T, Marin ML, Veith FJ, et al. Endovascular aortounifemoral grafts and femorofemoral bypass for bilateral limb-threatening ischemia. J Vasc Surg 1996;24:984–987.
40. Marin ML, Veith FJ. Preliminary experience with endovascular stented grafts for limb-threatening aortoiliac occlusive disease. In: Yao JST, Pearce WH, eds. The ischemic extremity: Advances in treatment. Norwalk, CT: Appleton & Lange, 1995.
41. Marin ML, Veith FJ, Sanchez LA, et al. Endovascular aortoiliac grafts in combination with standard infrainguinal arterial bypasses in the management of limb-threatening ischemia: Preliminary report. J Vasc Surg 1995;22:316–325.
42. Marin ML, Veith FJ. Clinical application of endovascular grafts in aortoiliac occlusive disease and vascular trauma. Cardiovasc Surg 1995;3(2):115–120.
43. Marin ML, Veith FJ. The role of stented grafts in the management of failed arterial reconstructions. Semin Vasc Surg 1994;7:188–194.
44. Marin ML, Veith FJ, Cynamon J, et al. Initial experience with transluminally placed endovascular grafts for the treatment of complex vascular lesions. Ann Surg 1995;222:449–469.
45. Brewster DC, Perler BA, Robison JG, Darling RC. Aortofemoral graft for multilevel occlusive disease: Predictors of success and need for distal bypass. Arch Surg 1982;117:1593–1600.
46. Malone JM, Moore WS, Goldstone J. The natural history of bilateral aortofemoral bypass grafts for ischemia of the lower extremities. Arch Surg 1975;110:1300–1306.
47. Baird RJ, Feldman P, Miles JT, et al. Subsequent downstream repair after aorta-iliac and aorta-femoral bypass operations. Surgery 1977;82:785–793.
48. Crawford ES, Manning LG, Kelly TF. "Redo" surgery after operations for aneurysm and occlusion of the abdominal aorta. Surgery 1977;81:41–52.
49. Szilagyi DE, Elliott JP, Smith RF, et al. A thirty-year survery of the reconstructive surgical treatment of aortoiliac occlusive disease. J Vasc Surg 1986;3:421–436.
50. Brewster DC. Clinical and anatomical considerations for surgery in aortoiliac disease and results of surgical treatment. Circulation 1991;83(Suppl I):I-42–I-52.
51. Kempczinski RF. Lower extremity arterial emboli from ulcerating atherosclerotic plaques. JAMA 1979;241:807–810.
52. Malone JM, Moore WS, Goldstone J. Life expectancy following aortofemoral arterial grafting. Surgery 1977;81:551–555.
53. Baron J, Bertrand M, Barre E, et al. Combined epidural and general anesthesia versus general anesthesia for abdominal aortic surgery. Anesthesiology 1991;75:611–618.
54. Christopherson R, Beattie C, Frank SM, et al. Perioperative morbidity in patients randomized to epidural or general anesthesia for lower extremity vascular surgery. Anesthesiology 1993;79:422–434.
55. Yeager MP, Glass DD, Neff RK, Brinck-Johnsen T. Epidural anesthesia and analgesia in high-risk surgical patients. Anesthesiology 1987;66:729–736.
56. Brewster DC, Meier GH, Darling RC, et al. Reoperation for aortofemoral graft limb occlusion: Optimal methods and long-term results. J Vasc Surg 1987;5:303–374.
57. Queral LA, Whitehouse WM Jr, Flinn WR, et al. Pelvic hemodynamics after aortoiliac reconstruction. Surgery 1979;86: 799–809.
58. Brief DK, Brener B, Alpert J, Parsonnet V. Crossover femorofemoral grafts followed up five years or more: An analysis. Arch Surg 1975;110:1294–1299.
59. Flanigan P, Pratt DG, Goodreau JJ, et al. Hemodynamic and angiographic guidelines in selection of patients for femorofemoral bypass. Arch Surg 1978;113:1257–1262.
60. Livesay JJ, Atkinson JB, Baker JD, et al. Late results of extraanatomic bypass. Arch Surg 1979;114:1260–1267.
61. Rutherford RB, Patt A, Pearce WH. Extra-anatomic bypass: A closer view. J Vasc Surg 1987;5:437–446.
62. LoGerfo FW, Johnson WC, Corson JD, et al. A comparison of the late patency rates of axillobilateral femoral and axillounilateral femoral grafts. Surgery 1977;81:33–40.

63. Ray LI, O'Connor JB, Davis CC, et al. Axillofemoral bypass: A critical reappraisal of its role in the management of aortoiliac occlusive disease. Am J Surg 1979;138:117–128.
64. Eugene J, Goldstone J, Moore WS. Fifteen-year experience with subcutaneous bypass grafts for lower extremity ischemia. Ann Surg 1976;286:177–183.
65. Chang JB. Current state of extraanatomic bypasses. Am J Surg 1986;152:202–205.
66. Ascer E, Veith FJ, Gupta SK, et al. Comparison of axillounifemoral and axillobifemoral bypass operations. Surgery 1985;97: 169–175.
67. Harris EJ, Taylor LM, McConnell DB, et al. Clinical results of axillobifemoral bypass using externally supported polytetrafluoroethylene. J Vasc Surg 1990;12:416–420.
68. Schneider JR, Besso SR, Walsh DB, et al. Femorofemoral versus aortobifemoral bypass: Outcome and hemodynamic results. J Vasc Surg 1994;19:43–57.
69. Self SB, Richardson JD, Klamer TW, et al. Utility of femorofemoral bypass: Comparison of results with indications for operation. Am Surg 1991;57:602–606.
70. Schneider JR, McDaniel MD, Walsh DB, et al. Axillofemoral bypass: Outcome and hemodynamic results in high-risk patients. J Vasc Surg 1992;15:952–963.
71. Taylor L, Moneta GL, McConnell D, et al. Axillofemoral grafting with externally supported polytetrafluoroethylene. Arch Surg 1994;129:588–595.
72. Calvelo A, Collado G, Masegosa A, et al. Value of secondary repair in axillo-femoral bypass grafting. (Abstract). J Cardiovasc Surg 1988;29(Suppl):10.
73. Donaldson MC, Louras JC, Bucknam CA. Axillofemoral bypass: A tool with a limited role. J Vasc Surg 1986;3:757–763.
74. Strecker EP, Hagen B, Liermann D, et al. Iliac and femoropopliteal vascular occlusive disease treated with flexible tantalum stents. Cardiovasc Intervent Radiol 1993;16:158–164.
75. Liermann D, Strecker EP, Peters J. The Strecker stent: Indications and results in iliac and femoropopliteal arteries. Cardiovasc Intervent Radiol 1992;15:298–305.
76. Vorwerk D, Guenther RW, Schumann K, et al. Primary stent placement for chronic iliac artery occlusions: Follow-up results in 103 patients. Radiology 1995;194:745–749.
77. Henriksen LO, Jorgensen B, Holstein PE, et al. Percutaneous transluminal angioplasty of infrarenal arteries in intermittent claudication. Acta Chir Scand 1988;154:573–576.
78. Johnston KW. Reanalysis of the results of balloon angioplasty of the femoral and popliteal arteries. Radiology 1992;183:767–771.
79. Sapoval MR, Chatellier G, Long AL, et al. Self-expandable stents for the treatment of iliac artery obstructive lesions: Long-term success and prognostic factors. AJR 1996;166:1173–1179.

Chapter 15

Endoluminal Grafting for Femoropopliteal Occlusive Disease

Edward B. Diethrich

Femoropopliteal (FP) occlusive disease remains one of the most difficult vascular maladies to treat. Although balloon angioplasty, atherectomy, and stenting are successful in providing acute results in over 90% of patients with FP disease, long-term patency with percutaneous intervention is elusive. Classic FP bypass grafting results in long-term patency in roughly two thirds of patients, and, as yet, success rates of the aforementioned endoluminal techniques are not able to equal these statistics (1–3).

Unfortunately, early results indicate that patients with FP disease do not always experience the benefits we usually associate with percutaneous intervention. Generally, we expect less invasive techniques to result in shorter hospital stays and a more rapid return to normal activities. In patients with FP disease, however, these benefits are not realized as often as they are in patients with disease in other anatomic regions. It is unfortunate that percutaneous intervention has not been more successful in these patients. Not only have they required more invasive treatment, but they have also been deprived of one of the major benefits of percutaneous revascularization—early intervention.

Our search for the reasons why less invasive methods of treatment for FP disease have failed to provide good long-term results has led us to firmly place the blame on a biologic phenomenon called "restenosis." It is now clear that rapid, cellular proliferation contributes to anastomotic stenosis, the most common cause of FP graft failure. Part of the struggle to understand the concept of the myoproliferative response to injury caused by angioplasty and the subsequent intimal hyperplasia has led us to realize that new devices and techniques may be important in discovering ways to improve long-term luminal patency.

Many investigators now believe that one way to discourage cellular growth in the intima is to provide a new artificial intraluminal surface that resists the proliferation of cells. This idea has prompted the development of a new device commonly referred to as an endoluminal graft (ELG) or covered stent. Lining the diseased arterial wall with biomaterials that exclude contact with blood may protect the lumen from the cellular overgrowth that results in constriction and restenosis. In addition, placement of the device effectively eliminates the conventional anastomotic junction—a site of inherent flow disturbances. The incidence of graft thrombosis may also be decreased.

The concept of intraluminal resurfacing is not a particularly new one, although it is certainly true that an abundance of new biomaterials and devices has encouraged the development of a number of different ELGs. Dotter's 1969 proposal for a transcatheter endarterial tube graft paved the way for modern ELG designs nearly 30 years ago (4), but the first attempt at endoluminal grafting was not realized until 1983. At that time, Dotter (5) deployed loosely wound nitinol coils in canine limbs; exclusion of abdominal aortic aneurysms using tightly wound nitinol coils was also demonstrated in a dog model by another investigator in the same year (6). Later, a nitinol stent covered with polyurethane tubing was implanted in sheep aortas (7), and Russian investigators used an ELG to treat iliac occlusive disease in a clinical setting (8). Not until the beginning of this decade did Parodi and colleagues (9) begin using a Dacron tube graft-Palmaz stent device for aneurysm exclusion in clinical applications.

Since the first clinical experiences with endoluminal grafting, rapid development of the technique has ensued. Approximately 24 ELG configurations have been designed and are being investigated around the world.

In Europe, a modified endograft is about to be tested in clinical trials; the results achieved with a similar prototype used in the superficial femoral artery have been favorable. The system comprises design changes that have evolved from a variety of experiences.

The device to be tested in the European trials will be assembled and packaged for immediate delivery through a 9F introducer sheath. Deployment in the superficial femoral artery does not require the use of an introducer sheath; this feature is optional. Polytetrafluoroethylene (PTFE) tubing will come in 5-cm increments from 5 cm to 30 cm long, and the expansion capability will range from 4 mm to 9 mm. Nitinol stents will be preinserted in the ends of the endograft. No serial dilations are required throughout the length of the graft because the endograft is mounted on a newly designed balloon that is the same length as the graft. The new delivery balloons have a very low profile after deflation and are easy to retract. Silicone sleeves are affixed to the catheter at the endograft terminals and secure the endograft to the catheter. These sleeves also facilitate adequate balloon inflation and release of the endograft.

At the Arizona Heart Institute, we have developed a novel ELG to reline the iliac and FP arteries (10–14). In our prototype design, which may be used in limb arteries of any length, a small-diameter, radially expandable tube graft combines with Palmaz stents for percutaneous delivery and deployment over occlusive lesions or aneurysms.

ELG DESIGN USED AT THE ARIZONA HEART INSTITUTE

Constructed of an ultrathin PTFE tube graft material, the Arizona Heart Institute's ELG device allows radial expansion via balloon dilation. The extruded material is available in lengths up to 40 cm; the tube grafts are supplied in unpredilated diameters of either 2 mm or 3 mm. The 2-mm tubes, which expand to a maximum of 8 mm, are used in the infrainguinal arteries; the 3-mm models (expanding to 15 mm) are appropriate for iliac applications. Grafts are currently available in incremental 5-cm lengths up to 40 cm and may be customized as needed for each lesion at the time of the procedure (Fig. 15.1).

A standard Palmaz stent (Johnson & Johnson Interventional Systems, Warren, NJ) is used for fixation of the ELG. The dimensions of the native artery are used to determine the selection of stents for distal and proximal attachment. Two different stent lengths are commonly used; the 29-mm or the 20-mm stents offer expansion ranges from 4 mm to 9 mm (models P294 and P204,

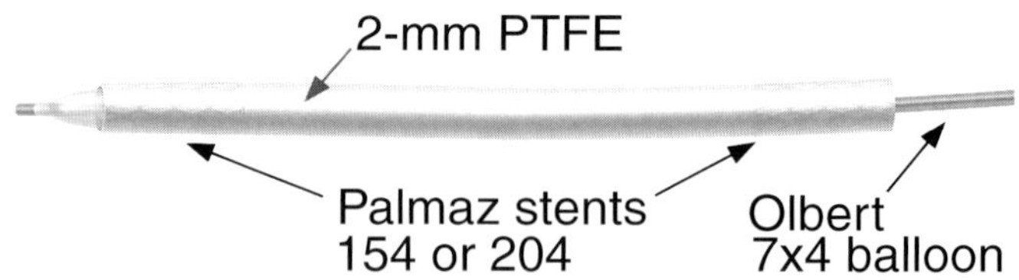

Figure 15.1. Arizona Heart Institute's (AHI) prototype ELG device. (PTFE = polytetrafluoroethylene).

respectively) and are frequently chosen for proximal deployment in the superficial femoral artery. The 15-mm model, P154, is often used for distal fixation. Whenever possible, shorter stents are chosen for distal fixation because they generally allow greater flexion and extension of the graft with movement.

ELG ASSEMBLY TECHNIQUES

A radiopaque-tipped 5F, 65-cm Digiflex angiographic catheter (Medi-tech, Inc, Watertown, MA) allows fluoroscopic measurement of the lesion and helps ensure proper sizing of the graft. The catheter is placed in the most distal portion of the lesion and then pulled back to the desired proximal location. Direct measurement determines the length of the catheter; both intravascular ultrasound and the angioscope are used to discern where the arterial pathologic process begins and ends. After the 2-mm tube is trimmed, the graft is filled with heparin from a syringe: the adapter is held at one end of the ELG during this process, and the other end is held closed. Appropriate-sized stents are then introduced into both ends of the graft and positioned flush with the end of the tube. Identifying markers (silver clips or gold wires) are placed on the ends. It is not necessary to join the stents to the graft with sutures, because friction created by the low-profile PTFE provides adequate stability.

A 7-mm × 4-cm Olbert angioplasty balloon catheter (Meadox Surgimed, Oakland, NJ) is used, and the ELG is threaded onto it in such a fashion that the distal stent rides on the tip of the balloon. The graft sleeve and proximal stent track up the catheter shaft, and percutaneous delivery through a 9F sheath allows delivery of the device (Fig. 15.2).

ELG DEPLOYMENT TECHNIQUES

Complete predilation of occlusive lesions is performed after standard antegrade percutaneous access is obtained through the common femoral artery and 5000 units of heparin have been administered. It is important that dilation of the entire length of the lesion is achieved. After placement of the wire, a long sheath (45 cm or 65 cm) with a radiopaque marker (Cook Inc, Bloomington, IN) is advanced through the lesion, and the ELG is introduced to the level of the radiopaque marker inside the sheath at the predetermined location. Retraction of the sheath is performed at this time, and proper positioning

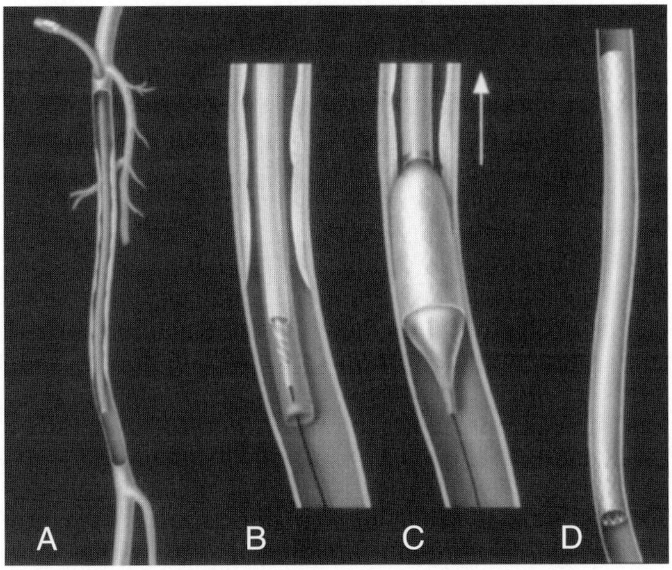

Figure 15.2. (**A**) The delivery sheath is positioned across the lesion. (**B**) The sheath is retracted to expose the ELG's distal stent which is (**C**) expanded against the arterial wall by balloon dilation. The balloon is deflated and sequentially reinflated to dilate the entire length of the graft (*arrow*) and the proximal stent. (**D**) The final appearance of the expanded ELG.

of the ELG is confirmed. The distal stent is expanded using 10 atmospheres of pressure. The entire length of the PTFE graft, from the distal to the proximal stent, is sequentially dilated with the delivery balloon. Retracting the Olbert balloon into the proximal stent is not a problem at this point because there is no profile associated with these balloons.

Additional dilations of both stents and the graft sleeve may be necessary; the control angiogram should be used to determine whether further dilation is required. High-pressure (17 atmospheres), 10-cm Blue Max balloons (Medi-tech, Inc) of appropriate diameter are best suited for final dilatations—an overdilation of 10% is adequate to offset PTFE recoil.

Modifications in the approach may be indicated in certain situations. When using a percutaneous antegrade femoral approach, a small incision may be made in a diseased common femoral artery to facilitate endarterectomy and patch graft angioplasty of the common femoral and profunda femoris arteries. If the occlusion is wire-resistant, recanalization of the superficial femoral artery may incorporate laser technology. After the occlusion has been crossed and the distal superficial femoral artery has been dilated, the ELG can be delivered.

Superficial femoral artery flush occlusions may be treated by an open common femoral arteriotomy. A retrograde popliteal approach may be appropriate in some circumstances.

Protamine is administered in those procedures in which the sheath is to be immediately removed. When the sheath is to be removed later, the patient is transferred to the intensive care unit, and the sheath is removed after the activated coagulation time has returned to normal.

Additional therapy with antibiotics or anticoagulation is not routinely used unless there is a specific indication. When the runoff status is marginal as in trifurcation disease, or if thrombus is observed during the initial procedure, heparin therapy is initiated and warfarin is administered.

IMAGING IN ELG DELIVERY

Proper stent deployment and optimal dilatation of the PTFE graft may be determined by several techniques. Intravascular ultrasound (IVUS) imaging is a newer technique that has demonstrated benefit in coronary angioplasty and stenting. Its use in peripheral vascular procedures indicates that it has practical application and clinical value. IVUS has been used after percutaneous transluminal angioplasty, intravascular stenting, atherectomy, and, more recently, in the deployment of ELGs (Fig. 15.3) (15–18).

IVUS has proved valuable in assessment after intervention and in diagnostic evaluation of arterial disease. The former constitutes its primary use; IVUS provides information about the success of intervention and the need for additional treatment. As a complement to angiography—which provides the endovascular surgeon with details of collateral circulation, vessel contour, quality of flow, and inflow and outflow—IVUS allows appreciation of the vessel wall and lumen and distinguishes soft plaque from calcification (19). Imaging with IVUS allows detection of intimal dissection flaps, thrombus formation, and ulceration; the luminal diameter and cross-sectional area may also be measured (20).

In some instances, IVUS allows detection of lesions

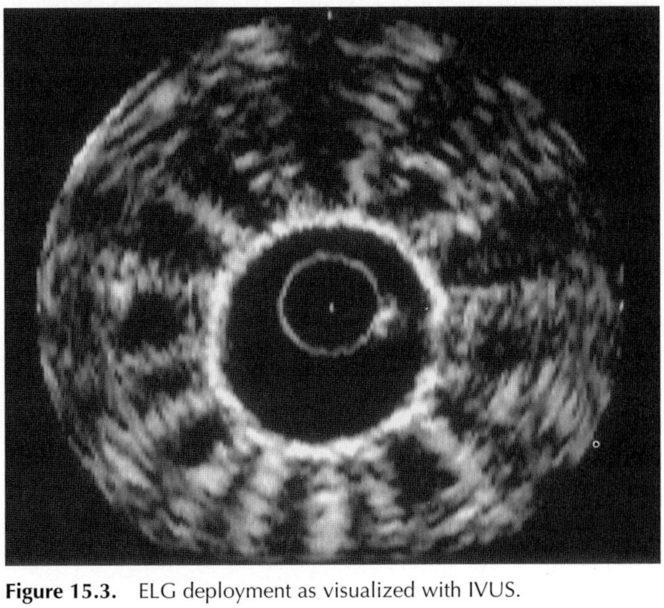

Figure 15.3. ELG deployment as visualized with IVUS.

missed on conventional arteriography (21). At the Arizona Heart Institute, we use IVUS in conjunction with angioscopy and control angiography to optimize visualization in the peripheral arteries.

RESULTS OF INITIAL EXPERIENCE

We have recently published our initial results with the use of ELGs as a treatment for FP occlusions (14). Patients were evaluated for ELG placement based on the following criteria: High potential for angioplasty failure (e.g., long complex aneurysmal/stenotic disease or occlusions); complications of angioplasty (e.g., dissection, vessel rupture); recurrent disease after angioplasty in which stents had been deployed; and aneurysmal disease. We did not evaluate patients with symptomatic FP disease based on their appropriateness for treatment with conventional bypass grafting. Instead, we examined the feasibility of endovascular intervention on an individual basis to determine whether we could appropriately treat the patient using these percutaneous techniques.

Fifty patients with 53 affected limbs participated in this study. The lesions were located in 47 native arteries, 5 FP grafts, and 1 femorotibial vein graft. There were 37 occlusions, 14 stenoses averaging 86% ± 10%, and 2 long combined stenotic-aneurysmal lesions (18 cm and 40 cm, respectively). Four of the FP grafts had distal anastomotic stenoses. In one graft, stents had been placed on two previous occasions. The fifth FP graft was chronically occluded. The femorotibial vein graft, which had Wallstents lining its entire length, occluded when the patient discontinued warfarin.

Access via a percutaneous procedure was achieved in 47 (89%) of the procedures. In six patients, popliteal access was gained. This enabled wire passage through two occlusions. Popliteal access was used in the remaining four patients to introduce the ELG. Bleeding in three obese patients required the use of sutures at the common femoral artery puncture sites.

The femoral bifurcation was surgically opened in seven cases for the following reasons:

- Three patients with FP grafts experienced graft thrombectomy.
- Three patients required patch angioplasty for occlusions at the superficial femoral artery origin.
- A balloon rupture occurred in the seventh patient during deployment of the proximal ELG stent.

In the last patient, introduction of a new stent was necessary to achieve an adequate result, and exposure of the superficial femoral artery was required.

Successful treatment in 51 limbs (48 patients) was achieved through implantation of 55 ELGs. The acute success rate for the procedure was 96%. Two patients were intraoperatively converted to FP bypasses, one because of a native vessel dissection and one due to a twisted ELG. A total of nine covered stents, 2 to 4 cm long, were placed for the treatment of focal disease or anastomotic stenosis, and two ELGs were placed in the same limb in three patients during the initial procedure. A second 22-cm ELG was used in the vein graft of a FP graft patient to treat proximal stenosis that was threatening the viability of an ELG implanted 8 months earlier. The mean ELG length was 23.6 ± 19.4 cm (range, 2 cm to 40 cm).

Complications related to the procedure occurred in 18 patients, and most of these were mild in nature. During ELG implantation, wire placement resulted in a distal dissection; a covered 29-mm stent was used to repair the defect. During the deployment of one graft, redilation was necessary to combat its initial collapse. A prototype ELG deployed in a 14F sheath was used early in our series, and, in two patients, evacuation of a hematoma and arterial repair were required. In seven other patients, hematoma or edema at the access site required no intervention. The completion angiogram in one patient revealed a small midgraft leak, and reevaluation at 2-week follow-up showed a pseudoaneurysm. We repaired this leak by implanting a covered P294 stent at the site.

Thrombosis of the ELG was seen in six patients; two were related to compression of the device after sheath removal at the popliteal access site. In each case, the occluded area was dilated with a balloon, and heparin was administered for a 24-hour period. Two other cases of graft thrombosis were treated with redilation alone. A patient whose FP graft had been clamped proximally developed a thrombosis in the ELG, and it was treated with open thrombectomy. Although deployment of a Wallstent over an area of recoil within the ELG in the last patient was successful, conventional bypass grafting 3 days after ELG implantation was necessary to achieve patency.

Short-term follow-up during the 30 days after the procedures revealed that two of the patients who had undergone successful recanalization of an acute thrombosis had suffered reocclusions. During this period, two additional cases of graft thrombosis were also seen. One case of thrombosis responded to lysing with a thrombolytic agent, but two other patients required treatment with FP bypass grafting because of angioplasty-induced injuries in their distal vessels. One patient required replacement of a fixation stent that was damaged during a fall.

Complications were evaluated in follow-up at 8.3 ± 5.5 months (range 1 to 20 months). In one patient, a covered stent at the distal anastomosis of an FP graft occluded after 2 months because of poor distal runoff; no further treatment was desired by the patient. Three months after implantation, another patient experienced an occlusion resulting from untreated disease distal to the ELG. In this case, balloon dilation of the distal lesion resulted in a wire dissection of the artery, and it was necessary to perform an FP below-knee vein bypass. At 6 months, a reocclusion in one of the acute thrombosis patients required an FP bypass.

Four other patients with ELG thrombosis were successfully treated with endoluminal techniques. The results of the life-table analysis of primary and secondary patencies were 72% and 84%, respectively.

DISCUSSION

Despite the success of percutaneous endoluminal intervention in a number of vessels, revascularization of occluded FP vessels by surgical endarterectomy is still considered the "gold standard." The results of our work with the ELG must be held to the same exacting standards to determine whether percutaneous interventions are truly a viable alternative to open surgical procedures. ELGs may offer a significantly improved mode of treatment for some patients. In those with FP graft occlusions, the removal of an end-to-side anastomosis may improve hemodynamics and successfully treat flow reductions resulting from distal or proximal disease progression or the anastomotic narrowing associated with intimal hyperplasia.

In covering a stent with an occlusive material, primary consideration must be given to the graft material. It should be thin-walled, flexible, biocompatible, and durable, and should possess controlled radial compliance and resistance to cellular adherence or migration in its lumen. PTFE is currently the most commonly used material in prosthetic construction because it meets all the structural requirements for use as an arterial conduit. In addition, it has a well-documented ability to provide long-term patency. It seems ideally suited for use as an ELG component (22,23). In longitudinal studies of FP bypass grafting, microporous PTFE graft lumens do not completely endothelialize and are not a site of intimal hyperplasia (24,25).

The inhibition of neointimal formation in expanded PTFE is directly related to the electronegativity and porosity of the material (26). The results of experiments using subcutaneous PTFE grafts in animals indicate that an internodal distance of at least 60 microns is necessary to promote optimal tissue ingrowth (27,28). Infiltration is not possible when this measure is less than 30 microns.

Although the latter results are preclinical, our own observations in the clinical setting indicate a similar phenomenon (Fig. 15.4). The PTFE sleeves used in our investigation have an internodal distance of 20 to 30 microns and a porosity of approximately 80%.

Covering the thromboembolic material of the lumen with the PTFE tube may offer protection against embolic consequences. The ELG can be deployed directly after wire passage and dilation, and this results in compression

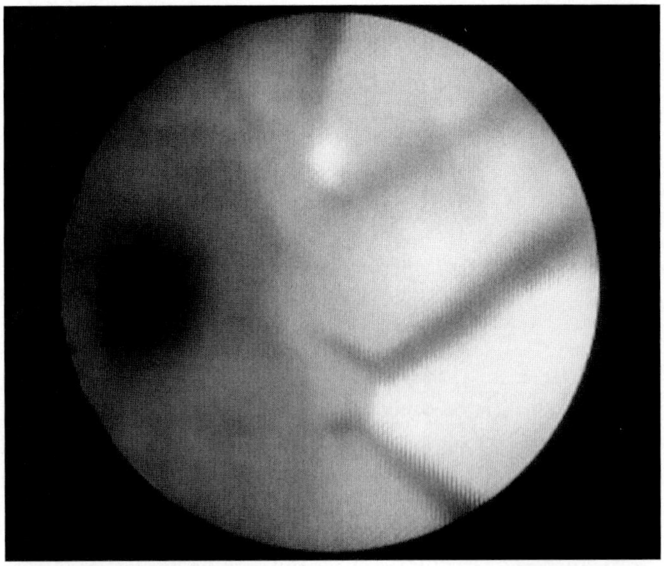

Figure 15.4. Angioscopic appearance of a covered stent 14 months after deployment. There is no evidence of intimal hyperplasia.

of thrombus against the arterial wall. Thrombolysis before ELG deployment is probably unnecessary when the procedure comprises adequate dilation and expansion of the graft. In our series, we used the ELG to treat long occlusive lesions, but did not encounter any embolic complications.

The profile of the graft is an important consideration in its successful use. The use of large-diameter grafts with the fixation anchors sutured to the graft has resulted in a device that must be folded into its delivery system and introduced through large sheaths (29,30). Low-profile ELGs can be constructed by using small-diameter, radially expandable PTFE tubes. These require a 9F sheath for lower limb applications, and it is conceivable that smaller devices may be possible in the future.

Determining the appropriate fixation device in current ELG configurations involves evaluation of its characteristics. We chose the Palmaz stent based primarily on its expansion capability and the availability of a variety of sizes. In addition, we appreciated the device's proven record of clinical safety and efficacy.

Our main complaint about the Palmaz stent has been its vulnerability to outside mechanical forces. Although we have experienced only one case in which this was problematic —a patient fell and deformed the stent such that a thrombosis resulted at the site—the potential for future complications of this sort certainly exists.

The search for an ideal fixation device is fraught with contradiction. Although clearly we are looking for a device that is sturdy enough to withstand significant crushing, we also desire a stent that possesses enough longitudinal flexibility to allow its successful use at points of flexion. It remains to be seen whether the new flexible stent designs will result in an improved ELG fixation device (Fig. 15.5).

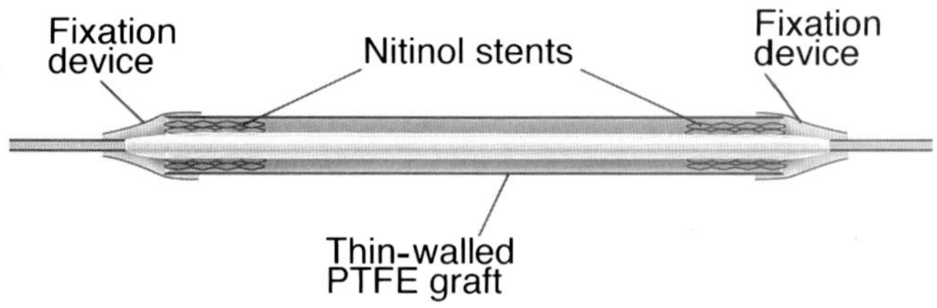

Figure 15.5. New ELG design being used in clinical trials; nitinol stents and thin-walled PTFE are incorporated into this design. This prototype features a long single-expansion balloon. (PTFE = polytetrafluoroethylene).

ELG design is not the only critical point in the successful use of these devices. The techniques used in delivery and deployment are extremely important; individual evaluation of the appropriate size of each ELG is crucial. Positioning of the fixation stents is also a critical step in the placement procedure. Minimizing the area of the stent outside the PTFE tube is important as well, and this technique-related detail was not apparent until later in our series when we began to see intimal hyperplasia on exposed stent surfaces. Those portions of the stent protected by PTFE were not affected, and it became clear that the key was deployment of the stent in a manner that lined it up perfectly with the tube (31).

The prevention of direct contact between the uncovered stent and the arterial wall reduces the incidence of intimal hyperplasia. Interrogating the entire vessel length and sizing the ELG so that its anchor points are on normal tissue are also critical in preventing intimal hyperplasia.

In our experience, graft expansion has been an important factor related to the success of our technique; a PTFE graft in the FP position must be at least 5 mm to 6 mm in diameter to achieve sufficient flow to prevent thrombosis. Early in our series, two patients who had discontinued their warfarin developed thrombosis within 30 days of the procedure because the ELG was not sufficiently expanded. As a result of these experiences, the composition of the PTFE graft material has been altered to reduce recoil to approximately 10% to 14%. A satisfactory luminal diameter can now be achieved by overdilating the entire length of the ELG by just 10%.

Predilation of calcium deposits may prevent cracking and an uneven surface under the ELG. Preparing the artery for ELG deployment by predilating may smooth the luminal contour and remodel calcified plaque so that it does not threaten the integrity of the tube material. Removal of long atherosclerotic segments before endovascular grafting may reduce the bulk of the atherosclerotic material and, consequently, the potential for graft compression (32,33). Our own experience in patients with existing Wallstents indicates that complete dilation is necessary before ELG deployment.

The use of PTFE in endoluminal aneurysm exclusion does not allow the same support an intact vessel provides; conventional PTFE bypass grafts have failed in some circumstances (34). Although it has been demonstrated that the use of PTFE with proximal and distal fixation in long lesions is a successful means of graft placement (30), concern about compression of graft segments without internal support has led some investigators to implant long PTFE tube grafts sutured to stents (35). However, this technique may reduce flexibility and increase the potential for thrombogenicity. We believe arterial pressure alone should be sufficient to achieve patency, but this requires adequate dilation of the ELG during deployment to ensure optimal luminal flow and pressure.

Complex lesions, angioplasty failure, and restenosis were among the complications we were confronted with in this initial series. Clearly, these conditions pose a challenge for any endoluminal technique. Although 50% of the complications were minor, the problem of graft thrombosis was vexatious. To circumvent this problem, we tried removing the sheath immediately after protamine administration to safeguard graft inflow; this solution, although effective in some circumstances, is not a universal remedy. We were somewhat surprised by the prevalence of thrombosis. In our experience (unpublished data) as well as that of others, endovascular grafting in the aortoiliac segment has been free of this event (29,36,37). However, evidence indicates with the Cragg (Endopro) device that interventions below the inguinal ligament may be more subject to thrombotic complications (37).

We are just now ascending a mountain of knowledge and understanding pertaining to endoluminal grafting techniques, whereas results would certainly indicate we have crested a similar headland of enlightenment regarding endarterectomy, the gold standard of therapy. Just as it has taken time to assimilate an understanding of the optimal procedure and technique in the latter discipline, so it is with the former. Although the early results of ELG placement for the treatment of FP occlusive disease reveal some nagging complications, endovascular grafting clearly is a conceptually attractive technique with significant potential. Rather than halt our ascent midway, I believe we should continue our collaboration with device manufacturers to achieve optimal ELG configuration and develop improved delivery and deployment techniques. Such a unified effort seems certain to bring us closer to the pinnacle of endoluminal grafting technology.

References

1. Capek P, McLean GK, Berkowitz HD. Femoropopliteal angioplasty: Factors influencing long-term success. Circulation 1991;83(Suppl 1):I-70–I-80.
2. Johnston KW. Femoral and popliteal arteries: Reanalysis of results of balloon angioplasty. Radiology 1992;183:767–771.
3. Moore WS. Therapeutic options for femoropopliteal occlusive disease. Circulation 1991;83(Suppl 1):I-91–I-93.

4. Dotter CT. Transluminally placed coil-spring endarterial tube grafts: Long-term patency in canine popliteal artery. Invest Radiol 1969;4:329–332.
5. Dotter CT, Buschmann RW, McKinney MK. Transluminal expandable nitinol coil stent grafting: Preliminary report. Radiology 1983:147:259–260.
6. Cragg AH, Lund G, Rysavy JA, et al. Percutaneous arterial grafting. Radiology 1984;150:45–49.
7. Balko A, Piasecki GJ, Shah DM, et al. Transluminal placement of intraluminal polyurethane prosthesis for abdominal aortic aneurysm. J Surg Res 1986;150:305–309.
8. Volodos NL, Shekhanin VE, Karpovich IP, et al. Self-fixing synthetic prosthesis for endoprosthetics of the vessels. Vestn Khir (Russia) 1986;137:123–125.
9. Parodi JC, Palmaz JC, Barone HD. Transfemoral intraluminal graft implantation for abdominal aortic aneurysms. Ann Vasc Surg 1991;5:491–499.
10. Diethrich EB, Papazoglou CO, Lundquist P, et al. Early experience with aneurysm exclusion devices and endoluminal bypass prostheses. (Abstract). J Intervent Cardiol 1994;7:108–109.
11. Diethrich EB, Papazoglou CO, Rodriguez-Lopez J, Lopez-Galarza L. Endoluminal grafts for percutaneous aneurysm exclusion and intraluminal bypass. (Abstract). Circulation 1994;90: I-206.
12. Papazoglou C, Lopez-Galarza L, Rodriguez-Lopez J, et al. Endoluminal grafting: The Arizona Heart Institute experience. (Abstract). J Endovasc Surg 1995;2:89–90.
13. Papazoglou C, Diethrich EB, Lopez-Galarza L, et al. Percutaneous endoluminal grafting for iliofemoral aneurysmal and occlusive disease. (Abstract). J Vasc Intervent Radiol 1995;6:40.
14. Diethrich EB, Papazoglou K. Endoluminal grafting for aneurysmal and occlusive disease in the femoropopliteal arteries: Early experience. J Endovasc Surg 1995;2:225–239.
15. Gussenhoven EJ, van der Lugt A, Pasterkamp G, et al. Intravascular ultrasound predictors of outcome after peripheral balloon angioplasty. Eur J Vasc Endovasc Surg 1995;10:279–288.
16. Cavaye DM, Diethrich EB, Santiago OJ, et al. Intravascular ultrasound imaging: An essential component of angioplasty assessment and vascular stent deployment. Int Angiol 1993;12:212–220.
17. Katzen BT, Benenati JF, Becker GJ, Zemel G. Role of intravascular ultrasound in peripheral atherectomy and stent deployment. (Abstract). Circulation 1991;84:II-542.
18. White RA, Scoccianti M, Back M, et al. Innovations in vascular imaging: Arteriography, three-dimensional CT scans, and two- and three-dimensional intravascular ultrasound evaluation of an abdominal aortic aneurysm. Ann Vasc Surg 1994;8:285–289.
19. The SHK, Gussenhoven EJ, Zhong Y, et al. Effect of balloon angioplasty on femoral artery evaluated with intravascular ultrasound imaging. Circulation 1992;86:483–493.
20. Kopchock GE, White RA, Guthrie C, et al. Intraluminal vascular ultrasound: Preliminary report of dimensional and morphological accuracy. Ann Vasc Surg 1990;4:291–296.
21. Scoccianti M, Verbin CS, Kopchock GE, et al. Intravascular ultrasound guidance for peripheral vascular interventions. J Endovasc Surg 1994;1:71–80.
22. McAuley CE, Steed DL, Webster MW. Seven-year follow-up of expanded polytetrafluoroethylene (PTFE) femoropopliteal bypass grafts. Ann Surg 1984;199:57–60.
23. Ascer E, Veith FJ, Gupta SK, et al. Six year experience with expanded polytetrafluoroethylene arterial grafts for limb salvage. J Cardiovasc Surg 1985;26:468–472.
24. Sanchez LA, Gupta SK, Veith FJ, et al. A ten year experience with one hundred fifty failing or threatened vein and polytetrafluoroethylene arterial bypass grafts. J Vasc Surg 1991;14:729–738.
25. Veith FJ, Gupta SK, Ascer E, et al. Six-year prospective multicenter randomized comparison of autologous saphenous vein and expanded polytetrafluoroethylene grafts in infrainguinal arterial reconstructions. J Vasc Surg 1986;3:104–114.
26. Griesler HP. Interactions of the blood/material interface. Ann Vasc Surg 1990;4:98–103.
27. Campbell CD, Goldfarb D, Roe R. A small arterial substitute: Expanded microporous polytetrafluoroethylene: Patency versus porosity. Ann Surg 1975;182:138–143.
28. Golden MA, Hanson SR, Kirkman TR, et al. Healing of polytetrafluoroethylene arterial grafts is influenced by graft porosity. J Vasc Surg 1990;11:838–845.
29. Parodi JC. Endovascular repair of abdominal aortic aneurysms and other arterial lesions. J Vasc Surg 1995;21:549–557.
30. Marin ML, Veith FJ, Cynamon J, et al. Transfemoral endovascular stented graft treatment of aortoiliac and femoropopliteal occlusive disease for limb salvage. Am J Surg 1994;168:156–162.
31. Diethrich EB. Commentary: Polymeric covers and coats for metallic stents: Microporous PTFE and the inhibition of intimal hyperplasia. J Endovasc Surg 1995;2:266–271.
32. Bergeron P. Stenting and endoluminal grafting of femoral and popliteal arteries. (Abstract). J Endovasc Surg 1995;2:197–198.
33. Bray A. Superficial femoral endarterectomy with intra-arterial PTFE grafting. J Endovasc Surg 1995;2:297–301.
34. Campbell CD, Brooks DH, Webster MW, et al. Aneurysm formation in expanded polytetrafluoroethylene prostheses. Surgery 1976;79:491–493.
35. Cragg AH, Dake MD. Percutaneous femoropopliteal graft placement. J Vasc Intervent Radiol 1993;4:455–463.
36. Sanchez LA, Marin ML, Veith FJ, et al. Placement of endovascular stented grafts via remote access sites: A new approach to the treatment of failed aortoiliofemoral reconstructions. Ann Vasc Surg 1995;9:1–8.
37. Henry M, Amor M, Ethevenot G, et al. Initial experience with the Cragg Endopro System I in the interventional treatment of peripheral vascular disease. J Endovasc Surg 1994;1:31–43.

Chapter 16

Endovascular Stented Grafts: An Independent View

John J. Bergan

As indicated elsewhere in this book, the introduction of vascular grafts by the endoluminal route has the potential to revolutionize vascular surgery in general and clinical practice in particular. Today, its use is especially attractive for high-risk patients who would die in an open operation. Predictably, the technique will go through evolutionary phases and will find a defined place in the clinical armamentarium.

Vascular surgeons have been disappointed in the past by promises made by new therapeutic innovations. One has only to remember a few of these disappointments, that is, the demise of the nylon graft and the disappearance of the homograft. Who among us remembers F. John Lewis' experiments with ivalon sponge as an aortic replacement? Whatever became of Sawyers' gas endarterectomy? All these and many others reached clinical application and then quietly disappeared.

The shelves of hospital operating suites have groaned under the weight of vascular inventions that caught the fancy of staff surgeons. Many expensive purchases, such as the laser, eventually found their way into the dumpster and into the dust bin of recorded time.

With these and many other disappointments in mind, relevant questions arise. What is the future of the stented graft, and what does one observer of the vascular scene predict for its fate?

ADVANTAGES OF MINIMAL INVASION

The patient's perception is that minimal invasion in surgery is equated with excellence. The prospect of less pain, decreased hospital confinement, and rapid return to productive activity is of great importance to the patient. The patient expects perfection in the surgical result, of course. That expectation has been fueled by dramatically successful advances in surgery, as chronicled in the press in the recent past.

Surgeons realistically expect less than perfection in surgical outcomes, but the patient does not. The patient reads the newspaper or The *New Yorker* and learns of dramatic advances in medicine such as conquering some forms of cancer and the apparently routine success of single- and multiple-organ transplants. He or she extrapolates from this. For the patient, exclusion of an aortic aneurysm by the transfemoral technique must be just as successful as laparoscopic cholecystectomy. The patient does not know of the unexpected increase in the number of common bile duct injuries induced by this new technique, nor does the patient anticipate the particular complications that will attend transfemoral endoluminal stent-graft placement.

Surgeons, for the most part, do not know the metabolic advantages of minimal invasion. However, two examples in vascular surgery are informative. The first is the prospective, randomized Veteran's Administration study, which compared percutaneous transluminal angioplasty (PTA) with arterial reconstruction (1). Review of the late morbidity of that multicenter clinical trial reveals an unexpected high incidence of adverse systemic effects in surgical patients (2). The randomization of more than 130 cases into each treatment group was without bias with regard to the important factors of age, sex, race, blood pressure, smoking history, history of cardiac failure, stroke, and the incidence of renal, hepatic, and pulmonary disease, or alcoholism. Therefore, when the natural history of progression of cardiac and renal dysfunction and mortality significantly differed on follow-up, this was a surprising finding.

Within 2 years, 17 deaths occurred in the surgery group and 15 in the PTA group. After another 4 years, 25 more deaths occurred in the surgical group and only 12 in the PTA group. A significant deterioration in renal function was noted in nine of the surgical patients and in none of the PTA patients. Even myocardial infarctions were greater on follow-up in surgical patients so that by 6 years after initiation of treatment, 18 in the surgery group and only 11 in the PTA group had one or more myocardial infarctions. Furthermore, twice as many patients (n = 19) in the surgical group had experienced congestive heart failure compared with the number of patients in the PTA group (n = 8). These data were unexpected and were made believable by the prospective nature of the randomized study.

In an entirely separate study designed to assess the efficacy of thrombolytic therapy as the initial treatment of acute peripheral arterial occlusion compared with operative revascularization, Ouriel and coworkers (3) demonstrated the advantages of minimal invasion. In this study, after patients met the defined entry criteria, 57 patients were randomized to intra-arterial thrombolytic therapy and 57 to operative revascularization.

The curious phenomenon of increased cumulative survival in patients randomized to thrombolysis was unexplained. Eight-four percent of the patients undergoing thrombolytic therapy survived 12 months compared with 58% who received surgical intervention (*P* = .01).

BIOLOGIC ADVANTAGES OF MINIMAL INVASION

The unexpected findings of decreased morbidity and mortality in the minimal-intervention groups previously mentioned have aroused disbelief in some observers. In commentary at the meeting of the Pacific Coast Surgical Society, Taylor voiced his skepticism but said, "In the future, randomized treatment trials for arterial disease should include an analysis of late deaths." In written commentary, Strandness said, "Unfortunately, I'm not sure what conclusions can be drawn from this study." Both commentators criticized the analysis by treatment given rather than intention to treat. Yet the thesis of the entire paper was that the unexpected finding of increased morbidity and mortality in the surgical group could be uncovered only if patients having surgery were the ones included in that treatment arm.

It is to Ouriel's credit that he followed the admonition of Taylor (perhaps unknowingly) and did report the late mortality of the minimal-intervention group. In doing so, he has provided additional knowledge apart from that which was intended in setting up and carrying out the study (3).

In searching for an explanation for the increased mortality and morbidity, authors have postulated that it may be the adrenomedullary response to the maximal stress caused by surgery (4) or that it may be the increased catecholamine release during surgical stress (5). Perhaps the answers are provided via the neuroendocrine response in general (6) or via the increased demand for myocardial oxygen caused by tachycardia, hypertension, anemia, sympathomimetic drugs, or the discontinuation of beta blockers (7).

The most important neuroendocrine response to stress is an increase in plasma epinephrine concentration (4). In man, the acute deleterious effects of such high concentrations of epinephrine include serious cardiac arrhythmias and lactic acidosis. Actual myocardial ischemia and even myocardial necrosis are reported (8). Additional findings of renal damage, pulmonary insufficiency, visceral congestion, and intravascular blood redistribution also have been observed (4). The actual mechanism of plasma catecholamine increase during such surgery is apparently an adrenergic response to afferent pain stimuli not blocked by anesthesia (5). As such, this would occur only in a surgical group of patients, not in those subjected to endovascular stent-graft placement.

Furthermore, plasma levels of norepinephrine and epinephrine as measured by radioenzymatic assay increase during operation in patients receiving inhalation anesthesia (8). Levels of norepinephrine and epinephrine continue to be elevated in the postoperative recovery period in such patients. However, these responses to elevated norepinephrine and epinephrine levels are not observed in patients receiving low and high spinal anesthesia (9).

LIMITED INITIAL USE

As with all new surgical procedures, including closed operations of heart valves, open cardiac surgery using a living, related individual as the blood oxygenator, and the early trials of cardiopulmonary bypass, the transfemoral intraluminal stent-graft has been tried on critically ill patients. Despite this, in Parodi's initial report of five patients with aortic stented grafts, no deaths occurred (10). A more realistic expectation was encountered by Chuter, who used bifurcated endovascular grafts in 15 patients at four centers (personal communication), and Parodi, whose subsequent experience is described in Chapter 8. Failure was encountered in several of these patients.

In these early days of treating somewhat complicated patients, the technique is undoubtedly going to have a difficult time. The procedures are complex and differ so much from standard that comparisons are flawed. The risk for wound infection is greater, and the risk for hematoma is much greater, especially groin hematoma. Difficulties exist in prevention of graft irregularities at anchoring sites and of kinks in the graft body and limbs. Perigraft leaks occur at the proximal and distal anastomoses. Nevertheless, the procedure is compared with surgical aortofemoral reconstruction.

The results of such aortoiliac reconstruction and prosthetic graft replacement are acceptable today. These include an operative mortality rate under 5%, a median survival of just under 10 years with more than 25% of patients surviving for more than 15 years. Although late patency rates decrease to 75% and 70% after 10 and 15 years, respectively, secondary interventions hold the major amputation rates to well under 5% early on (11).

As the initial phases of development pass and as specific new graft materials are introduced, along with improved instrumentation, the procedure will be offered to increasing numbers of patients. Thus, its limitations will be learned.

At the present time and with the present techniques, approximately 10% of patients with aortic aneurysms may be considered for endovascular stented graft placement, but with improved delivery systems, it is anticipated that from 70% to 80% will be candidates.

INCREASED SURGICAL SKILLS AND CREDENTIALING

The particular problems related to the stented graft, such as failure of proximal or distal anchoring and production of distal embolization, predictably will not be entirely solved but may come under control. As this happens, surgeons will need to be instructed in the new techniques. Safe practices must be taught and practice obtained in hand-eye coordination to develop skills that can be used with external imaging devices. These very special skills, currently the property of interventional radiologists, need to be learned by vascular surgeons.

The problem of endovascular surgery credentialing has been discussed, and guidelines have been drawn up and published (12). With this in mind, it is interesting to note in the history of vascular surgery that the first carotid endarterectomy performed at the Johns Hopkins Hospital as described by G. Rainey Williams (13) reflects on our current litigation-conscious times. As Williams described that event, "no one on the operating team had ever seen a carotid endarterectomy, and there certainly were no animal models." He goes on to say, "In the current litigiousness society and with detailed credentialing, one wonders whether this procedure would even be considered."

The subject of credentialing brings up the problem of territories and territoriality. Radiologists and cardiologists have become exceptionally skilled at endovascular techniques. Their skills are not easily learned by surgeons, and they presently dominate the control of interventions using radiologic techniques. Their knowledge of x-ray physics, digital imaging enhancement, and even catheter nomenclature is not easily learned. This tends to put stented graft placement in the radiologic or heart station arenas, which is not attractive to surgeons who are dramatically concerned with the possibility of graft infection.

ROLE OF IMAGING TECHNIQUES

White (14,15) has proposed that endovascular imaging techniques provide an answer to stent-graft surgeons. His suggestion is to extend knowledge derived from angioscopy to intravascular ultrasonography. This would allow the entire technique of stented graft placement to be confined within the operating theater and its sterile atmosphere. The intravascular ultrasound computerized processing system displays real-time images of vessels and possibly allows greater precision in placement of the stents than can be achieved by external imaging techniques today.

Already, the combination of intravascular ultrasonography and angioscopy has proved to be effective in inspection of arteries after various treatments. Thus, residual stenoses of artery or graft, presence or absence of complications of plaque dissection, intimal flaps, and vessel perforation have been immediately detected and treated. White's solution has been criticized because of equipment costs. However, these are a fraction of the expenses accumulated in setting up an operating theater with full angiography capabilities.

CONCLUSIONS

Early in evaluating the experience of stented graft placement, we speculated on the natural history of the procedure saying the following:

> Predictably, it will be offered at first to patients who are at prohibitive risk for conventional aortic surgery. As experience grows, it will be offered to patients who are good surgical risks, even those with aneurysms smaller than the ones conventionally requiring surgical repair. During this time, complications will occur. As every interventional procedure has its own complications, new problems will arise. Opposition to the procedure will be mounted.

Each of those predictions has come true, and in light of progress summarized in this book, the final prediction is undoubtedly going to be fulfilled. This prediction was "Now that this initial barrier is broached, new applications are predictable. Such change is inevitable."

References

1. Wilson SE, Wolf GL, Cross AP. Percutaneous transluminal angioplasty versus operation for peripheral arteriosclerosis. J Vasc Surg 1989;9:1–9.
2. Bergan JJ, Wilson SE, Wolf G, Deupree RH. Unexpected, late cardiovascular effects of surgery for peripheral artery disease. Arch Surg 1992;127:1119–1124.
3. Ouriel J, Shortell CX, DeWeese JA, et al. A comparison of thrombolytic therapy with operative revascularization in the initial treatment of acute peripheral arterial ischemia. J Vasc Surg 1994;19:1021–1030.
4. Wortsman J, Frank S, Cryer PE. Adrenomedullary response to maximal stress in humans. Am J Med 1984;77:779–784.
5. Halter JB, Pflug AE, Porte D Jr. Mechanism of plasma cate-

cholamine increased during surgical stress in man. J Clin Endocrinol Metab 1977;45:936–944.
6. Udelsman R, Norton JA, Jelenich SE, et al. Responses of the hypothalamic-pituitary-adrenal and renin-angiotensin axes and the sympathetic system during controlled surgical and anesthetic stress. J Clin Endocrinol Metab 1987;64:986–994.
7. Mangano DT. Perioperative cardiac morbidity. Anesthesiology 1990;72:153–184.
8. Horak A, Raine R, Opie LH, Lloyd EA. Severe myocardial ischemia induced by intravenous adrenaline. Br Med J 1983;286:519.
9. Pflug AE, Halter JB. Effect of spinal anesthesia on adrenergic tone and the neuroendocrine responses to surgical stress in humans. Anesthesiology 1981;55:120–126.
10. Parodi JC, Palmaz JC, Barone HD. Transfemoral intraluminal graft implantation for abdominal aortic aneurysms. Ann Vasc Surg 1991;5:491–499.
11. Nevelsteen A, Wouters L, Suy R. Aortofemoral Dacron reconstruction for aortoiliac occlusive disease: A 25-year survey. Eur J Vasc Surg 1991;5:179–186.
12. White RA, Fogarty TJ, Baker WH, et al. Endovascular surgery credentialing and training for vascular surgeons. J Vasc Surg 1993;17:1095–1102.
13. Williams GR. The first carotid endarterectomy at the Johns Hopkins Hospital. Arch Surg 1994;129:454–455.
14. White GH, White RA, eds. Angioscopy: Vascular and coronary applications. Chicago: Mosby-Year Book Medical Publishers, 1989.
15. Cavaye DM, White GH, eds. Intravascular ultrasound. New York: Raven Press, 1993.

Chapter 17

Development of a Program of Endoluminal Aneurysm Repair: Experience in 250 Patients

Geoffrey H. White, James May, and Weiyun Yu

Endoluminal graft repair of aortic aneurysms has now been in trial within our vascular surgery unit for 6 years. We commenced a clinical program of endovascular grafting in May 1992, initially using custom-made stent-grafts and modifications of the Parodi graft (1,2). During the intervening period, device development has rapidly progressed with the emergence of a variety of balloon-deployed (3–5) and self-expanding endovascular grafts (6–10), most of which feature the combination of a vascular stent with a prosthetic vascular graft.

This chapter reports the overall experience of the unit, up to late 1997, which totals 252 clinical cases of endoluminal graft implantation (Tables 17.1 and 17.2). We have seen endoluminal repair go through various stages—from pioneering work involving the use of improvised, home-made graft-stent combinations, to first-generation experimental devices produced by commercial firms, to the current group of second-generation devices.

A large number of improvements have been made in the technology of the devices, in surgical techniques, and in knowledge of the complications and intermediate-term outcome (Table 17.3). Long-term outcome beyond 3 to 4 years of follow-up remains unknown.

Special interests of the unit in recent years have included the relevant merits of endovascular repair versus conventional open repair, the relative qualities of different devices, the diagnosis and management of endoleaks, and the occurrence of technical problems and other adverse events associated with these procedures. These aspects have been explored in several comparative studies, which form the basis of this chapter.

DEVELOPMENT OF THE PROGRAM

Our initial human clinical experience was with coated stents (conventional Palmaz arterial stents coated with thin-walled PTFE graft material; n = 2) and with modifications of Parodi stent/graft devices in several configurations (n = 10) (1,11). Subsequently, we have also gained experience with the use of the EVT device (Endovascular Technologies, Menlo Park, CA; n = 19) (12,13), the Chuter bifurcated graft (n = 1) (7), the Mintec Stentor and Endopro grafts (n = 24) (8), the Boston Scientific Vanguard graft (n = 48) (14), BSC Passager graft (n = 9), Bard tube graft (n = 5) (15), World Medical Talent graft (n = 10) (16), Medtronic Aneurx graft (n = 10) (17), Baxter tube graft (n = 2), and with the White-Yu Endovascular GAD graft (n = 102) (18). Additional cases were done as redo secondary endoluminal graft implantations to seal endoleaks. The configuration of grafts used is presented in Table 17.2.

Early Experience

Our initial cases were done with modifications of the Parodi device. In our hands, these were complicated by a high incidence of stent and graft displacement during the implantation procedure because of lack of columnar support within the system that we used. The prototype White-Yu Endovascular GAD graft evolved after we recognized that an alternative method of endoluminal graft fixation was required. This system was developed by members of our vascular unit, and clinical trials of this graft commenced in 1993 after bench model testing and animal implants (2).

Table 17.1. Experience With a Series of Devices in 250 Patients

Device	Number of patients
Self-expanding devices	
EVT Endovascular Grafting System	19
Chuter Bifurcated Graft (Cook/Meadox)	
Mintec Stentor (bifurcated graft)	1
Mintec Endopro (peripheral vessels)	15
Bard graft (tube graft)	9
BSC Vanguard (bifurcated graft)	5
BSC Passager (tube graft)	48
Talent graft (bifurcated aortic + iliac tube grafts)	9
Aneurx graft (Medtronic, bifurcated)	10
Hemobahn (Gore, peripheral occlusive device)	10
Balloon-expandable devices	
Parodi stent/graft	2
Coated Stent (Palmaz/PTFE)	10
White-Yu GAD Graft (Sydney)	110
Baxter GAD Graft (tube aortic)	2

Table 17.2. Type and Configuration of Endoluminal Grafts Used in 250 Patients[a]

	Tube	Tapered aortoiliac aortofemoral	Bifurcated	Peripheral
Coated stent (Palmaz)	—	—	—	2 (superficial femoral, subclavian
Parodi	3	6	1	—
EVT	12	2	5	—
Chuter	—	—	1	—
Mintec graft	—	—	15	9 (iliac, femoral)
Vanguard	1 (thoracic)	—	47	9 (iliac, femoral)
Talent	—	—	5	5 (iliac)
Bard	5	—	—	—
Aneurx	—	—	10	—
Hemobahn	—	—	—	2
White/Yu	42 AAA[b]; 4 thoracic	23	27	14 (iliac, popliteal)

[a] From May 1992 to December 1997.
[b] Forty-two procedures in 37 patients; 5 patients had procedures redone to seal early or late endoleaks.

The White-Yu device features the interweaving of metallic wireforms into the wall of a conventional polyester vascular graft fabric. The wireforms, which we called graft attachment devices (GADs), were intermittently spaced in such a fashion as to provide flexibility as well as adequate support along the whole length of the graft wall.

Recent modifications to the White-Yu graft include an improved sheath design (18F internal diameter) and delivery system, which have been developed in conjunction with a commercial company (Baxter Healthcare Corporation, Irvine, CA), resulting in a new iteration of this graft system. An international multicenter clinical trial of the Baxter endovascular tube graft commenced in December 1996; this study includes a concurrent control group of patients undergoing open repair of abdominal aortic aneurysms (AAAs). Trials of the aortoiliac and bifurcated Baxter endovascular grafts will commence in 1998.

SYDNEY EXPERIENCE WITH THE WHITE-YU ENDOVASCULAR GRAFT

Variations of the White-Yu graft design have been used in the treatment of AAAs, thoracic aneurysms, aortic dissections, false aneurysms, iliac aneurysms, and popliteal aneurysms (18). This section of the chapter includes the experience in the first 102 patients, treated over a period of 3½ years.

Endovascular Graft Design

We reasoned that because vascular stents were primarily developed for treatment of occlusive arterial disease, their use for attachment or fixation of graft materials to the aortic wall may have been a compromise in design development. The White-Yu endovascular GAD graft is constructed of a conventional woven polyester prosthesis, which incorporates a patented intrinsic wireform graft attachment device system along the body of the graft. The wireforms were constructed of stainless steel early in the series and more recently have been constructed of annealed Elgiloy.[1] Wireforms are interwoven through the polyester graft matrix to provide attachment to the vessel wall and radial support after endovascular deployment of the graft. The graft is deployed by balloon expansion using large vascular balloons with diameters of 10 to 27 mm.

The White-Yu endovascular graft was specifically designed for flexibility, thus allowing conformation of the wireform-supported graft to the aneurysm morphology. The system is modular; graft diameters of 8 to 30 mm and lengths of 5 to 20 cm have been used in the clinical series. Two or more grafts can be used in an overlapping fashion (or "trombone" technique), which allows customization of the graft length intraoperatively (19).

The graft system was made in the following versions:

1. Tube, straight graft for AAAs.
2. Tube graft variations used for thoracic aorta, iliac, or peripheral aneurysms.

[1] Elgiloy is a lightweight wire made from a corrosion-resistant alloy with excellent fatigue-resistant characteristics. It has an extensive clinical history as a long-term implantable material, including use in cardiac bioprostheses (e.g., porcine and pericardial mitral and aortic valves).

Table 17.3. Improvements in Endoluminal Grafts Over 6 Years

Availability of off-the-shelf graft devices
Decrease in sheath sizes
Improved hemostatic systems
Fully supported grafts
Improved radiopaque markers
Availability of extension cuffs
Better graft fabrics
Stronger, more reliable stent support
Improved columnar support
Development of secondary procedures to treat endoleaks and other complications

3. Aortoiliac graft (tapered or step-down grafts used in conjunction with occlusion of the contralateral iliac artery and restoration of flow to that limb by conventional femorofemoral crossover grafting).
4. Bifurcated graft version.

Patients and Methods

One hundred and two patients had endoluminal graft treatment of aortic aneurysms or dissections (n = 88) and iliac (n = 13) or popliteal (n = 1) aneurysms. The technique was used only in the elective setting, except for one patient who had emergency endoluminal repair of acute symptomatic AAA. The mean age of the patients was 72 years (range 36 to 87 years); there were 94 males and 8 females. Informed consent was obtained from each patient after detailed discussion of the techniques, risks, and possible complications. Preoperative imaging consisted of contrast-enhanced computed tomography (CT) scan and calibrated aortogram in all patients. Duplex scanning was selectively used for sizing the iliac and femoral vessels in cases in which the diameter had not been well demonstrated on the radiologic studies.

The grafts were deployed by balloon expansion after transfemoral sheath access. When femoral artery access was not feasible owing to atherosclerotic disease or severe tortuosity of the femoral or iliac vessels, access was gained by the retroperitoneal approach to the common iliac artery. In all patients, a completion on-table aortogram was obtained before removal of the access sheath and closure of the femoral arteriotomy. The angiographic cine-loop images were carefully inspected by frame-by-frame review for any evidence of graft complication or blood flow within the aneurysm sac, which would be evidence of an incomplete seal between the endoluminal graft and the aortic wall (endoleak) (20,21).

Outcome Assessment and Follow-up

Procedure "success" was defined as deployment of a patent endoluminal graft at the intended site, without endoleak, without graft displacement, and without conversion to open repair. In addition to completion angiograms, all patients were studied by contrast-enhanced CT of the aorta before discharge from the hospital to confirm that an endoleak was not present. Many patients also had duplex scan imaging of the aorta and graft in situ. Patients had clinical examination and progress CT scans at 6, 12, and 18 months and then annually. The mean follow-up interval for this trial was 21 months (range 2 to 45 months) postoperatively.

Results

The perioperative mortality rate (within 30 days of surgery) for the 101 elective endoluminal repair patients was 2%. The one patient treated for acute symptoms died on the third postoperative day despite a successful graft implant procedure. The mean procedural time was 102 minutes for iliac aneurysms, 111 minutes for tube grafts and 193 minutes for aortoiliac or bifurcated repair of AAAs. The results for each anatomical site and type of graft are presented in Table 17.4.

For *tube graft* treatment of AAAs, 35 of the 37 (95%) grafts were implanted in satisfactory position, with the two unsuccessful cases being due to graft misplacement and balloon trauma to the aortic wall. Primary endoleaks occurred in 5 patients (21).

For *aortoiliac grafts,* the overall treatment success rate was 95% (19 of 20). In this group, the single failure was caused by graft thrombosis due to inadequate anticoagulation at the time of operation.

For *bifurcated grafts,* technical device failure occurred with the early bifurcated graft design in our initial three cases. Improvement in graft design resulted in successful bifurcated endograft deployment in all but one of the subsequent cases. Early thrombosis occurred in one patient who then had distal revascularization by an axillofemoral graft. The four failed cases were successfully converted to open repair. One patient in this group has required late conversion to open surgery for management of a secondary endoleak which occurred at an interval of 8 months after the implantation procedure.

Table 17.4. Results Obtained With the White-Yu Endovascular Graft in 105 Patients

	Number	Success	% With 1-degree endoleak	Conversion
Tube graft (AAA)	37	30 (81%)	5 (14%)	2 (5%)
Aortoiliac	23	22 (96%)	0	1 (4%)
Bifurcated	27	12 (44%)	0	4 (15%)[a]
Thoracic	4	4 (100%)	0	0
Peripheral	14	13 (93%)	0	0

[a] Plus one patient treated by axillofemoral bypass.

All but one of the *iliac* and *popliteal graft* cases were successful. The failed case occurred when sheath access could not be gained to the common iliac artery in a patient with tortuous, calcified iliac vessels. The procedure was abandoned, and the patient was later treated by a percutaneous stent-graft using a more flexible sheath.

Perioperative Complications

Immediate conversion to open repair. Immediate conversion was required in seven patients (6.9%) (Table 17.5). The indications for conversion to open repair were: Graft implanted over renal arteries (n = 1), rupture of aorta (n = 1), technical problems with bifurcated grafts (n = 4), and graft thrombosis (n = 1). Conversion was required in 2 of 37 tube graft procedures (5.4%), 1 of 20 aortoiliac procedures (5%), 4 of 27 bifurcated procedures (14.8%), and in no cases of thoracic or iliac aneurysm endoluminal repair.

Primary endoleak. Defined by contrast enhancement outside the graft but within the aneurysm sac on arteriography or early CT scan, endoleaks were detected in 5 of the 102 patients (4.9%), with all cases occurring in tube graft patients and during the first 15 months of prototype clinical trial. Endoleaks were strongly associated with short distal aneurysmal necks (less than 15 mm) in tube graft cases.

Vascular complications. Two patients required surgical repair of a perforated iliac artery caused by access sheath trauma. A single case of embolus to the distal femoral artery occurred 2 weeks after graft implantation, but there was no incidence of microembolization, as reported in other series (5,22). Approximately 15% of patients have reported temporary neuralgia of the medial or lateral cutaneous nerves of the thigh. This seems to have been associated with stretching of the femoral and iliac arteries by the larger 24F sheaths initially used (perhaps resulting in trauma or ischemia of a femoral nerve branch), because this complication has not been seen since we changed to more flexible 18F sheaths. When all minor complications are included (wound, femoral artery dissection, and so on), the overall incidence of local and vascular complications was 33% (13). Medical and other perioperative complications have been documented in recent publications (13,23).

Table 17.5. Indications for Conversion to Open Repair in 7 of 102 Cases of Endoluminal Aneurysm Grafting

Tube grafts	Aortoiliac grafts	Bifurcated grafts
Misplacement of endoluminal graft above the renal arteries (n = 1) Rupture of the aortic proximal neck (n = 1)	Acute graft thrombosis (n = 1)	Failed graft deployment (n = 3) Acute graft displaceement (n = 1)

Late Complications

Late conversion to open repair. Late conversion was required in an additional five patients. (Three of these conversions were required for management of persistent endoleaks, and two were in patients who suffered aneurysm rupture due to secondary endoleaks [21].) Conversion to open repair was associated with technical expertise and patient selection: In the first 15 months of experience with this graft, immediate conversion to open repair was required for 6 of 47 cases (12.7%) and late conversion in 3 (6.4%), whereas just 1 of 55 cases (1.8%) required conversion to open repair during the next 30 months.

Late secondary endoleak. This occurred in six patients with tube grafts and in two patients with a bifurcated graft in follow-up to date. An endoleak has recurred in one patient who had been previously treated for a primary endoleak. Three patients were managed by secondary endograft implantation in an attempt to obtain seal of the leak. An endoleak has not been detected on postprocedure imaging for any aortoiliac, thoracic, or iliac graft case.

Follow-up imaging studies in tube grafts and aortoiliac grafts have shown no evidence of late graft thrombosis or stenosis to date (100% patency). One patient with a bifurcated graft has had thrombosis of one iliac limb at 12 months follow-up, requiring treatment by a femorofemoral crossover graft, and another patient developed stenosis within the graft limb, which was successfully treated by balloon dilatation. Migration of the proximal end of two grafts was observed on 2-year follow-up imaging in two patients. Diminishment in aneurysm size has been consistently observed in successfully treated cases (24).

SUMMARY

This section presents the results obtained in the initial 102 consecutive patients treated with the White-Yu endovascular GAD graft, to the point at which a new iteration of the graft has entered prospective clinical trials. During the course of this early experience, improvements in patient selection, surgical techniques, and equipment have resulted in a lower incidence of conversion to open repair and a lower incidence of endoleak. The operative time was longer and the early success rate lower for AAAs with complex morphology, but improvements in the bifurcated design were instrumental in reducing complications in the latter half of the series.

Disadvantages of the prototype system were mainly related to the large diameter and relative inflexibility of the initial sheaths used. These features have been addressed with the development of the new commercial version.

Late failures were due to inadequate graft seal resulting in endoleaks; these appear to be due to hemodynamic stress on the distal graft attachment zone in patients who had tube grafts implanted, particularly in patients with short distal aneurysm necks. We believe that tube grafts should not be used in patients with a distal neck less than 15 to 20 mm long; in such cases, bifurcated or aortoiliac grafts are now preferred.

THE BAXTER ENDOVASCULAR GRAFT

The Baxter endovascular graft is an improved version of the White-Yu prototype graft, with most modifications being in the delivery system. Minor device alterations to the implant have been carried out, and the deployment balloons have also been modified specifically for endoluminal graft applications.

The delivery system is composed of two primary units—the introducer sheath unit and the endovascular graft delivery unit.

The introducer sheath unit contains a tapered introducer trocar for initial placement and a passive hemostatic valve system to prevent blood loss from the access sheath. The endovascular graft delivery unit contains the prosthetic device (tube or bifurcated) mounted over a deployment balloon and housed within a loader. The loader interlocks with the introducer sheath unit, facilitating easy transfer of the device into the introducer sheath for deployment into the aorta. The study is a multicenter, prospective, nonrandomized, concurrent controlled clinical trial. Endoluminal repair patients from the United States, Europe, and Australia are being enrolled concurrently with conventional open surgical repair patients from the same clinical sites.

ENDOLUMINAL GRAFTING OF THORACIC ANEURYSMS AND DISSECTIONS

We have used endoluminal grafts to treat thoracic aneurysms in six patients. In two, the aneurysm was arising from the false channel of an aortic dissection. Endoluminal graft procedures within the thoracic aorta have certain inherent problems and dangers (25,26).

In our limited experience, all thoracic repairs to date have been successful, with no incidence of paraplegia, endoleak, or graft displacement. The use of balloon-expanded grafts in this site is limited by the unavailability of balloons greater than 30 mm in diameter.

EXPERIENCE WITH OTHER GRAFTS

As presented in Tables 17.1 and 17.2, we have in trials many other forms of endovascular grafts. The incidence of conversion to open repair and early (primary) endoleak for each graft are presented in Table 17.6. The grafts are listed in chronologic order of our initial use of each graft type, which allows the observation that the incidence of early conversion to open repair has markedly decreased as our experience grew and as the devices were improved. Note that late conversions to open repair for management of late endoleak or other causes of graft failure are not included in this table.

Conversely, the incidence of primary endoleak has increased, although approximately 50% were retrograde endoleaks from lumbar or mesenteric arteries. These differences may reflect variations in patient selection as well as differences in device efficacy.

Device Technical Complications

During trials of the various grafts, we became interested in the concept of prospectively monitoring the incidence of intraprocedural device-related technical problems as a method of comparing different devices in addition to the usual outcome parameters (27). Current techniques for endoluminal grafting of AAAs use either balloon-expandable (BE) or self-expanding (SE) devices. There are theoretical advantages and disadvantages for each type of deployment. In an initial study of these problems, we aimed to prospectively evaluate the incidence of technical problems and deployment-related complications during a series of consecutive patients undergoing endoluminal grafting of AAAs with SE devices.

Data regarding the occurrence of intraoperative technical difficulties and device complications were prospectively recorded for 41 consecutive AAA cases done with SE endovascular graft devices (Vanguard bifurcated graft [n = 34], Talent bifurcated endoluminal graft [n = 3], Bard endoluminal tube graft [n = 4]), in addition to the usual records of patient outcome and procedure complications (27). All technical difficulties and device-related complications that occurred during the procedure were labeled Device Technical Complications (DTC); these included any intraoperative problem with guidewires,

Table 17.6. Incidence of Conversion to Open Repair and Early Endoleak According to Type of Graft

	Total cases	Conversion to open repair	Primary endoleak
Parodi	10	6	1
EVT	19	2	1
White/Yu	110	5	5
Chuter	1	0	0
Mintec graft	15	1	1
Vanguard	48	0	11
Talent	10	0	2
Bard	5	0	1
Aneurx	10	1	2

access sheaths, balloons, graft delivery systems, and endoluminal graft components. We also determined the effect of patient selection factors or surgeon error on the incidence of DTCs.

A total of 91 DTCs were documented in 34 of the 41 SE cases (83%) (range 0 to 10 per patient). Despite these problems, grafts were implanted in all patients, and no patient required conversion to open repair. Patient anatomy contributed to the occurrence of DTCs in 10 cases and surgeon error in 1 case. Procedure success, defined as exclusion of the aneurysm sac by the endograft, without evidence of endoleak, was achieved in 33 patients (80%). Primary endoleaks occurred in seven patients (perigraft endoleak [n = 5], retrograde flow from patent inferior mesenteric artery [n = 2]). Two deaths (5%) occurred in the perioperative 30-day period.

This study showed a very high incidence of DTCs with the designs of SE endoluminal grafts that were tested. We concluded that, in addition to conventional prospective studies of surgical outcome, careful recording and analysis of DTCs may have a valuable role in the assessment of new technologies, similar to the role of critical incident reporting in anesthesia practice.

Comparison of Outcome Parameters for Different Endoluminal Grafts

In 1997, we compared the procedure outcome parameters and early follow-up results for our series of AAA patients treated with SE endoluminal grafts (n = 83) up to that time with results for a group of patients who had been treated with BE devices (n = 88) (27). Excluded from this analysis were patients who had endoluminal graft treatment of iliac or thoracic aneurysms and the 10 early AAA patients treated with improvised devices manufactured on a side bench in the operating room from graft and stent components (modified Parodi graft).

Primary endoleak was defined as incomplete endograft seal resulting in incomplete exclusion of the aneurysm sac, detected within the 30-day perioperative period by imaging studies (completion angiography and predischarge contrast-enhanced spiral CT scan) (21). Late endoleak was defined as any similar occurrence after the 30-day interval in a patient who had previously been demonstrated to have no endoleak. Procedure success was defined as complete exclusion of the aneurysm sac, with no endoleaks seen on angiogram or CT scan and no mortality within the perioperative 30-day interval.

The procedure outcome parameters for these groups are shown in Table 17.7. The major failure mode in the perioperative period for the total series of endoluminal graft procedures was primary perigraft endoleak, which occurred in nine patients in the SE group compared with five patients in the BE group. Primary endoleaks sealed spontaneously within 6 months in three patients in the SE group (but recurred at 12 months in one of these), and in one patient in the BE group. Endoleaks that did not seal spontaneously within 6 months were treated by active management in the majority of others. Retrograde endoleaks from a patent inferior mesenteric artery occurred in two patients in the SE group and one in the BE group.

Overall technical success (graft deployed with complete exclusion of the AAA sac) was achieved in 73 of 88 (83%) in the BE group versus 66 of 83 (79%) patients in the SE group. No graft infection or late embolization in either group occurred.

Table 17.7. Comparison of Early Outcome Parameters for 2 Groups of Patients Treated With Balloon-Expandable and Self-Expanding Endoluminal Grafts (1993–1997)

	Balloon-expandable (n = 88)	Self-expanding (n = 83)	*P* value
Graft deployment	85 (96%)	81 (97%)	ns
Conversion to open surgery	8 (9%)	2 (2.4%)	ns
Distal embolization	1 (1.2%)	4 (4.8%)	ns
Primary endoleak	5 (5.7%)	9 (10.8%)	ns
Retrograde endoleak (IMA)	1 (1.1%)	2 (2.4%)	ns
Number successful[a]	73 (83%)	66 (79%)	ns
Late endoleak	7 (7.9%)	2 (2.4%)[b]	ns

[a] Success is defined as an aneurysm sac completely sealed by endoluminal graft (i.e., no conversion for open primary endoleak or retrograde endoleak).
[b] Short follow-up interval.

The Pathology of Healing of AAAs After Endoluminal Repair

The prospect of excluded AAAs reestablishing communication with the circulation as a result of the natural progression of the aneurysmal process into the AAA neck is a continuing cause for concern. In this situation, expansion of the proximal neck around a prosthesis of fixed dimensions would encourage distal migration of the prosthesis. It is also known that the integrity of union between an artery and a prosthetic graft, which are sutured together at conventional open operation, depends on the continued strength of the nonabsorbable suture material. Little is known about the healing process between an endoluminally placed prosthesis and the aortic wall. We reported the results of autopsy 7 months after successful endoluminal repair of AAAs (28), and the histopathology of transluminally placed endovascular grafts in occlusive disease has been reported by Marin and associates (29).

To further our understanding of this area of concern, we performed a combined study of the histopathology of the human reaction to the placement of endoluminal AAA prostheses and the changes in morphology of AAAs after treatment with such prostheses. The pathology was studied in two autopsies of patients who died 4 and 9 months, after successful endoluminal AAA repairs. This was compared with the histopathology of the aorta of an age-matched normal control patient.

Histopathology Technique

Tissue sections were taken throughout both of the formalin fixed grafts (3) and embedded in paraffin. Initial evaluation of the sections followed histologic staining with hematoxylin and eosin. Further evaluation was performed using connective tissue staining with trichrome. The tissue was also examined immunohistochemically by antibodies to a smooth muscle actin (Sigma, St Louis, MO) immunochemicals monoclonal anti-alpha smooth muscle actin), macrophages (CD68) and T cells (CE43). Endothelial cells were sought using the endothelial cell antibody Sigma Anti-Human von Willebrand Factor. Secondary antibodies used were biotinylated anti-mouse IgG and biotinylated anti-rabbit IgG. Positive and negative control tissues were used. The detection system used was the avidin biotin peroxidase system. A counterstain with hematoxylin was performed to visualize cell structure.

Sections taken from both abdominal aortas above the level of the endografts showed microscopic evidence of advanced atheromatous disease (complicated atheroma with calcification, ulceration and thrombus formation) when compared with normal aortic histology of age-matched aorta.

Microscopic examination showed that most of the interior of the stent-graft was patchily covered with a smooth jelly-like lining composed mostly of fibrin, with enmeshed leukocytes and lacking an endothelial surface (7). The interior was similar to that described by Clowes and colleagues (30) and by Florey and associates (31) as a "pseudointima." The fibrinous clot covering the proximal metal component of the prosthesis was 1 to 2 mm thick, whereas the distal part of the Dacron graft was more patchily covered by 1-mm thick fibrinous lining.

The proximal parts of the prostheses were completely covered with fibrin, whereas the remainder were patchily covered by longitudinally oriented patches of fibrin. The proximal pseudointima had an average thickness of 2 mm, whereas the more distal pseudointima was somewhat thinner, averaging 1 mm. No evidence of endothelial cells was seen. Along the remainder of the prosthesis there was no evidence of healing cells such as fibroblasts or inflammatory cells between the thrombus and the surface of the graft.

The aneurysm wall showed little structure. The aneurysmal wall consisted of intima and adventitia only. Few medial smooth muscle cells could be seen. A layer of laminated thrombus, varying in thickness according to the extent of dilation of the aneurysm, was present. In contrast, the common iliac artery, which contained an extension of the Dacron graft, showed more healing response with a strong fibrous reaction to the prosthesis. The wall was thinned, atheromatous, and calcified, with thrombus narrowing the lumen. However, granulation tissue was growing through the thrombus, and acute and chronic inflammatory cells were present. There was no evidence of smooth muscle hyperplasia.

The overall amount of inflammation was low, with only a small area of moderate inflammation in one section of aneurysm wall. This inflammatory infiltrate consisted of lymphocytes and plasma cells with occasional macrophages and multinucleated giant cells in relation to the calcified atheroma. This contrasts with the extensive inflammatory infiltrate usually seen in advanced untreated aneurysm walls. Inflammatory cells did not extend into the thrombus but remained within the residual aneurysmal wall.

Placement and Mechanical Stability of the Prosthesis

In the autopsies of both cases, the prosthesis appears to have remained stable at the site of original endoluminal placement. The aortic wall at the site of implantation in both cases did not show evidence of aneurysmal degeneration.

Healing Responses to the Prostheses

The internal surface of both grafts showed a fibrinous internal lining with no endothelial cells seen. Because endothelialization is incomplete in open reconstructions of the aorta that have been in widespread use for many decades, endothelialization of endoluminal prostheses, although theoretically desirable, does not appear to be essential to their stability in practice. A lack of smooth muscle hyperplasia was observed despite the time course from procedure to presentation.

Changes in AAA Morphology After Endovascular Grafting

The fate of the sac and neck of the AAA was observed by contrast-enhanced CT in a group of 47 patients who had undergone endoluminal AAA repair 18 months (or more) previously. The CT scans were performed within 10 days of the operation and at 6, 12, and 18 months after operation. Several parameters were examined and published in two reports (32,33). Of particular interest were the following:

- The presence or absence of endoleak into the aneurysmal sac.
- The maximum transverse diameter (MTD) of the AAA.
- The dimensions of the proximal neck of the aneurysm.

Results

For the purpose of analysis, patients were divided into two groups:

Group 1—Those in whom there was a decrease in the MTD of the aneurysmal sac (n = 39).
Group 2—Those in whom there was an increase in the MTD of the aneurysmal sac (n = 8).

Endoleak into the Aneurysmal Sac

No instances of endoleak were observed in any of the 39 patients is group 1. Endoleaks were observed in five of the eight patients in group 2. In three of these, endoleaks were noted within the first 10 days after operation. Endoleaks were not observed in the other two patients until 12 months or more after the operation.

Maximum Transverse Diameter

The MTD in all patients in group 1 diminished. This diminution was progressive with the mean decrease being 0.44 cm at 6 months, 0.80 cm at 12 months, and 1.15 cm at 18 months. Patients in group 2 fell into two subgroups—those with endoleaks and those without. The three patients with perioperative endoleaks all experienced an increase in MTD at 6 months after operation. Successful secondary endoluminal repair was accompanied by diminution in MTD in two of these. In the third patient, persistent endoleak was noted after secondary endoluminal repair; this patient was converted to open AAA repair. In one of the two successful secondary endoluminal repairs, recurrent endoleaks were noted 6 months later. This was accompanied by an increase in MTD, which culminated in rupture of the AAA. This patient recovered after repair despite having an ejection fraction of 13%. In the remaining two patients in this sub-group, endoleaks did not become apparent until 12 months or more after operation. Both endoleaks were preceded by an increase in MTD, which exceeded the preoperative figure. Both had also experienced an initial decrease in MTD at 6 and 12 months, before the increase. In the second subgroup, no endoleak was noted despite an increase in MTD. This increase did not persist beyond 6 months in two of the three patients.

Changes in Proximal Neck

In all the patients in group 1, the diameter of the proximal neck increased. The magnitude of this increase was about 10%. Examination of the mean proximal neck diameter at 6, 12, and 18 months did not reveal any progressive increase. In group 2, a similar increase in diameter of the proximal neck was observed at the same intervals after operation. No decrease in the length of the proximal neck was observed in group 1 or group 2.

Discussion

We have postulated that the paradoxical increase in the diameter of the proximal neck, concurrent with a decrease in the diameter of the sac of the AAA is due to deliberate oversizing of the endograft. Such oversizing is commonly practiced to avoid migration of the endograft. This theory is supported by the fact that the increase in the diameter of the neck has not been progressive and the magnitude of the increase in diameter is similar to that by which the endograft was oversized. However, the changes in the neck of the AAA may be due to progression of aneurysmal degeneration into the wall of the aorta between the aneurysm and the renal arteries.

It is well known that this process is part of the natural history of untreated AAAs. If this were so, one might expect to see evidence of aneurysmal degeneration in autopsy specimens obtained from patients with implanted endografts who have died from other causes. We have previously recorded one such patient (28), and two other patients are reported in this chapter. In all three patients, there was good apposition between the endograft and the neck of the AAA. In addition, no evidence of aneurysmal degeneration was noted in the neck of the AAA. It should be emphasized that the length of follow-up was less than 1 year in all three patients.

Distal migration of the proximal attachment device of an endograft has been independently reported by Ivancev (34) and Chuter (35). Both investigators were using endografts that were anchored by modified Gianturco stents. In some of these patients, the migration occurred in endografts that had been implanted for more than 1 year. We have now also observed this late migration of endografts in patients in whom the endoluminal method was used for AAA repair.

A COMPARATIVE STUDY OF OUTCOME FOR MATCHED GROUPS OF PATIENTS UNDERGOING ENDOLUMINAL REPAIR AND OPEN REPAIR OF AAAs

There have been two main schools of thought with respect to trials of endoluminal grafts. One school of thought was that endoluminal repairs should be offered only to patients who are not medically fit to undergo conventional open repair; the other was that endoluminal repairs should also be available for medically fit patients under strict investigational protocols, so that a true comparison with conventional open repair becomes available, including the long-term effects on the morphology of AAAs as well as the more immediate factors presented herein. Our policy has been to offer the technique to both groups.

In this early study, we analyzed the outcome of two

matched groups of AAA patients (36). One group had undergone endoluminal repair (n = 28), and the other was an historical control group who had open repair (n = 27). The patients were retrospectively selected on the basis that they were medically and anatomically suitable for either of the two methods of treatment. High-risk patients were excluded, as were patients with aneurysms that did not fulfill the criteria for endoluminal grafting. Patients in both groups were comparable in relation to age, sex, risk factors, and dimensions and morphology of AAAs.

Despite the exclusion criteria, comorbidity remained significant in the included groups, with more than 50% of both groups having ischemic heart disease or chronic airways limitation. The key to the comparative analysis of outcome in this study is the assumption that the patients were suitable for both treatments.

At the time of this study, the initial enthusiasm for endoluminal grafting had been tempered somewhat by the realization that these cases can be extremely complex and that complications are frequent. In a previous study, we had shown that the results of endoluminal grafting of aneurysms were dependent on the morphology of the aneurysm, with a higher incidence of complications being encountered in patients who had more complex aneurysms or tortuous iliac arteries used for transluminal access (37).

The factors analyzed for each group were operating time, operative blood loss, type and dimensions of graft used, duration of intensive care unit (ICU) admission required, duration of hospital stay, and incidence and severity of complications. Blood loss was calculated from records of cell-saver salvage, suction bottles, and sponge weighing. Duration of hospital stay was calculated from the day before operation. These data were collected prospectively for the endoluminal group and retrospectively for open repair.

Morbidity and mortality were analyzed for both groups according to the recommendations of the Ad Hoc Committee on Reporting Standards for arterial aneurysms, with complications being listed in two categories as either local/vascular or remote/systemic (38,39).

Table 17.8. A Comparative Study of Outcome for Matched Groups of Patients Undergoing Endoluminal Repair and Open Repair[a]

	GP I (n = 27)	GP II (n = 28)	P value[b,c]
Hospital stay (mean)	12.4 days	11.1 days	ns
Blood loss (mean)	1422 mL	873 mL	<.05
Duration of operation (mean)	2.6 hr	3.1 hr	ns
Duration of ICU admission (mean)	1.8 days	0.7 days	<.05
Local/vascular complications	4 (15%)	7 (25%)	ns
Systemic/remote complications	10 (37%)	8 (29%)	ns
Deaths <30 days	1 (3.7%)	0	ns
Deaths late	0	3 (11%)	ns

[a] The outcome of AAA repair for a group of 27 patients undergoing open repair (Gp I) compared with a matched group of 28 patients undergoing endoluminal repair (Gp II). The comparison is of mean values; complications are listed by number of patients.
[b] Unpaired t-test.
[c] Mann-Whitney test.

Results: Comparison of Outcomes

The mean values for operation time, blood loss, ICU care stay, and hospital stay for open and endoluminal repair, were 2.6 versus 3.1 hours, 1422 versus 873 mL, 1.8 versus 0.7 days, and 12.4 versus 11.1 days ($P<.05$), respectively (Table 17.8). Local/vascular complications occurred in 15% of patients in the open group compared with 25% in the endoluminal group, whereas remote/systemic complications occurred in 37% and 29%, respectively. Five of 28 patients in the endoluminal group had complications requiring early operative repair (n = 3) or late revision (n = 2). When comparison was made on an intention-to-treat basis, the incidence of local/vascular complications was significantly greater for endoluminal repair (P .047).

Despite the patients' mean age of 69 years, the mortality rates in both groups (endoluminal 0%; open 3.7%) compared favorably with those reported recently for elective aneurysm repair (36) and were consistent with the selection of patients deemed to be fit for surgery. The significant reduction in ICU stay for the endoluminal group is impressive and supports the opinion that endoluminal repair is less invasive and less stressful to the patient.

Complications

The complications for each group are presented in Table 17.9.

The following local/vascular complications were encountered with endoluminal repair:

- Iliac artery trauma ([n = 2] both requiring immediate operative repair).
- Incomplete exclusion of the aneurysm sac (n = 2).
- Graft overlap of the renal artery orifice (n = 1) with horseshoe kidney.
- Groin wound complications (n = 4; a total of five major and four minor complications in seven patients).

The following remote/systemic complications occurred in eight patients:

- Myocardial infarction (n = 1).
- Atrial fibrillation (n = 2).

Table 17.9. A Comparative Study of Outcome for Matched Groups of Patients Undergoing Endoluminal Repair and Open Repair

Group I (n = 27), open repair, no complications in 15 (56%)	Group II (n = 28), endoluminal repair, no complications in 16 (57%)
Local/vascular	Local-vascular
Impaired perfusion of one kidney (renal artery damage)	Iliac artery trauma (n = 2)
Occlusion of left common iliac artery	Incomplete exclusion of aneurysm sac (n = 2)
Occlusion of left common femoral artery	Occlusion renal artery
Abdominal wound dehiscence	Groin incision wound infection
Remote/systemic	Groin incision wound seroma/lymphocele (n = 3)
Deep venous thrombosis, pulmonary embolus	Remote/systemic
Cardiac failure, pulmonary edema (n = 2)	Urinary retention (n = 2)
First-degree heart block	Myocardial infarction
Renal failure (n = 3)	Cerebrovascular accident
Transient cerebral ischemic attacks	Atrial fibrillation (n = 2)
Myocardial infarction	Lung consolidation
Vocal cord injury, respiratory failure	Acute renal failure

- Cerebrovascular accident (n = 1).
- Lobar atelectasis (n = 1).
- Urinary retention due to prostatic disease (n = 2).
- Acute renal failure (n = 1) due to graft overlap of the renal arteries, as described above.

Exposure to contrast did not result in renal impairment in any patient. Sixteen patients (57%) in this group were free of complications.

Five of 28 patients in the endoluminal group had complications requiring early operative repair (n = 3) or late revision (n = 2). All were associated with potentially correctable technical factors. Thin flexible access and delivery sheaths with smaller external diameters are required to reduce the incidence of iliac artery trauma during the procedure. Improved imaging systems are required in the operating room to reduce the chances of misplacement of the endoluminal graft or of failing to detect leak of contrast from the distal or proximal ends of the graft denoting incomplete exclusion of the aneurysm sac, which would be best managed in many cases by insertion of a second overlapping endograft.

Commentary

From this study, we concluded that there was significantly less blood loss and shorter ICU stay with endoluminal repair. The incidence of local/vascular complications had a tendency to be higher for endoluminal compared with the standard open method (and was significantly greater if failed procedures were included). On the other hand, in this early experience with prototype devices, patients who were medically suitable for open surgery did not derive benefit from the less invasive endoluminal technique with respect to duration of operation, length of hospital stay, or perioperative morbidity and mortality. This study pointed to a need for improved devices and access systems, but because those with endoluminal repair did not have a worse outcome, a randomized study was justified in the group of patients who are not unfit for open repair.

There is little argument that endoluminal repair may legitimately be offered to selected high-risk patients who have favorable AAA morphology, but the use of the endoluminal technique for patients who are suitable for conventional surgical repair remains contentious. This decision ultimately depends on long-term results.

Blood loss in this study was significantly less for the endoluminal group (mean 873 mL versus 1422 mL, mean 700 mL versus 1200 mL; $P<.05$), especially for the majority of patients who had successful transfemoral endoluminal graft placement, whose mean loss was only 410 mL. In these patients, bleeding occurred only by leakage from the sheath and hemostatic valve, particularly during exchange of various items of equipment (guidewires, angiography catheters, sheaths, and graft devices). The prototype hemostatic valve system used for most of these patients was unsatisfactory and was modified to reduce blood leakage. There is also an obligatory loss when the femoral artery is flushed at the completion of the case to avoid embolization of any thrombus or debris that may have accumulated during the procedure.

Additional bleeding from traumatized iliac arteries in two patients contributed to the overall mean blood loss of 850 mL.

A STUDY OF ENDOLUMINAL REPAIR VERSUS NO TREATMENT FOR SMALL AAAs

In a further study (40), we examined the relative merits of endoluminal repair compared with conservative treatment in a retrospective study of 110 patients from our register. During the study period, 43 patients had endoluminal repair and 67 patients had no treatment for small AAAs.

Despite patient selection for different management in each group, close analysis revealed that the groups were similar with regard to comorbidities and risk factors, as well as age, sex, and size of aneurysm.

The endografts used were the modified Parodi (n = 2), the EVT (Endovascular Technologies) (n = 8), White-Yu (n = 27), the Mintec Stentor (n = 5), and the Chuter (n = 1). The graft configuration was tubular in 26, aortoiliac in 5, and bifurcated in 12. Surgical exposure of the access arteries was used in all patients. The size of the introducing sheath (internal diameter) was 24F in 37 patients and 18F in 6.

Sixty-seven patients with AAAs of 5 cm or less were conservatively managed with correction of risk factors, hypertension, cardiovascular disease, and lipids. Patient characteristics and comorbidities were similar to those of the endoluminal group (Table 17.10). Patients in this group were followed up at intervals of 6 or 12 months by clinical examination and contrast-enhanced CT. The duration of follow-up ranged from 1 to 51 months with a mean of 18 months. Patients with AAAs growing larger than 5 cm in diameter were offered intervention.

Results

A comparison of the results in terms of AAA rupture, systemic complications, and death is given in Table 17.11.

Endoluminal Repair Group

Endoluminal repair failed in 6 of 43 patients (14%), requiring conversion to open repair. Four of these failures occurred at the original operation, and two were secondary conversions. Two of the primary conversions were due to access problems early in our experience, whereas the remaining two were due to inability to deploy a bifurcated graft. Local vascular complications occurred in 11 patients (36,41). Systemic/remote complications occurred in 10 patients. Renal insufficiency was the most worrying of these complications, occurring in four patients. Two patients with myocardial infarction and three with atrial fibrillation recovered without incident. There were two perioperative deaths in the endoluminal repair group (4.7%).

In follow-up of the endoluminal repair group, all grafts remained functional with no incidence of thrombosis (other than one limb of a bifurcated graft), embolism, or infection. The mean maximum diameter of the AAA sac after successful endoluminal repair showed a progressive decrease from 4.41 to 3.92 cm at 12 months. One late death occurred from spontaneous rupture of the esophagus 8 months after successful endoluminal AAA repair.

Table 17.10. A Study of Endoluminal Repair versus No Treatment for Small AAAs

Characteristics of patients	No treatment	Endoluminal repair
Number of patients	67	43
Mean age (years)	71.4	69.6
Sex (M/F)	58/9	40/3
Maximum diameter AAA (mean cm)	4.08	4.41
Ischemic heart disease (%)	34 (51)	19 (44)
Hypertension (%)	19 (28)	8 (19)
Pulmonary disease (%)	9 (13)	4 (9)
Diabetes mellitus (%)	5 (7)	4 (9)
Chronic renal failure (%)	0 (0)	1 (2)
Transient ischemic attack (%)	1 (1)	2 (5)

Table 17.11. A Study of Endoluminal Repair versus No Treatment for Small AAAs

	Endoluminal repair	No treatment
Aneurysm rupture	0	1
Systemic complications		
Renal insufficiency		
Obstructive		
Prerenal	2	0
Postrenal	1	0
Contrast medium-induced	1	0
Cardiac		
Myocardial infarction	2	2
Cardiac arrhythmia	4	0
Transient ischemic attack	0	1
Deaths		
Perioperative	2	0
Late	1	2
Total	3	2

No-Treatment Group

Sixty-seven patients with AAAs 5 cm or less were managed conservatively with correction of risk factors. The mean maximum diameter of the AAAs in this group at presentation was 4.08 cm. The growth rate of the AAA was an increase in mean maximum diameter of 0.52 cm per year. During the 4-year study period, 21 (31%) of the 67 AAAs enlarged beyond 5 cm in maximum diameter. Eleven of the 21 patients whose AAAs exceeded 5 cm in diameter were operated on. Seven AAAs were repaired using the endoluminal method and four by the conventional open method.

One death occurred from aneurysm rupture in the no-treatment group (1.5%). There was one non–aneurysm-related death from myocardial infarction. Morbidity was limited to two nonfatal episodes of myocardial infarction and one transient ischemic attack.

Discussion

In assessing the relative merits of two methods of treatment, a concurrent comparison has a useful place. It provides an approximate estimate of the risks and benefits of each method. Such an estimate allows a decision as to whether a randomized trial of two methods is justified. Although the randomized trial remains the gold standard for comparing two methods of treatment, there are disadvantages in using this method from the outset. At the one extreme, one method may be so obviously superior that the patients randomized to the other method will clearly be disadvantaged and the trial may have to be suspended. At the other extreme, one method may be so hazardous as to render the randomization of patients unethical. In this study, the mortality in the group of patients with small AAAs treated by endoluminal repair (n = 3) was similar to the group managed by no treatment (n = 2). However, endoluminal repair had an unacceptably high rate of failure and conversion to open repair (14%) with a 25% vascular complication rate.

The latter results suggest that a randomized trial of endoluminal repair versus no treatment will become justified in the subset of patients with small AAAs of 5 cm or less, when the mortality of endoluminal repair can be reduced and the incidence of complications can be reduced by further improvements in endoluminal technology.

It is probable that the results of this study were skewed by initial high mortality and complication rates associated with learning the endoluminal method during the early phase. In fact, both perioperative deaths and four of six conversions to open repair occurred during the early phase of our experience with endoluminal graft techniques, up to the end of 1994. During 1995 and 1996, only two patients required conversion to open repair. Furthermore, five of the six conversions to open repair occurred in cases in which a new device was being put in trial for the first or second time only (and four of the six occurred with devices that are now superseded). Poor-quality imaging equipment may also play a role. Five of six conversions to open repair occurred during a period when we were using an older image intensifier, whereas only one has occurred since state-of-the-art imaging equipment was installed.

It is also our experience that procedural complications are related to access sheath size. Up until the latter stages of this study, all of our endoluminal cases were done with sheaths of 24F internal diameter (ID)/29F external diameter (ED). Recent devices may be introduced via sheaths of 18F ID/21F ED, and these newer access sheaths are more flexible. This is likely to reduce the incidence of vascular complications and improve outcome.

Availability of smaller sheaths will also increase the proportion of patients who are candidates for endoluminal repair. These factors suggest that morbidity and mortality are strongly related to experience, technique, and specific device design.

One of the premises on which this study was based was the nexus between small AAAs and the presence of a proximal neck of sufficient length to enable endoluminal repair to be undertaken. It is therefore of interest that an AAA diameter of 5 cm or less is no guarantee of suitability for endoluminal repair. Indeed, Hopkinson and colleagues (42) found little correlation between AAA diameter and suitability of the proximal neck for endoluminal repair. They found 19 of 44 patients (43%) with AAAs 4.5 to 5.4 cm in diameter to be unsuitable. They also found that shortening and widening of the proximal neck increases with aneurysm size but only after the aneurysm expands beyond 7 cm in diameter. It is also of interest to note that in the present study only 26 of 43 (60%) AAAs were suitable for tube graft repair despite their small size.

A limitation in this study was the inability to conclude whether the risk of complication is greater with one or the other device because of the small number of patients in whom some of the devices were used. Analysis did not reveal any difference in the complication rate between the largest group of 27 patients with the White-Yu prosthesis and the remaining group of 16 patients in whom various prostheses were used.

The evolution of these devices also makes it impossible at this time to form definite conclusions regarding the role of endoluminal repair for small aneurysms. In the present study, the endoluminal repair group has gained an asset in having their aneurysms repaired at a cost of early morbidity after operation. However, any proposal to screen for aneurysms with a view to treating them while they are small enough to be suitable for endoluminal repair would seem premature because (1) the long-term outcome of endoluminal repair is not known, (2) the current morbidity of this method is high even with small aneurysms; and (3) a significant number of small aneurysms will be unsuitable despite their size. Improvements in results and devices during the last few years suggest that there will be a role for a prospective randomized study in the future.

DURABILITY OF ENDOLUMINAL REPAIR: A CONCURRENT COMPARISON WITH OPEN REPAIR

To determine the durability of endoluminal repair, we did a life-table analysis of the long-term outcome of endoluminal repair compared with open repair of AAAs treated concurrently by the same surgeons during a defined interval (43). The study included 362 consecutive patients with AAAs who underwent repair between May 1992 and May 1996. Fifty-three patients who had open operations for ruptured AAAs, plus two patients who had endoluminal repair of false AAAs and four patients who had secondary endoluminal repair of AAAs, were excluded, leaving 303 patients having elective repair of true AAAs in the study.

The elective operations were conventional open repair

in 195 patients (151 male, 44 female; mean age 69) and endoluminal repair in 108 patients (100 male, 8 female; mean age 70). The decision to perform endoluminal repair was based on comorbidities precluding open repair (n = 48) and patient choice (n = 60).

Graft configuration in the open repair group was tubular (n = 180) and bifurcated (n = 15); and in the endoluminal repair group tubular (n = 48), aortoiliac/femoral (n = 25), and bifurcated (n = 35). All procedures were performed in the operating room, and radiographic guidance was used in the endoluminal repair group.

Follow-up was by interview, examination, and telephone. In addition, contrast-enhanced CT was performed within the first 10 days after operation; 6 months and 12 months after operation; and annually thereafter in endoluminal repair patients.

Outcome measures were successful exclusion of the aneurysm sac from the general circulation and survival. Data were analyzed by the life-table method. Other outcome measures were length of hospital stay, length of ICU stay, and operative blood loss.

There was no significant difference between the perioperative mortality rate for open repair—11 deaths in 195 patients (5.6%) and endoluminal repair—6 deaths in 108 patients (5.6%). Three of the six deaths in the latter group occurred in patients with successful endoluminal repair, and three occurred in 18 patients with failed endoluminal repair who were converted to open repair. Similarly, there was no significant difference in survival between the endoluminal and open repair groups when analyzed by the log rank test (*P* .14). Graft failure, however, was significantly higher in the endoluminal repair group compared with that of the open repair group (Fisher's Exact test $P<.001$). Success in the endoluminal repair group was defined as continuing graft function without endoleak or conversion to open repair. The Kaplan-Meier survival curve for graft failure times for the endoluminal group revealed a 3-year graft success probability of 70%.

This study suggests that endoluminal repair is safe, sharing the same perioperative mortality as open repair, despite 44% of endoluminal repair patients being rejected as unfit for open repair. Conventional open repair is the most reliable method of successfully treating AAAs. The endoluminal method, however, results in shorter length of hospital stay, shorter length of ICU stay, and less blood loss than does the open method. Patients opting for the endoluminal method of repair should be made aware that the minimally invasive technique carries the disadvantage of a higher failure rate.

Discussion

This was the first study that attempted to compare the outcome of endoluminal treatment of AAAs with that of a concurrent control group undergoing open repair in consecutive patients. It must be acknowledged at the outset that the two groups, although comparable in relation to age, sex, and size of aneurysms, markedly differed with regard to comorbidities. These conditions were sufficiently severe in 44% of the patients in the endoluminal group to lead to rejection for open repair at major vascular centers. Despite this, the perioperative mortality rate was identical for both groups. It should also be pointed out that the mortality rate for the endoluminal group includes not only those dying after successful endoluminal repair, but also those dying as a result of conversion from failed endoluminal repair to open repair.

Similarly, no significant difference was found in survival between the two groups. One interpretation of these results is that the two groups are equally healthy. However, it seems more likely that the less healthy endoluminal group has undergone a procedure of lesser magnitude, which has resulted in the same outcome as the healthier open repair group who has undergone a major open procedure.

There is general acceptance that the gold standard for comparing two methods of treatment is a concurrent randomized study. However, with such a vast difference between the two methods and the level of discomfort experienced by patients in each group, it has become apparent that a randomized trial is impractical. There is virtually no way of preventing patients from withdrawing from a randomized study if they fail to draw the minimally invasive method that many seek. If the trial happens to be a multicenter one, there are similar problems with preventing patients from registering with several centers in a sequential manner until they draw the method of their choice.

The limitations of this study must also be addressed. The retrospective analysis of the data for the open repair group may have underestimated the instance of complications for this group. Regular follow-up and use of contrast-enhanced CT in the endoluminal repair group make it highly unlikely that any failures or complications would be overlooked. Those in the open repair group were not subjected to such scrutiny. Only one failure from aortoenteric fistula was encountered.

It is possible that false aneurysms may have been overlooked. The chance of undetected failure from this cause, however, is probably small because the incidence of false aneurysm is low in the first 5 years after open repair of AAAs.

Although the open repair and endoluminal repair operations were performed by the same group of surgeons, the endoluminal repair group carried the burden of the learning curve for a new technique. This is well illustrated by the fall in the primary conversion rate from 20% in the first half of the study to 8% in the second half. All patients undergoing elective endoluminal repair from the outset of the program in May 1992 were included to present a "worst case scenario" for endoluminal repair to avoid any unrealistic expectations for the procedure that might arise from a selected series. For the same reason, the life-table curve for endoluminal repair includes the outcome for primary repair only. The improved results that can be achieved with successful secondary endoluminal repair, thus avoiding conversion to open repair, have been excluded.

The significance of endoleaks deserves further comment. In the authors' experience, primary endoleaks usually result from errors of judgment in case selection, errors of technique, or device faults. Secondary endoleaks usually result from endoluminal device failure. In this regard, it is reassuring to see that the failure rate due to device failure is limited to one per year in a group of 95 successful endoluminal repairs in 108 patients.

Endoluminal repair shared the same mortality rate as open repair in this study, despite almost 50% of the endoluminal repair patients being at high-risk and unfit for open repair. If successful repair of AAAs was the only criterion, the inescapable conclusion would be that open repair is the most reliable method of achieving this end. However, other factors must be taken into consideration in assessing the relative merits of the two methods of aneurysm repair.

This study shows that the endoluminal method may result in a shorter length of hospital stay, a shorter length of ICU stay, and less blood loss than the open method. Despite these advantages, it must be emphasized that the endoluminal method is not free of major morbidity and mortality.

References

1. May J, White GH, Yu W, et al. Treatment of complex abdominal aortic aneurysms by a combination of endoluminal and extraluminal aorto-bifemoral grafts. J Vasc Surg 1994;19:924–933.
2. White GH, Yu W, May J, et al. A new nonstented balloon-expandable graft for straight or bifurcated endoluminal bypass. J Endovasc Surg 1994;1:16–24.
3. Parodi JC, Palmaz JC, Barone HD. Transfemoral intraluminal graft implantation for abdominal aortic aneurysms. Ann Vasc Surg 1991;5:491–499.
4. Marin ML, Veith F, Cynamon J, et al. Initial experience with transluminally placed endovascular grafts for the treatment of complex vascular lesions. Ann Surg 1995;222:449–465.
5. Parodi JC. Endovascular repair of abdominal aortic aneurysms and other arterial lesions. J Vasc Surg 1995;21:549–557.
6. Yusuf SW, Whitaker SC, Wenham PW, et al. Early results of endovascular abdominal aortic aneurysm repair. (Abstract). J Endovasc Surg 1996;3:98–99.
7. Chuter TAM, Wendt G, Hopkinson BR, et al. Transfemoral insertion of a bifurcated endovascular graft for aortic aneurysm repair: The first 22 patients. Cardiovasc Surg 1995;3:121–128.
8. Miahle C, Amicabile C. Traitement endovasculaire des aneurismes de l'aorte sous-renale par l'endoprosthèese Stentor. Série préliminaire. J Mal Vasc 1995;20:290–295.
9. Blum U, Langer M, Spillner G, et al. Abdominal aortic aneurysms: preliminary technical and clinical results with transfemoral placement of endovascular self-expanding stent-grafts. Radiology 1996;198:25–31.
10. Dake M, Miller DC, Semba CP, et al. Transluminal placement of endovascular stent-grafts for the treatment of descending thoracic aortic aneurysms. N Engl J Med 1994;331:1729–1734.
11. May J, White GH, Waugh R, et al. Transluminal placement of a prosthetic graft-stent device for treatment of subclavian artery aneurysm. J Vasc Surg 1993;18:1056–1059.
12. Moore WS, Rutherford R. Transfemoral endovascular repair of abdominal aortic aneurysm: Results of the North American EVT phase 1 trial. J Vasc Surg 1996;23:543–553.
13. May J, White GH, Yu W, et al. Early experience with the Sydney and EVT prostheses for endoluminal treatment of abdominal aortic aneurysms. J Endovasc Surg 1995;2:240–247.
14. Blum U, Voshage G, Lammer J, et al. Abdominal aortic aneurysms: Endovascular self-expanding stent-grafts. N Engl J Med 1997; 336:13–20.
15. May J, Bell PRF, Gaines P. Early experience with endovascular AAA repair with the Bard device. (Abstract). J Endovasc Surg 1998;5:94.
16. Fry PD, Machan L, Doyle DL, Morris C. The Talent graft for endoluminal aortic aneurysm repair: Initial experience. (Abstract). J Endovasc Surg 1997;4(Suppl 1):I11–I12.
17. Biasi GM, Piglionica MR, Meregaglia D, et al. Infrarenal AAA endoluminal repair using the Medtronic AneuRx stent graft: A European one-year multicenter experience. J Endovasc Surg 1998;5(Suppl 1):I4–I5.
18. White GH, Yu W, May J, et al. Three year experience with the White-Yu endovascular GAD graft in aortic and iliac aneurysms. J Endovasc Surg 1997;4:124–136.
19. Yu W, White GH, May J, Stephen MS. Endoluminal repair of abdominal aortic aneurysms using the trombone technique. Asian J Surg 1996;19:37–40.
20. White GH, Yu W, May J. "Endoleak"—A proposed new terminology to describe incomplete aneurysm exclusion by an endoluminal graft. (Letter). J Endovasc Surg 1996;3:124–125.
21. White GH, Yu W, May J, et al. Endoleak as a complication of endoluminal grafting of abdominal aortic aneurysms: Classification, incidence, diagnosis and management. J Endovasc Surg 1997; 4:152–168.
22. Marin ML, Veith F, Cynamon J, et al. Initial experience with transluminally placed endovascular grafts for the treatment of complex vascular lesions. Ann Surg 1995;222:449–465.
23. May J, White GH, Yu W, et al. Surgical management of complications following endoluminal grafting of abdominal aortic aneurysms. Eur J Vasc Endovasc Surg 1995;10:51–59.
24. May J, White GH, Yu W, et al. A prospective study of changes in morphology and dimensions of abdominal aortic aneurysms following endoluminal repair: A preliminary report. J Endovasc Surg 1995;2:343–347.
25. Dake M, Miller DC, Semba CP, et al. Transluminal placement of endovascular stent-grafts for the treatment of descending thoracic aortic aneurysms. N Engl J Med 1994;331:1729–1734.
26. Dorros G, Cohn JM. Adenosine-induced transient cardiac asystole enhances precise deployment of stent-grafts in the thoracic or abdominal aorta. J Endovasc Surg 1996;3:270–272.
27. White GH, May J, Waugh RC, et al. High incidence of intraoperative technical difficulties and device complications with self-expanding stent-grafts used for endovascular repair of abdominal aortic aneurysms. J Vasc Surg 1998 (in press).
28. McGahan TJ, Berry GA, McGahan SL, et al. Results of autopsy 7 months after successful endoluminal treatment of an infrarenal abdominal aortic aneurysm. J Endovasc Surg 1995;2:348–355.
29. Marin M, Veith FJ, Cynamon J, et al. Human transluminally placed endovascular stented grafts: Preliminary histopathologic analysis of healing grafts in aortoiliac and femoral artery occlusive disease. J Vasc Surg 1995;595–604.
30. Clowes AW, Kirkman TR, Reidy MA. Mechanisms of arterial graft healing: Rapid transmural capillary ingrowth provides a source of intimal endothelium and smooth muscle in porous PTFE prosthesis. Am J Pathol 1986;123:220–230.
31. Florey HW, Greer SJ, Kiser J, et al. The development of pseudointima lining fabric grafts of the aorta. Br J Exp Pathol 1962;43:655–660.
32. May J, White GH, Yu W, et al. A prospective study of changes in morphology and dimensions of abdominal aortic aneurysms following endoluminal repair: A preliminary report. J Endovasc Surg 1995;2:343–347.

33. May J, White GH, Yu W, et al. A prospective study of anatomico-pathological changes in abdominal aortic aneurysms following endoluminal repair: Is the aneurysmal process reversed? Eur J Vasc Endovasc Surg 1996;12:11–17.
34. Ivancev K. Malmo experience with Ivancev prosthesis. Presented at Endovascular Aortic Aneurysm Surgery Symposium, Nottingham, April 1996.
35. Chuter TAM. Experience with Chuter bifurcated prosthesis. Presented at Endovascular Aortic Aneurysm Surgery Symposium, Nottingham, April 1996.
36. White GH, May J, McGahan T, et al. Historical control comparison of outcome for endoluminal versus open repair of abdominal aortic aneurysms. J Vasc Surg 1996;23:201–212.
37. May J, White GH, Yu W, et al. Results of endoluminal grafting of abdominal aortic aneurysms are dependent on aneurysm morphology. Ann Vasc Surg 1996;10:254–261.
38. Rutherford RB, Flanigan DP, Gupta SK, et al. Suggested standards for reports dealing with lower extremity ischemia. J Vasc Surg 1986;4:80–94.
39. Johnson KW, Rutherford RB, Tilson DM, et al. Suggested standards for reporting on arterial aneurysms. J Vasc Surg 1991;13:444–450.
40. May, J, White GH, Yu W, et al. Endoluminal repair versus no treatment for small abdominal aortic aneurysms. Eur J Vasc Endovasc Surg 1998 (in press).
41. May J, White GH, Yu W, et al. Surgical management of complications following endoluminal grafting of abdominal aortic aneurysms. Eur J Vasc Endovasc Surg 1995;10:51–59.
42. Hopkinson BR, Armon MP, Latief K, et al. Influence of abdominal aortic aneurysm size on the feasibility of endovascular repair. J Endovasc Surg (in press).
43. May J, White GH, Yu W, et al. Concurrent comparison of endoluminal versus open repair in the treatment of abdominal aortic aneurysms: Analysis of 303 patients by life table method. J Vasc Surg 1998 (in press).

Index

Page numbers in *italics* denote figures; those followed by a "t" denote tables.